METHODS IN MOLECULAR

Series Editor
John M. Walker
School of Life and Medical Sciences
University of Hertfordshire
Hatfield, Hertfordshire, UK

For further volumes:
http://www.springer.com/series/7651

For over 35 years, biological scientists have come to rely on the research protocols and methodologies in the critically acclaimed *Methods in Molecular Biology* series. The series was the first to introduce the step-by-step protocols approach that has become the standard in all biomedical protocol publishing. Each protocol is provided in readily-reproducible step-by-step fashion, opening with an introductory overview, a list of the materials and reagents needed to complete the experiment, and followed by a detailed procedure that is supported with a helpful notes section offering tips and tricks of the trade as well as troubleshooting advice. These hallmark features were introduced by series editor Dr. John Walker and constitute the key ingredient in each and every volume of the *Methods in Molecular Biology* series. Tested and trusted, comprehensive and reliable, all protocols from the series are indexed in PubMed.

Histidine Phosphorylation

Methods and Protocols

Edited by

Claire E. Eyers

Department of Biochemistry, Centre for Proteome Research, Institute of Integrative Biology, University of Liverpool, Liverpool, UK

Editor
Claire E. Eyers
Department of Biochemistry
Centre for Proteome Research
Institute of Integrative Biology
University of Liverpool
Liverpool, UK

ISSN 1064-3745 ISSN 1940-6029 (electronic)
Methods in Molecular Biology
ISBN 978-1-4939-9886-9 ISBN 978-1-4939-9884-5 (eBook)
https://doi.org/10.1007/978-1-4939-9884-5

This Humana imprint is published by the registered company Springer Science+Business Media, LLC, part of Springer Nature.
The registered company address is: 233 Spring Street, New York, NY 10013, U.S.A.

Preface

Phosphorylation of proteins was first reported over 100 years ago. Since then, this type of protein modification has been demonstrated to critically regulate numerous aspects of protein function, controlling the cellular signaling networks that govern cell growth, differentiation, and the physiological responses to external stimuli. Although phosphorylation of histidine was first discovered in bovine mitochondria in 1962, the biological roles of phosphohistidine have been slow to be uncovered. Current literature primarily describes this modification in the context of two-component signaling systems in prokaryotes, fungi, and plants, or as a phosphoenzyme intermediate during catalysis. Only relatively recently have there been studies demonstrating its involvement in mammalian signaling, which is in marked contrast to investigations of other phosphorylated amino acids. The rate of progress in defining and understanding histidine phosphorylation in all organisms has been severely hampered by the fact that this modification is relatively unstable; experimental conditions for its analysis therefore need to be extremely carefully controlled. For the same reason, key tools that biochemists typically rely on to help elucidate functional roles for phosphorylation, i.e., phosphospecific antibodies, have only recently been generated.

Histidine Phosphorylation details methodologies for the characterization of histidine phosphorylation, and the enzymes that regulate this modification, in both in vitro and cell-based systems. More comprehensive background review articles on histidine kinases and phosphatases are also presented. As is customary in the *Methods in Molecular Biology* book series, each chapter introduces the techniques in the context of this modification and describes in a clear format the reagents, equipment, and methods required for its investigation. The experience of the authors in studying this labile modification is evident in the *Notes* sections for each chapter and should not be overlooked! Whether you are interested in expressing recombinant histidine kinases and establishing an assay to quantify histidine phosphorylation, or seeking to understand the dynamics and localization of histidine phosphorylated proteins in cells, the techniques described in this book are an ideal starting point to advance the field of *Histidine Phosphorylation*, particularly the nascent evaluation of its roles in mammalian cell signaling.

Liverpool, UK *Claire E. Eyers*

Contents

Contributors

KEVIN ADAM • *Molecular and Cell Biology Laboratory, Salk Institute for Biological Studies, La Jolla, CA, USA*

PAUL V. ATTWOOD • *School of Molecular Sciences, The University of Western Australia, Crawley, WA, Australia*

AMY M. BARRIOS • *Department of Medicinal Chemistry, University of Utah, Salt Lake City, UT, USA*

ELIZABETH M. BOON • *Department of Chemistry, Stony Brook University, Stony Brook, NY, USA; Institute of Chemical Biology and Drug Discovery, Stony Brook, NY, USA*

ALEKSANDRA E. BURY • *Molecular Microbial Physiology Group, Swammerdam Institute for Life Sciences, University of Amsterdam, Amsterdam, The Netherlands*

ALEJANDRO BUSCHIAZZO • *Laboratory of Molecular and Structural Microbiology, Institut Pasteur de Montevideo, Montevideo, Uruguay; Département de Microbiologie, Institut Pasteur, Paris, France*

DOMINIC P. BYRNE • *Department of Biochemistry, Institute of Integrative Biology, University of Liverpool, Liverpool, UK*

OSCAR MARTIN CAMACHO • *Department of Functional and Comparative Genomics, Institute of Integrative Biology, University of Liverpool, Liverpool, UK*

PATRICIA CASINO • *Departament de Bioquímica i Biología Molecular, Universitat de València, Burjassot, Spain; Estructura de Recerca Interdisciplinar en Biotecnologia i Biomedicina (ERI BIOTECMED), Universitat de València, Burjassot, Spain*

W. SETH CHILDERS • *Department of Chemistry, University of Pittsburgh, Pittsburgh, PA, USA; Chevron Science Center, University of Pittsburgh, Pittsburgh, PA, USA*

ALICE K. M. CLUBBS COLDRON • *Department of Biochemistry, Institute of Integrative Biology, University of Liverpool, Liverpool, UK*

SAMUEL W. DUVALL • *Department of Chemistry, University of Pittsburgh, Pittsburgh, PA, USA*

CLAIRE E. EYERS • *Department of Biochemistry, Centre for Proteome Research, Institute of Integrative Biology, University of Liverpool, Liverpool, UK*

PATRICK A. EYERS • *Department of Biochemistry, Institute of Integrative Biology, University of Liverpool, Liverpool, UK*

JONATHAN T. FISCHER • *Department of Chemistry, Stony Brook University, Stony Brook, NY, USA*

GEMMA HARDMAN • *Department of Biochemistry, Centre for Proteome Research, Institute of Integrative Biology, University of Liverpool, Liverpool, UK*

ILANA HECKLER • *Department of Chemistry, Stony Brook University, Stony Brook, NY, USA*

JAN HEJÁTKO • *Central European Institute of Technology and National Centre for Biomolecular Research, Faculty of Science, Masaryk University, Brno, Czech Republic*

KLAAS J. HELLINGWERF • *Molecular Microbial Physiology Group, Swammerdam Institute for Life Sciences, University of Amsterdam, Amsterdam, The Netherlands*

TONY HUNTER • *Molecular and Cell Biology Laboratory, Salk Institute for Biological Studies, La Jolla, CA, USA*

ANDREW R. JONES • *Department of Functional and Comparative Genomics, Institute of Integrative Biology, University of Liverpool, Liverpool, UK*

RAJASREE KALAGIRI • *Molecular and Cell Biology Laboratory, Salk Institute for Biological Studies, La Jolla, CA, USA*
KIMBERLY A. KOWALLIS • *Department of Chemistry, University of Pittsburgh, Pittsburgh, PA, USA*
NATALIE LUHTALA • *Molecular and Cell Biology Laboratory, Salk Institute for Biological Studies, La Jolla, CA, USA*
ALBERTO MARINA • *Instituto de Biomedicina de Valencia, Consejo Superior de Investigaciones Científicas (IBV-CSIC), Valencia, Spain; CIBER de Enfermedades Raras (CIBERER-ISCIII), Madrid, Spain*
BRANDON S. MCCULLOUGH • *Department of Medicinal Chemistry, University of Utah, Salt Lake City, UT, USA*
CRISTINA MIDEROS-MORA • *Instituto de Biomedicina de Valencia, Consejo Superior de Investigaciones Científicas (IBV-CSIC), Valencia, Spain; Facultad de Ciencias de la Salud Eugenio Espejo, Universidad Tecnológica Equinoccial, Quito, Ecuador*
LAURA MIGUEL-ROMERO • *Instituto de Biomedicina de Valencia, Consejo Superior de Investigaciones Científicas (IBV-CSIC), Valencia, Spain*
BLANKA PEKÁROVÁ • *Central European Institute of Technology and National Centre for Biomolecular Research, Faculty of Science, Masaryk University, Brno, Czech Republic*
DANIEL J. RIGDEN • *Department of Biochemistry, Institute of Integrative Biology, University of Liverpool, Liverpool, UK*
AGNIESZKA SZMITKOWSKA • *Central European Institute of Technology and National Centre for Biomolecular Research, Faculty of Science, Masaryk University, Brno, Czech Republic*
LAUREN J. TOMLINSON • *Department of Biochemistry, Centre for Proteome Research, Institute of Integrative Biology, University of Liverpool, Liverpool, UK*
FELIPE TRAJTENBERG • *Laboratory of Molecular and Structural Microbiology, Institut Pasteur de Montevideo, Montevideo, Uruguay*
WEI ZHAO • *Department of Chemistry, University of Pittsburgh, Pittsburgh, PA, USA*

Chapter 1

Protein Dynamics in Phosphoryl-Transfer Signaling Mediated by Two-Component Systems

Felipe Trajtenberg and Alejandro Buschiazzo

Abstract

The ability to perceive the environment, an essential attribute in living organisms, is linked to the evolution of signaling proteins that recognize specific signals and execute predetermined responses. Such proteins constitute concerted systems that can be as simple as a unique protein, able to recognize a ligand and exert a phenotypic change, or extremely complex pathways engaging dozens of different proteins which act in coordination with feedback loops and signal modulation. To understand how cells sense their surroundings and mount specific adaptive responses, we need to decipher the molecular workings of signal recognition, internalization, transfer, and conversion into chemical changes inside the cell. Protein allostery and dynamics play a central role. Here, we review recent progress on the study of two-component systems, important signaling machineries of prokaryotes and lower eukaryotes. Such systems implicate a sensory histidine kinase and a separate response regulator protein. Both components exploit protein flexibility to effect specific conformational rearrangements, modulating protein–protein interactions, and ultimately transmitting information accurately. Recent work has revealed how histidine kinases switch between discrete functional states according to the presence or absence of the signal, shifting key amino acid positions that define their catalytic activity. In concert with the cognate response regulator's allosteric changes, the phosphoryl-transfer flow during the signaling process is exquisitely fine-tuned for proper specificity, efficiency and directionality.

Key words Bacterial signaling, Protein phosphorylation, Allostery, Histidine kinase, Response regulator

1 Introduction

A vast number of organisms use two-component systems (TCSs) as an efficient means of sensing, transmitting and processing information, ultimately ensuring cellular homeostasis. Almost ubiquitous in bacteria, also present in fungi and plants, the simplest TCSs work by a concerted action of two protein components: a sensory histidine kinase (HK) and a response regulator (RR). These two components communicate by transferring a phosphoryl group from a particular histidine residue on the HK to a specific aspartate on the RR (Fig. 1a). This phosphotransfer reaction can take place once the

Claire E. Eyers (ed.), *Histidine Phosphorylation: Methods and Protocols*, Methods in Molecular Biology, vol. 2077,
https://doi.org/10.1007/978-1-4939-9884-5_1,

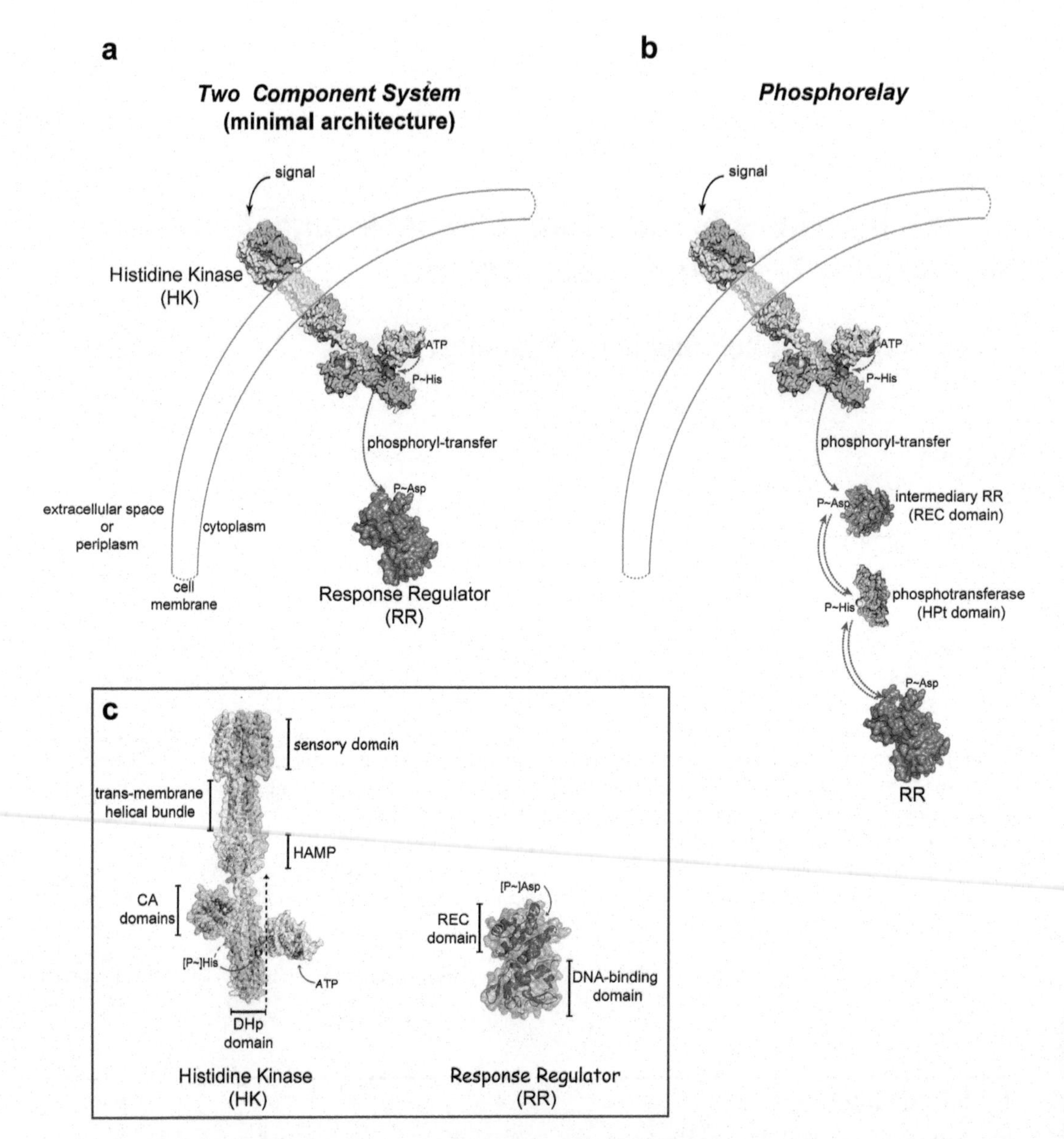

Fig. 1 Two-component system pathways and protein domain architecture. (**a**) Minimal organization of a canonical TCS comprising a sensory protein (histidine kinase HK) and an output effector one (response regulator RR). HKs are typically dimeric (each protomer distinguished with a different color), and although they can be cytoplasmic soluble proteins, a transmembrane representative has been chosen for this illustration. The phosphoryl-transfer is typically irreversible (P-$His_{HK} \rightarrow Asp_{RR}$) in simple TCSs. (**b**) A phosphorelay pathway is schematized, showing the role of intermediary proteins that receive and transfer the phosphoryl group, often through reversible reactions (double-headed arrows). (**c**) Modular domain organization of HKs and RRs. Both protein components can have a larger or smaller number of domains than drawn in the figure, but the minimal architecture that defines each component always includes at least a DHp and CA domains in the HK, and a REC domain in the RR. HAMP (signaling domain found in *H*istidine kinases, *A*denylyl cyclases, *M*ethyl-accepting chemotaxis proteins, and *P*hosphatases); DHp (*D*imerization and *H*istidine *p*hosphotransfer domain); CA (*C*atalytic and *A*TP-binding domain); REC (*Rec*eiver domain)

His residue is phosphorylated via an autophosphorylation reaction catalyzed by the HK itself, using ATP as the phosphodonor substrate. The autokinase and phosphotransferase reactions are typically coupled, and correspond to a particular structure/functional state(s) of the HK defined as "switched on," or "kinase-active." At some point the phosphorylated RR (P-RR) species is eventually dephosphorylated by hydrolysis, rendering inorganic phosphate (P_i). The latter reaction is frequently accelerated by the specific HK partner when switched off. In this kinase-inactive state, the HK acts as a phosphatase, accelerating P-RR dephosphorylation in a specific way: only the cognate HK partner and not other HKs can catalyze the dephosphorylation [1]. HKs thus constitute a fascinating example of enzymes with paradoxical activities [2], in that they catalyze reactions with opposite outcomes: the phosphorylation of its specific substrate, as well as its dephosphorylation, according to the signaling state of the pathway. Overall, a fundamental aspect behind the cascade of TCSs-mediated posttranslational modifications, is that the P-RR active species, is functionally distinct from the inactive unphosphorylated species. The biological functions engaged by the phosphorylated form of the RR, constitute the very output of the signaling cascade, ultimately orchestrating an adaptive response. The reactions involved in such cascades are tightly regulated, ensuring an accurate transmission of information. Thus, the evolution of mechanisms that minimize signal-independent activations and futile cycles has been critically important. Futile cycles could arise by uncontrolled phosphoryl-transfer and dephosphorylation of the RR (both catalyzed by the same HK enzyme), which can be overcome by molecular means that separate such reactions efficiently in time.

A minimalistic two-component organization, with one HK and one RR (Fig. 1a), is frequently observed in many TCSs, involved in signaling pathways that respond to a broad range of signals such as cell quorum, osmolarity, temperature, or antibiotics, among many others [3]. However, TCSs have also evolved in some cases into more complex linear or branched cascade systems, called phosphorelays, where additional domains and/or proteins integrate the phosphoryl-transfer circuits between the upstream sensory HK (s) and the output RR(s) (Fig. 1b). Phosphorelay intermediary proteins comprise variants of the same type of domains found in simple HK:RR TCSs, often also including *H*istidine *P*hospho*t*ransfer (HPt) proteins that harbor a phosphorylatable histidine residue, albeit structurally different from HKs and unable to autophosphorylate. Nonetheless, the mechanistic workings of TCSs and phosphorelays are thought to be similar at the molecular level, given the structural resemblance of the different modules that participate in signal transmission.

More often than not, signal sensation triggers HK activation. The opposite holds in some cases, where the signal turns off an otherwise constitutive autokinase activity, such as in CheA-regulated chemotaxis [4], among several other pathways [5, 6]. HK "on/off" switching is one of the key elements in defining the outcome of the TCS pathway, especially considering that most HKs can act both as a kinase and as a phosphatase of their cognate RR partners. Signal sensing is known to modulate the on/off switching transition in HKs [7], a key step in TCS signaling that this review will focus on. Recent progress on understanding the mechanisms whereby activation switching is also engaged in controlling the specific association between HKs and RRs, will also be elaborated. We shall see that HK and HK:RR functional transitions rely on the modulation of dynamic features of the proteins. Additional signaling steps such as HK autophosphorylation [8, 9] and RR activation [10], are also critically dependent on protein flexibility, but will not be reviewed here.

2 Key Structural Features of TCS Proteins

Currently over 600 three-dimensional structures of separate HKs and RRs have been reported, most of them determined by X-ray crystallography, and also by NMR [11, 12]. Both TCS proteins display modular architectures, including domains belonging to a number of different classes which appear with varying frequencies in different HKs and RRs. However, few specific domains define whether a given protein is an HK or an RR (Fig. 1c). Response regulators always comprise at least one receiver domain (REC), which harbors the phosphorylatable aspartate residue within a conserved α/β Rossmann-like topology fold. RRs can be single-domain proteins, or may comprise additional domains, such as DNA-binding or enzymatic modules, among many others [13]. HKs are defined by two distinct domains (Fig. 1c): (1) a centrally localized DHp (*D*imerization and *H*istidine-*p*hosphotransfer) domain, which is typically an elongated all-helical module engaged in homodimerization and includes the phosphorylatable histidine residue; and (2) a CA (*C*atalytic and *A*TP-binding) domain, which is a globular α/β module belonging to the Bergerat fold of ATPases [14]. CA domains bind ATP and exert a slow ATPase activity. Many HKs comprise additional modular domains, for example sensory domains, transmembrane regions, HAMP (for *H*istidine kinase, *A*denylyl cyclase, *M*ethyl-accepting chemotaxis protein, and *P*hosphatase), PAS (for *P*er-*A*rnt-*S*im), or GAF (for c*G*MP-specific phosphodiesterases, *A*denylyl cyclases, and *F*hlA) domains (*see* [11] for a review). There are also more complex hybrid architectures that include REC, DHp, and CA domains in the same polypeptide [3, 15]. Such hybrid TCS proteins are classified as hybrid-RRs or hybrid-HKs if the REC domain is respectively placed N- or C-terminal to the HK modules [16].

Structures of full-length RRs have been solved in active and inactive conformations, including some in complex with their cognate DNA [17, 18]. In contrast, no structures of full-length transmembrane HKs have yet been determined. The most complete picture so far corresponds to the high-resolution images of a portion of NarQ from *E. coli* [19], comprising the periplasmic sensory domain, the TM region, and the first signal transmission domain (a HAMP) on the cytoplasmic portion of this HK. Comparison of apo and signal-bound crystal structures [19] revealed the workings of a signal-triggered activation mechanism. However, downstream signal transmission to the catalytic region (DHp and CA domains) remains hypothetical, as the entire intracytoplasmic region was lacking from the crystallized constructs. On the other hand, a few crystal structures of the whole catalytic portion of HKs have been determined (DHp + CA modules, sometimes with additional intracytoplasmic domains) [9, 20, 21], and in one case different conformations corresponding to distinct kinase/phosphatase functional states were captured [22]. By combining information from both sensory/transmembrane and catalytic intracytoplasmic snapshots, a common picture starts to arise which will be elaborated further below.

3 Classification of HKs in Families

HKs were discovered in the 1980s [23] as enzymes that catalyze autophosphorylation on a conserved His residue, using ATP as the phosphodonor substrate [24]. The Bergerat fold of HK CA domains, different from Walker ATPases, is shared with other, distantly related, slow ATPases (DNA gyrase, *H*sp90, and Mut*L*) constituting the so-called GHKL superfamily [25].

Early sequence alignments revealed the essential features distinguishing HK subclasses [26]. In particular, those comparative studies identified a marked correlation of particular HK classes to RR classes, confirming that these signaling systems tend to function as two-component pairs or dyads [27]. Current HK classifications have been simplified to a fewer number of groups, notably using hidden Markov profile approaches [28], with all the HK sequences comprised within the single Pfam clan His_Kinase_A (CL0025). The classification of HKs in different families [29, 30] appears to have relevant mechanistic implications.

Within the His_Kinase_A clan four different families currently classify HKs according to DHp domain sequences: (1) HisKA (PFAM family PF00512), (2) HisKA_2 (PF07568), (3) HisKA_3 (PF07730), and (4) HWE_HK (PF07536). Additionally, HK can be classified according to their CA domain: HATPase_c (PF02518), HATPase_c2 (PF13581), and HATPase_c5 (PF14501), the latter including examples of HKs with novel secondary structure elements [31] compared to the most populated family, the HATPase_c.

HisKA (covering ~80% of all DHp-containing HKs) and HisKA_3 (~15%) DHp domains, as all DHp sequences, are predicted to contain two antiparallel α-helices forming a helical hairpin that, when dimerized, form a 4-helical bundle. Very few examples of monomeric HKs have been reported, corresponding exclusively to members of the peculiar HisKA_2/HWE_HK group [32]. DHp helix α1 tends to be longer than α2; hence, the N-terminal portion of the typically dimeric DHp results in two parallel helices, with predicted tendency to form a left-handed coiled-coil. The conservation of hydrophobic residues in positions *a* and *d* within the coiled-coil heptad repeat pattern (or equivalent hydrophobic-core positions in undecad repeats identified in certain HKs), is far from ideal: those key positions are often occupied by polar residues instead, which reduce the energetic stability of an otherwise tight helix-to-helix binding. Early sequence alignment studies and prediction of expected helical packing behaviors [33, 34] proposed a regulatory function for such marginally stable coiled-coils in the helical domain that precedes the His-containing phosphorylation site in HKs.

4 HK Conformational Switch: The Signal Regulates Kinase/Phosphatase Activities

The proposed regulatory function of the coiled-coil region in the DHp domain of HKs was observed directly for the first time in the crystal structures of DesK from B. subtilis [22] (Fig. 2), a HK that belongs to the HisKA_3 family. The kinase-off phosphatase state comprises a well-formed left-handed coiled-coil arrangement, whereas the kinase-on states, prephosphorylation and postphosphorylation, reveal a disrupted coiled-coil whereby the DHp helices α1 of both protomers dissociate from each other [22, 35]. These crystallographic studies comparing wild-type species and point mutants that trap functionally relevant conformational states, showcased a successful approach to study such dynamic systems [22, 35, 36]. In summary, HK active states are associated with higher interdomain flexibility and a disrupted DHp coiled-coil segment, which are in turn linked to asymmetric dimer assemblies. An inverse picture correlates to kinase-inactive, phosphatase states of HKs. Such a switching scheme based on HK flexibility regulation and substantial DHp/CA interdomain rearrangements is consistent with data gathered from several TCSs and HK families, using a variety of different approaches, like differential proteolysis sensitivity [37], cysteine engineering and cross-linking analyses [38, 39], hydrogen–deuterium exchange assessed by mass spectrometry [40], NMR [41], and other spectroscopic techniques like electron spin resonance [42], among others.

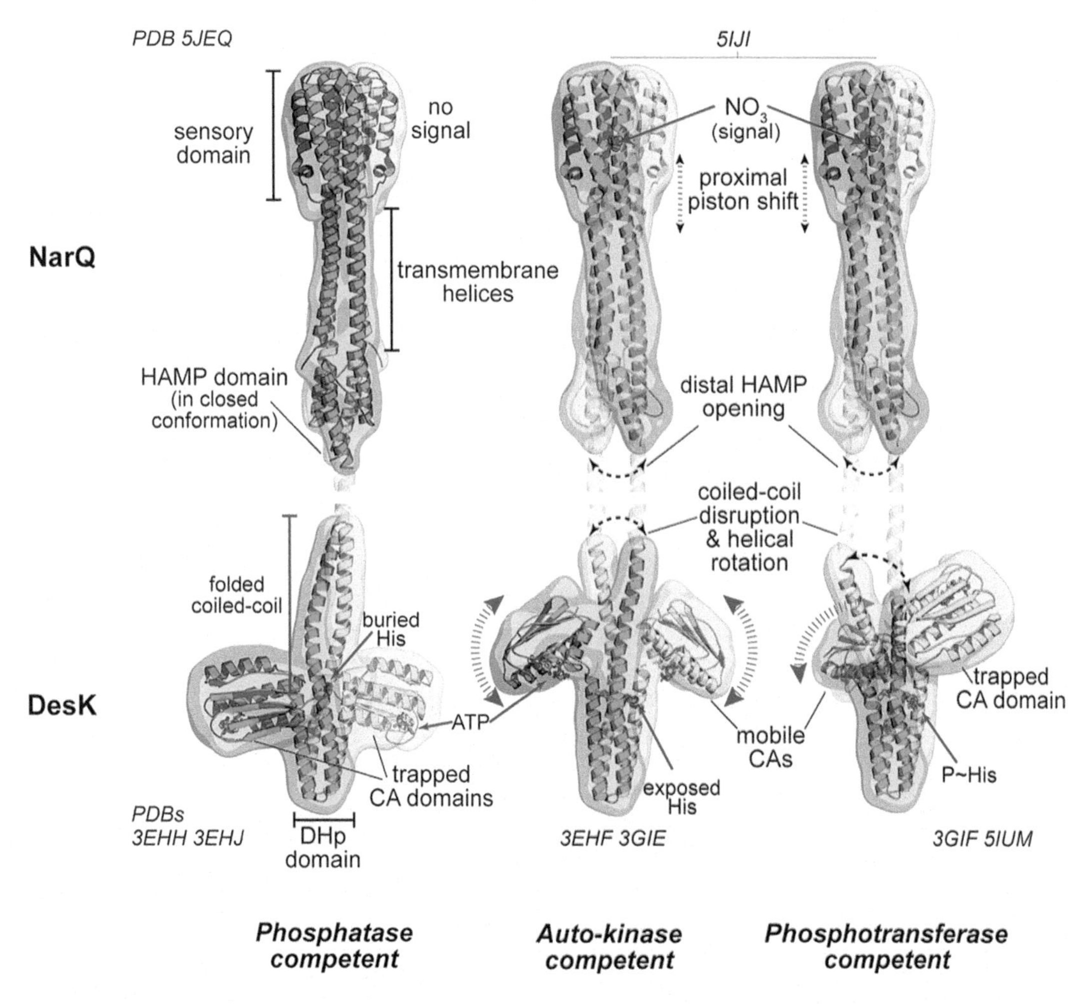

Fig. 2 Protein dynamics in histidine kinase activation switching. Integrating data from two different HisKA_3 histidine kinases, NarQ (top row) [19] and DesK (bottom) [22, 35], an overall picture of HK activation has been uncovered. The two protomers of the HKs' dimers are colored green and yellow. The structures are shown as cartoons with overlaid semitransparent solvent-accessible surfaces. The connection between both is disrupted to illustrate it is hypothetical, as no 3D structure of a full-length transmembrane HK has yet been determined. PDB accession codes are shown for each depicted structure. An elaborated description is given in the text, but note the general trend of symmetric to asymmetric organization when going from the phosphatase- to the phosphotransferase-competent states. DHp domain coiled-coil disruption and coupled helical rotational motion in the active state correlate with freely mobile CA domains, poised for autophosphorylation. Note the added regulatory effect of position rearrangement of the phosphorylatable His, swinging from a buried location within the DHp 4-helix bundle (phosphatase state) to a solvent-exposed one (kinase-active and phosphotransferase states)

Crystallization of the sensory and transmembrane portion of the nitrate-sensing HK NarQ [19], including the first intracytoplasmic signal-transmission module (a HAMP domain), was a major achievement leading to the currently only available 3D structure of a HK encompassing its transmembrane segment, detail which is key for understanding signal transduction. These crystal structures of apo and nitrate-bound NarQ consistently showcase the relevance of the DHp coiled-coil breaking/making mechanism (Fig. 2), and for the first time revealed that small signal-triggered piston-like movements within the sensory domain are amplified and transduced into a large scissoring motion of the HAMP module, leading directly to DHp domain coiled-coil disruption [19, 43].

The DHp conformational rearrangements that drive coiled-coil assembly and disruption are also coupled to CA domain mobility control. Indeed, major reorganization of the CA domains relative to the central helical DHp along the signaling cycle appears to be a universal feature in the regulation of all HKs, even though variations in the details of the effecting mechanism appear to exist. As mentioned above, in HisKA_3 HKs such as NarQ, DesK, or LiaS among others, the DHp helices rotate as a coupled movement to the coiled-coil folding/unfolding switch. Rotational rearrangements in HKs' signaling helices had been proposed in HK activation in quorum sensing [6] and light perception [44] pathways, but had yet to be verified by structural interrogation of both conformations. A first observation was achieved in B. subtilis DesK [22] and then confirmed in several other cases such as NarX [45] and NsaS [46]. Such findings pinpoint the crucial importance of the phosphorylatable histidine, which occupies a special skip position within the DHp coiled-coil heptad repeat pattern of HisKA_3 HKs, allowing for a dramatic rotational shift during activation. This rotational shift is maximal in the region surrounding the HK's phosphorylation site [22], resulting in exposure or burying of the key His to the solvent [35]. Consequently, this shift generates or obliterates a docking surface for the CA domains (Fig. 2). A large DHp-CA interface burying approximately 1200 Å^2, is observed in the phosphatase state, leading to a symmetrically open, more rigid, butterfly-shaped structure [22]. This rigid configuration minimizes the likelihood of CA-bound ATP moieties reaching the reactive His on the central DHp. In contrast, the kinase/phosphotransferase state, with substantially reduced or no interdomain interface, results in liberated CA domains, poised to engage in autophosphorylation reactions, or recruitment and phosphotransfer to the cognate RR. Such HK active states typically display asymmetric structures (Fig. 2).

In the case of the HisKA family of HKs, snapshots of the same kinase in different functional states are still missing. Nevertheless, from available structures [8, 9, 20], key conformational rearrangements appear to be shared between HisKA and HisKA_3 proteins. In HisKA HKs the coiled-coil sequence pattern is slightly different, suggesting that the phosphorylatable His will not swing between exposed and buried conformations as in HisKA_3 kinases. Nevertheless, comparing the structures of *Thermotoga maritima* HK853 [20, 47], and CpxA [9] and EnvZ [8] from *E. coli*, suggests that a transition from a symmetric to an asymmetric configuration is indeed a hallmark of HK activation.

Switching implies substantial protein flexibility, but in such a way as to populate distinct states: a form of modulated flexibility, enabling a single sequence to adopt at least two (or more) different 3D structures. This situation is different from a continuum (e.g., an unstructured protein), allowing for specific distinct states to be defined as either active or inactive. Certainly, each one of such alternative states can also display flexibility on their own, but not confounding active and inactive species. Experimental evidence supporting dynamic switching has mostly come from NMR studies of response regulators [10, 48], for which the inactive state appears to be much more conformationally dynamic compared to the Asp-phosphorylated species. The latter is the active form, competent to initiate a response, typically acquiring a more rigid, unique 3D structure. More work is needed in this area to understand HK activation switching [49], with a seemingly opposite scenario to RRs: more rigid inactive HKs switching to flexible and partially unfolded forms in the active state. NMR-based approaches face a challenging goal since HKs are typically larger proteins compared to their RR partners, and frequently comprise transmembrane segments.

Once the HK is activated, and in the presence of physiologic high cellular ATP concentrations, HKs catalyze the autophosphorylation reaction, which is typically an asymmetric process: one His in the dimer is more rapidly/efficiently phosphorylated than the other [50, 51]. This asymmetry is also consistent with the molecular features of crystal structures of HKs trapped in Michaelis complexes with ATP [8, 9] representing snapshots of the ATP γ-phosphoryl transfer reaction.

5 HK-RR Complexes

For historic reasons the two components, HKs and RRs, have largely been studied separately. Transiently associated protein–protein complexes, such as those formed by HK:RR binding, present huge technical challenges given the spatial/temporal resolution required to study their structural and dynamic features. The 3D structures of associated His- and Asp-containing TCS proteins in binary complexes have been determined mostly for phosphorelay pathways [52–57]. The His-containing proteins of such complexes are intermediary components with no autokinase activity (hence not bona fide HKs), functioning only as carriers of phosphoryl moieties between upstream and downstream Asp-containing receiver domains. The reason for this bias in structural determination of phosphorelay complexes is not clear, and is likely related to lower intrinsic flexibility of phosphotransfer proteins compared to

HKs. Although limited in number, experimental 3D structures of bona fide HK:RR complexes have provided extremely relevant information [35, 47, 58], notably showing that the reaction centers of the phosphotransferase and phosphatase reactions are assembled with residues from both protein partners [35].

Early on, it became clear that some sort of code dictating pairwise interactions between specific HKs and RRs ("specificity code") was biologically relevant, given that dozens of highly similar RRs are being coexpressed at any given time in the cell, and that cross talk among "noncognate" HKs and RRs is not usually observed [59–62]. Over the past decade, important contributions have allowed specificity determinants to be restricted to a small subset of defined amino acid positions in the HKs' DHp domain [63, 64], revealing that this interface is extremely plastic and tolerant to natural mutations [65]. Crystal structures of HK:RR complexes accurately explain those observations, revealing low surface complementarity and very few protein–protein contacts [35, 47, 58], just enough to retain affinity between the specific (or cognate) partners. The TCS DesK:DesR is currently the only system for which crystal structures of the HK:RR complex have been determined in both phosphatase and phosphotransferase states [35] using X-ray diffraction (Fig. 3). Structure-based point-mutants that trap DesK in the phosphatase state (DHp coiled-coil hyperstabilization), or that mimic its phosphotransferase form (a glutamate substituting the phosphorylatable histidine), were demonstrated to be important to stabilize the complex in solution [66]. The HK recognizes its cognate RR largely using the same interface in both states; the outcome of the pathway is dictated by the positioning of a few key residues from the HK partner in interaction with the RR's active site. Such subtle but decisive dissimilarity, explains why phosphotransfer and dephosphorylation reactions cannot be catalyzed simultaneously, inhibiting futile cycles [35, 67].

The phosphatase reaction. Besides the universal phosphotransferase activity that catalyzes phosphoryl-transfer from the P-HK to the inactive state of the RR, most DHp-containing HKs (i.e., excluding CheA/CheA-like HKs) are able to accelerate the dephosphorylation of their specific P-RR partner(s) when the HK itself is not kinase-active [67]. High resolution crystal structures representing snapshots of the phosphatase state just prior to dephosphorylation [35], uncovered a highly symmetric organization of the phosphorylated form of DesR (mimicked by covalently bound BeF_3^- to the reactive aspartate) in complex with DesK, a HisKA_3 HK. A balanced stoichiometry of two RR moieties bound to the two HK protomers, is coherent with a strongly symmetric butterfly-shaped HK dimer. This trend has also been observed HisKA HKs: the crystal structure of *T. maritima* HK853:RR468 complex reveals a strongly symmetric organization and was posited to be a snapshot of the phosphatase state [47].

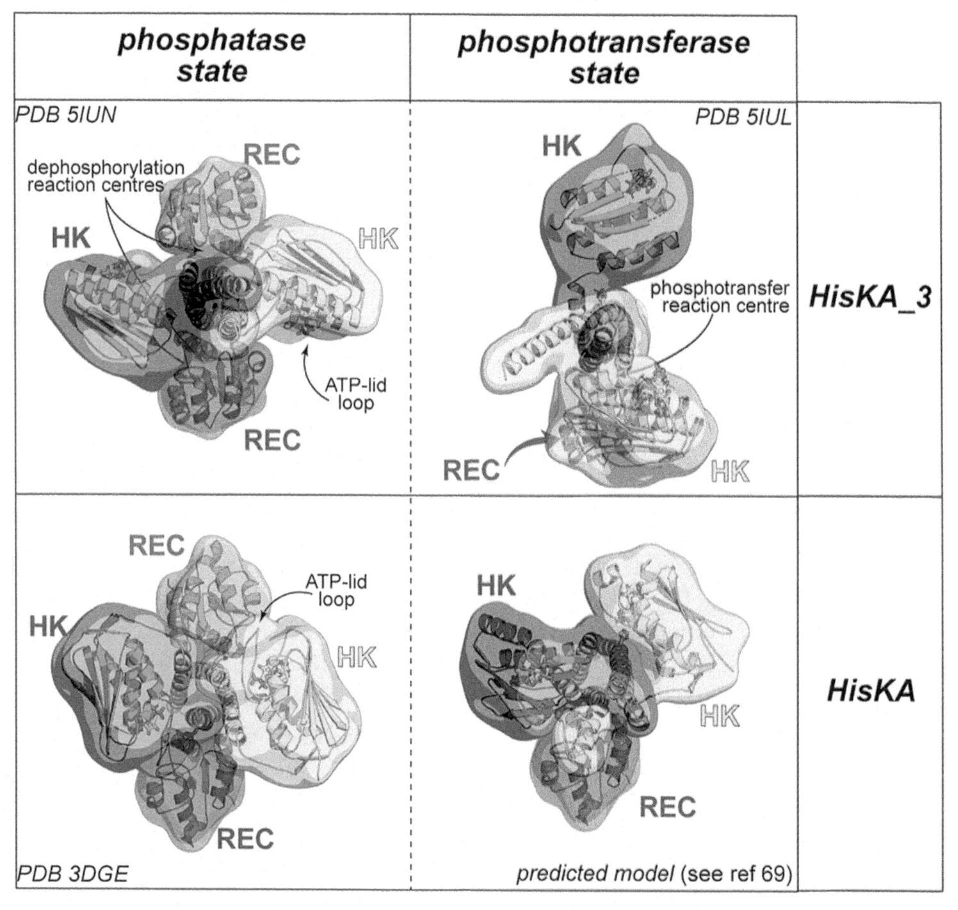

Fig. 3 3D structures of His-kinases in complex with their response regulators. Overall 3D configuration of TCS complexes, representing snapshots of the P-RR dephosphorylation reaction (phosphatase state) and the phosphoryl-transfer to RR (phosphotransferase state). Coloring scheme and model representations are the same as in Fig. 2, with the receiver domain (REC) of bound RRs colored orange and violet. The perspectives have been chosen by maximizing the structural superposition of the central DHp domain helices, which remains similar in all the panels. PDB accession codes or relevant references are indicated. (*Top row*) Both states have been captured in a HisKA_3 HK TCS (DesK:DesR). HK DHp coiled-coil disruption and CA domain reorganization, complex asymmetry are hallmarks of the phosphotransferase form. Subtle but critical repositioning of key amino acid residues in both HK and RR are not seen in this perspective (*see* ref. 35 for details). (*Bottom row*) The analysis of complexes engaging HKs that belong to the HisKA family allows identification of common features. The HisKA phosphotransferase state has not yet been observed but has been proposed on the basis of the autophosphorylating, active states of two different HisKA HKs

A glutamine, just over one helical turn C-terminal to the phosphorylatable His, is highly conserved in HisKA_3 HKs (in some kinases substituted by an Asn, or even a Thr in the equivalent position of HisKA HKs). The polar residue in this position is proposed to facilitate the correct positioning of a hydroxyl anion, which performs the nucleophilic attack on the Asp-bound phosphate of the RR [35, 68, 69]. The key role of residues other than the phosphorylatable histidine for the phosphatase function of the HK is consistent with early reports [70] demonstrating that the dephosphorylation reaction is not the reverse of phosphoryl-transfer, as later confirmed [67]. The HK is not backphosphorylated to take off the RR's phosphate. Instead, hydrolysis produces inorganic orthophosphate. Reversed phosphotransfer was initially proposed to be relevant in DHp-containing HKs such as EnvZ [71], but later proved to be the likely result of not using the full-length form of the HK under in vitro conditions [72]. Genuine reversed phosphoryl-transfer reactions have been reported in CheA-like pathways [73] and phosphorelay systems [74, 75], a biologically relevant reaction course, yet distinct from HK-mediated P-RR dephosphorylation (see below). Although the role of HKs acting as P-RR phosphatases is often difficult to quantify in vivo [76], the physiological importance of such activity is now widely accepted [1], ensuring robustness with respect to ATP and TCS protein concentrations [2, 77], and allowing for minimization of cross talk between different TCS pathways [59, 78]. The phosphatase activity of HKs has been shown to depend on the presence of ATP (or ADP) in some cases, a link that seems to hold for HisKA and not HisKA_3 HKs. The reason for this difference might be due to the ATP-lid, a loop of variable length in different HK sub-classes, juxtaposed to the ATP-binding pocket, and which often becomes ordered upon ATP binding and Mg^{2+} coordination. The ATP-lid makes direct contacts with the P-RR partner in HisKA-containing complexes [47], but not in HisKA_3-containing ones [35], consistent with the fact that HisKA_3 HKs have evolved to include substantially shorter ATP-lid loops in (Fig. 3).

The phosphotransferase reaction. In contrast to the phosphatase state, the phosphotransferase complex of DesK:DesR is strongly asymmetric, not only in the organization of the HK itself but also in the 1:2 RR:HK stoichiometry (Fig. 3). At the reaction center, the activation-switched rotational shift of the reactive histidine region places this residue perfectly in line for the RR's aspartate to perform a nucleophilic attack on the His-bound phosphate group. For reasons that are not yet clear, while one CA domain remains freely mobile on the side that engages in HK:RR association, the other CA remains bound to the DHp through a different, and smaller, interface than the one observed in the phosphatase complex. This has been confirmed in different crystal forms under variable crystal packing environments (PDBs 3GIF, 5IUM, 5IUJ), supporting its biological relevance.

A 3D structure corresponding to a HisKA HK in complex with its RR in the phosphotransferase state is not yet available. However, it is predicted to be asymmetric [79], as observed for the HisKA HKs CpxA and EnvZ captured in nucleotide-bound Michaelis complexes poised for autophosphorylation [8, 9]. It is clear that more complete and direct structural data of HisKA HK phosphotransferase complexes are needed. The rotational motion implicating the HK phosphorylation site and neighboring residues in HisKA_3 HKs will likely mean that it will not be possible to switch from phosphatase to phosphotransferase states due to the different positioning of the reactive His in the coiled-coil heptad repeat register of HisKA kinases. Albeit less dramatic, a change in the relative position of the His, modulating its precise location within the active site (closer or farther away from the RR aspartate), should not be ruled out.

6 Conclusions and Perspectives

Great progress is being achieved in the structural biology of TCSs. The importance of protein dynamics in populating distinct conformations, essential for the function of histidine kinases, is thus being highlighted. The controlled flexibility within and among their modular domains, underlies the regulation of their autokinase activity, as well as the way they interact with their cognate response regulator partners, dictating further catalytic roles as phosphotransferases or as phosphatases. Mechanistic insights about HK activation converge toward a scenario where the CA domains' mobility is regulated by a conformational rearrangement of the DHp domain, itself triggered by the sensory region through disruption or assembly of the connecting coiled-coil.

The molecular details of HK:RR complexes have also started to reveal reaction directionality features. As stated above, more often than not *bona fide* TCSs catalyze highly irreversible P-HK → RR phosphoryl-transfer reactions (Fig. 1a). Interestingly, phosphorelay pathways display many examples of reversible reactions from, and to, phosphorylatable Asp/His residues (Fig. 1b). This intriguing divergence may well find its molecular basis in the distance between the phosphorylatable histidine and aspartate residues and, directly correlated with this distance, in how symmetrically the RR-bound Mg^{2+} cation sits with respect to both reactive amino acid side chains. An unexpectedly large distance (>7.5 Å) was observed between the phosphorylatable residues in the DesK:DesR phosphotransferase complex [35], compatible with a nucleophilic substitution with significant dissociative character. The Mg^{2+} cation, essential for the phosphoryl-transfer reactions, is an appealing candidate to stabilize the transition state phosphoryl anion as it is being transferred: a symmetric Mg^{2+} position would allow for transfer to occur in both directions. In support for such hypothesis, available structures of phosphorelay TCS complexes reveal significantly shorter distances between the reactive His and Asp (<5.5 Å), anticipating an associative nucleophilic substitution correlated with a more symmetric position of the Mg^{2+} cation. Ongoing and future work shall verify whether this mechanism underlies directionality control.

Additional structures of HK:RR complexes in different signaling states, and from different HK and RR protein families, are needed to confirm and generalize the molecular mechanisms that govern TCS signaling in bacteria. For instance, the *T. maritima* HK853:RR468 complex was proposed to represent the phosphatase state of the pathway [47] displaying a highly symmetrical organization with a 2:2 stoichiometry and the CA domains rigidly packed against the central DHp. It is worth noting that the relative orientation of the CA domains is strikingly different compared to the phosphatase DesK:DesR complex (Fig. 3) which could reflect an authentic difference between HisKA and HisKA_3 families, with still unknown functional consequences. Further structural data, exploiting techniques in addition to X-ray crystallography, are required to complete the full picture of such dynamic protein systems. Comparison of the *T. maritima* phosphatase complex with other HisKA HK autophosphorylation snapshots reveal that the location of the CA domains in the former complex are intriguingly close to the position they will likely adopt in the kinase-active state [8, 9]. Furthermore, structural comparison of free HK853 [20] with the HK853:RR468 complex [47] discloses a more flexible coiled-coil region toward the N-terminus of the DHp $\alpha1$ helices in the complex, supporting the hypothesis that the complex captures the phosphotransferase state or a phosphotransferase-like intermediate form.

Cryo-electron microscopy and tomography, as well as ever more powerful NMR approaches, are anticipated to play key roles in structural determination of full-length HKs and HK:RR complexes with high resolution, and unveil detailed pictures of their dynamic behaviors.

In analogy with electric engineering, an input signal (an antibiotic, temperature, salt, quorum, etc.) triggers a deviation from cellular homeostasis due to physical or chemical effects on the cells' components and/or metabolic status. By means of a suitable TCS, that particular signal can be detected, transmitted, and processed into a "control" output, ultimately reestablishing homeostasis, a fascinating example of exquisite biological regulation. Suitability in this context means that (1) the particular TCS's HK is able to detect the specific signal (input sensitivity); (2) the HK is able to catalyze ATP-dependent autophosphorylation in a regulated manner, allosterically modulated by its sensory status (HK autokinase on/off switching); (3) the P-HK associates to the correct RR, selecting it out from dozens of simultaneously coexpressed RRs (specificity code/proper wire connectivity); (4) phosphatase vs. phosphotransferase activities are tightly regulated such that futile cycles and P_i loss are minimized (efficiency/

lossless signal transmission); (5) reversible vs. irreversible phosphotransfer reactions among reactive histidine- and aspartate-containing protein domains are properly modulated according to the needs of the pathway (signal transmission directionality); and (6) the P-RR output affinity is modulated to effect the output response, be it DNA-binding, protein–protein association, and/or enzymatic catalysis of downstream substrates (output device on/off switching). The complex puzzle of TCS mechanistic regulation is just starting to be solved. The complete understanding of how signals are efficiently transmitted by TCSs will be instrumental for pathway engineering and synthetic biology approaches.

Acknowledgments

This work was partially funded by grant # FCE 1_2017_1_136291 (ANII, Uruguay). We wish to thank Alberto Marina for discussions and useful suggestions.

References

1. Gao R, Stock AM (2017) Quantitative kinetic analyses of shutting off a two-component system. MBio 8(3):e00412–17
2. Hart Y, Alon U (2013) The utility of paradoxical components in biological circuits. Mol Cell 49(2):213–221
3. Stock AM, Robinson VL, Goudreau PN (2000) Two-component signal transduction. Annu Rev Biochem 69:183–215
4. Parkinson JS, Hazelbauer GL, Falke JJ (2015) Signaling and sensory adaptation in Escherichia coli chemoreceptors: 2015 update. Trends Microbiol 23(5):257–266
5. Dupre E, Lesne E, Guerin J, Lensink MF, Verger A, de Ruyck J, Brysbaert G, Vezin H, Locht C, Antoine R, Jacob-Dubuisson F (2015) Signal transduction by BvgS sensor kinase: binding of modulator nicotinate affects the conformation and dynamics of the entire periplasmic moiety. J Biol Chem 290 (38):23307–23319
6. Neiditch MB, Federle MJ, Pompeani AJ, Kelly RC, Swem DL, Jeffrey PD, Bassler BL, Hughson FM (2006) Ligand-induced asymmetry in histidine sensor kinase complex regulates quorum sensing. Cell 126(6):1095–1108
7. Zschiedrich CP, Keidel V, Szurmant H (2016) Molecular mechanisms of two-component signal transduction. J Mol Biol 428 (19):3752–3775
8. Casino P, Miguel-Romero L, Marina A (2014) Visualizing autophosphorylation in histidine kinases. Nat Commun 5:3258
9. Mechaly AE, Sassoon N, Betton JM, Alzari PM (2014) Segmental helical motions and dynamical asymmetry modulate histidine kinase autophosphorylation. PLoS Biol 12(1): e1001776
10. Pontiggia F, Pachov DV, Clarkson MW, Villali J, Hagan MF, Pande VS, Kern D (2015) Free energy landscape of activation in a signalling protein at atomic resolution. Nat Commun 6:7284
11. Bhate MP, Molnar KS, Goulian M, DeGrado WF (2015) Signal transduction in histidine kinases: insights from new structures. Structure 23(6):981–994
12. Gao R, Stock AM (2010) Molecular strategies for phosphorylation-mediated regulation of response regulator activity. Curr Opin Microbiol 13(2):160–167
13. Galperin MY (2006) Structural classification of bacterial response regulators: diversity of output domains and domain combinations. J Bacteriol 188(12):4169–4182
14. Bergerat A, de Massy B, Gadelle D, Varoutas PC, Nicolas A, Forterre P (1997) An atypical topoisomerase II from Archaea with implications for meiotic recombination. Nature 386 (6623):414–417
15. Cock PJ, Whitworth DE (2007) Evolution of prokaryotic two-component system signalling pathways: gene fusions and fissions. Mol Biol Evol 24(11):2355–2357
16. Wuichet K, Cantwell BJ, Zhulin IB (2010) Evolution and phyletic distribution of

two-component signal transduction systems. Curr Opin Microbiol 13(2):219–225

17. Lou YC, Weng TH, Li YC, Kao YF, Lin WF, Peng HL, Chou SH, Hsiao CD, Chen C (2015) Structure and dynamics of polymyxin-resistance-associated response regulator PmrA in complex with promoter DNA. Nat Commun 6:8838
18. Narayanan A, Kumar S, Evrard AN, Paul LN, Yernool DA (2014) An asymmetric heterodomain interface stabilizes a response regulator-DNA complex. Nat Commun 5:3282
19. Gushchin I, Melnikov I, Polovinkin V, Ishchenko A, Yuzhakova A, Buslaev P, Bourenkov G, Grudinin S, Round E, Balandin T, Borshchevskiy V, Willbold D, Leonard G, Büldt G, Popov A, Gordeliy V (2017) Mechanism of transmembrane signalling by sensor histidine kinases. Science 356 (6342):eaah6345
20. Marina A, Waldburger CD, Hendrickson WA (2005) Structure of the entire cytoplasmic portion of a sensor histidine-kinase protein. EMBO J 24(24):4247–4259
21. Rivera-Cancel G, Ko W-h, Tomchick DR, Correa F, Gardner KH (2014) Full-length structure of a monomeric histidine kinase reveals basis for sensory regulation. Proc Natl Acad Sci 111:17839–17844
22. Albanesi D, Martin M, Trajtenberg F, Mansilla MC, Haouz A, Alzari PM, de Mendoza D, Buschiazzo A (2009) Structural plasticity and catalysis regulation of a thermosensor histidine kinase. Proc Natl Acad Sci U S A 106 (38):16185–16190
23. Stock JB, Ninfa AJ, Stock AM (1989) Protein phosphorylation and regulation of adaptive responses in bacteria. Microbiol Rev 53 (4):450–490
24. Hess JF, Oosawa K, Matsumura P, Simon MI (1987) Protein phosphorylation is involved in bacterial chemotaxis. Proc Natl Acad Sci 84:7609–7613
25. Dutta R, Inouye M (2000) GHKL, an emergent ATPase/kinase superfamily. Trends Biochem Sci 25(1):24–28
26. Grebe TW, Stock JB (1999) The histidine protein kinase superfamily. Adv Microb Physiol 41:139–227
27. Parkinson JS, Kofoid EC (1992) Communication modules in bacterial signalling proteins. Annu Rev Genet 26:71–112
28. Finn RD, Coggill P, Eberhardt RY, Eddy SR, Mistry J, Mitchell AL, Potter SC, Punta M, Qureshi M, Sangrador-Vegas A, Salazar GA, Tate J, Bateman A (2016) The Pfam protein families database: towards a more sustainable future. Nucleic Acids Res 44(Database issue): D279–D285
29. Galperin MY, Nikolskaya AN (2007) Identification of sensory and signal-transducing domains in two-component signalling systems. Methods Enzymol 422:47–74
30. Ulrich LE, Zhulin IB (2010) The MiST2 database: a comprehensive genomics resource on microbial signal transduction. Nucleic Acids Res 38(Database issue):D401–D407
31. Srivastava SK, Rajasree K, Fasim A, Arakere G, Gopal B (2014) Influence of the AgrC-AgrA complex on the response time of Staphylococcus aureus quorum sensing. J Bacteriol 196 (15):2876–2888
32. Herrou J, Crosson S, Fiebig A (2017) Structure and function of HWE/HisKA2-family sensor histidine kinases. Curr Opin Microbiol 36:47–54
33. Anantharaman V, Balaji S, Aravind L (2006) The signalling helix: a common functional theme in diverse signalling proteins. Biol Direct 1:25
34. Singh M, Berger B, Kim PS, Berger JM, Cochran AG (1998) Computational learning reveals coiled coil-like motifs in histidine kinase linker domains. Proc Natl Acad Sci U S A 95 (6):2738–2743
35. Trajtenberg F, Imelio JA, Machado MR, Larrieux N, Marti MA, Obal G, Mechaly AE, Buschiazzo A (2016) Regulation of signalling directionality revealed by 3D snapshots of a kinase:regulator complex in action. eLife 5: e21422
36. Saita E, Abriata LA, Tsai YT, Trajtenberg F, Lemmin T, Buschiazzo A, Dal Peraro M, de Mendoza D, Albanesi D (2015) A coiled coil switch mediates cold sensing by the thermosensory protein DesK. Mol Microbiol 98 (2):258–271
37. Purcell EB, McDonald CA, Palfey BA, Crosson S (2010) An analysis of the solution structure and signalling mechanism of LovK, a sensor histidine kinase integrating light and redox signals. Biochemistry 49(31):6761–6770
38. Monzel C, Unden G (2015) Transmembrane signalling in the sensor kinase DcuS of Escherichia coli: a long-range piston-type displacement of transmembrane helix 2. Proc Natl Acad Sci U S A 112(35):11042–11047
39. Yusuf R, Nguyen TL, Heininger A, Lawrence RJ, Hall BA, Draheim RR (2018) In vivo crosslinking and transmembrane helix dynamics support a bidirectional non-piston model of signalling within *E. coli* EnvZ. bioRxiv. https://doi.org/10.1101/206888
40. Wang LC, Morgan LK, Godakumbura P, Kenney LJ, Anand GS (2012) The inner membrane

histidine kinase EnvZ senses osmolality via helix-coil transitions in the cytoplasm. EMBO J 31(11):2648–2659

41. Wang X, Vallurupalli P, Vu A, Lee K, Sun S, Bai WJ, Wu C, Zhou H, Shea JE, Kay LE, Dahlquist FW (2014) The linker between the dimerization and catalytic domains of the CheA histidine kinase propagates changes in structure and dynamics that are important for enzymatic activity. Biochemistry 53 (5):855–861
42. Bhatnagar J, Borbat PP, Pollard AM, Bilwes AM, Freed JH, Crane BR (2010) Structure of the ternary complex formed by a chemotaxis receptor signalling domain, the CheA histidine kinase, and the coupling protein CheW as determined by pulsed dipolar ESR spectroscopy. Biochemistry 49(18):3824–3841
43. Gushchin I, Gordeliy V (2018) Transmembrane signal transduction in two-component systems: piston, scissoring, or helical rotation? BioEssays 40(2):1700197. https://doi.org/10.1002/bies.201700197
44. Moglich A, Ayers RA, Moffat K (2009) Design and signalling mechanism of light-regulated histidine kinases. J Mol Biol 385 (5):1433–1444
45. Huynh TN, Noriega CE, Stewart V (2013) Missense substitutions reflecting regulatory control of transmitter phosphatase activity in two-component signalling. Mol Microbiol 88 (3):459–472
46. Bhate MP, Lemmin T, Kuenze G, Mensa B, Ganguly S, Peters J, Schmidt N, Pelton JG, Gross C, Meiler J, DeGrado WF (2018) Structure and function of the transmembrane domain of NsaS, an antibiotic sensing histidine kinase in S. aureus. J Am Chem Soc 140 (24):7471–7485
47. Casino P, Rubio V, Marina A (2009) Structural insight into partner specificity and phosphoryl transfer in two-component signal transduction. Cell 139(2):325–336
48. Volkman BF, Lipson D, Wemmer DE, Kern D (2001) Two-state allosteric behavior in a single-domain signalling protein. Science 291 (5512):2429–2433
49. Minato Y, Ueda T, Machiyama A, Iwai H, Shimada I (2017) Dynamic domain arrangement of CheA-CheY complex regulates bacterial thermotaxis, as revealed by NMR. Sci Rep 7 (1):16462
50. Jiang P, Peliska JA, Ninfa AJ (2000) Asymmetry in the autophosphorylation of the two-component regulatory system transmitter protein nitrogen regulator II of Escherichia coli. Biochemistry 39(17):5057–5065
51. Trajtenberg F, Grana M, Ruetalo N, Botti H, Buschiazzo A (2010) Structural and enzymatic insights into the ATP binding and autophosphorylation mechanism of a sensor histidine kinase. J Biol Chem 285(32):24892–24903
52. Bauer J, Reiss K, Veerabagu M, Heunemann M, Harter K, Stehle T (2013) Structure-function analysis of Arabidopsis thaliana histidine kinase AHK5 bound to its cognate phosphotransfer protein AHP1. Mol Plant 6(3):959–970
53. Bell CH, Porter SL, Strawson A, Stuart DI, Armitage JP (2010) Using structural information to change the phosphotransfer specificity of a two-component chemotaxis signalling complex. PLoS Biol 8(2):e1000306
54. Mo G, Zhou H, Kawamura T, Dahlquist FW (2012) Solution structure of a complex of the histidine autokinase CheA with its substrate CheY. Biochemistry 51(18):3786–3798
55. Varughese KI, Tsigelny I, Zhao H (2006) The crystal structure of beryllofluoride Spo0F in complex with the phosphotransferase Spo0B represents a phosphotransfer pretransition state. J Bacteriol 188(13):4970–4977
56. Willett JW, Herrou J, Briegel A, Rotskoff G, Crosson S (2015) Structural asymmetry in a conserved signalling system that regulates division, replication, and virulence of an intracellular pathogen. Proc Natl Acad Sci U S A 112 (28):E3709–E3718
57. Zhao X, Copeland DM, Soares AS, West AH (2008) Crystal structure of a complex between the phosphorelay protein YPD1 and the response regulator domain of SLN1 bound to a phosphoryl analog. J Mol Biol 375(4):1141–1151
58. Yamada S, Sugimoto H, Kobayashi M, Ohno A, Nakamura H, Shiro Y (2009) Structure of PAS-linked histidine kinase and the response regulator complex. Structure 17 (10):1333–1344
59. Capra EJ, Perchuk BS, Skerker JM, Laub MT (2012) Adaptive mutations that prevent crosstalk enable the expansion of paralogous signalling protein families. Cell 150(1):222–232
60. Villanueva M, Garcia B, Valle J, Rapun B, Ruiz de Los Mozos I, Solano C, Marti M, Penades JR, Toledo-Arana A, Lasa I (2018) Sensory deprivation in Staphylococcus aureus. Nat Commun 9(1):523
61. Willett JW, Tiwari N, Muller S, Hummels KR, Houtman JC, Fuentes EJ, Kirby JR (2013) Specificity residues determine binding affinity for two-component signal transduction systems. MBio 4(6):e00420–13
62. Skerker JM, Prasol MS, Perchuk BS, Biondi EG, Laub MT (2005) Two-component signal

transduction pathways regulating growth and cell cycle progression in a bacterium: a system-level analysis. PLoS Biol 3(10):e334

63. Podgornaia AI, Casino P, Marina A, Laub MT (2013) Structural basis of a rationally rewired protein-protein interface critical to bacterial signalling. Structure 21(9):1636–1647
64. Skerker JM, Perchuk BS, Siryaporn A, Lubin EA, Ashenberg O, Goulian M, Laub MT (2008) Rewiring the specificity of two-component signal transduction systems. Cell 133(6):1043–1054
65. Podgornaia AI, Laub MT (2015) Protein evolution. Pervasive degeneracy and epistasis in a protein-protein interface. Science 347 (6222):673–677
66. Imelio JA, Larrieux N, Mechaly AE, Trajtenberg F, Buschiazzo A (2017) Snapshots of the signalling complex DesK:DesR in different functional states using rational mutagenesis and X-ray crystallography. Bio-Protocol 7(16): e2510
67. Huynh TN, Stewart V (2011) Negative control in two-component signal transduction by transmitter phosphatase activity. Mol Microbiol 82(2):275–286
68. Pazy Y, Motaleb MA, Guarnieri MT, Charon NW, Zhao R, Silversmith RE (2010) Identical phosphatase mechanisms achieved through distinct modes of binding phosphoprotein substrate. Proc Natl Acad Sci U S A 107(5):1924–1929
69. Huynh TN, Noriega CE, Stewart V (2010) Conserved mechanism for sensor phosphatase control of two-component signalling revealed in the nitrate sensor NarX. Proc Natl Acad Sci U S A 107(49):21140–21145
70. Kamberov ES, Atkinson MR, Chandran P, Ninfa AJ (1994) Effect of mutations in Escherichia coli glnL (ntrB), encoding nitrogen regulator II (NRII or NtrB), on the phosphatase activity involved in bacterial nitrogen regulation. J Biol Chem 269(45):28294–28299
71. Dutta R, Inouye M (1996) Reverse phosphotransfer from OmpR to EnvZ in a kinase−/phosphatase+ mutant of EnvZ (EnvZ. N347D), a bifunctional signal transducer of Escherichia coli. J Biol Chem 271 (3):1424–1429
72. Hsing W, Silhavy TJ (1997) Function of conserved histidine-243 in phosphatase activity of EnvZ, the sensor for porin osmoregulation in Escherichia coli. J Bacteriol 179 (11):3729–3735
73. Tindall MJ, Porter SL, Maini PK, Armitage JP (2010) Modeling chemotaxis reveals the role of reversed phosphotransfer and a bi-functional kinase-phosphatase. PLoS Comput Biol 6(8): e1000896
74. Pena-Sandoval GR, Kwon O, Georgellis D (2005) Requirement of the receiver and phosphotransfer domains of ArcB for efficient dephosphorylation of phosphorylated ArcA in vivo. J Bacteriol 187(9):3267–3272
75. Uhl MA, Miller JF (1996) Central role of the BvgS receiver as a phosphorylated intermediate in a complex two-component phosphorelay. J Biol Chem 271(52):33176–33180
76. Kenney LJ (2010) How important is the phosphatase activity of sensor kinases? Curr Opin Microbiol 13(2):168–176
77. Batchelor E, Goulian M (2003) Robustness and the cycle of phosphorylation and dephosphorylation in a two-component regulatory system. Proc Natl Acad Sci U S A 100 (2):691–696
78. Siryaporn A, Goulian M (2008) Cross-talk suppression between the CpxA-CpxR and EnvZ-OmpR two-component systems in *E. coli*. Mol Microbiol 70(2):494–506
79. Mechaly AE, Soto Diaz S, Sassoon N, Buschiazzo A, Betton JM, Alzari PM (2017) Structural coupling between autokinase and phosphotransferase reactions in a bacterial histidine kinase. Structure 25(6):939–944

Chapter 2

A High-Throughput Strategy for Recombinant Protein Expression and Solubility Screen in *Escherichia coli*: A Case of Sensor Histidine Kinase

Agnieszka Szmitkowska, Blanka Pekárová, and Jan Hejátko

Abstract

Determining conditions optimal for host growth, maximal protein yield, and lysis buffer composition is of critical importance for the efficient purification of soluble and well-folded recombinant proteins suitable for functional and/or structural studies. Small-scale optimization of conditions for protein production and stability saves time, labor, and costs. Here we describe a protocol for quick protein production and solubility screen using TissueLyser II system from Qiagen enabling simultaneous processing of 96 protein samples, with application to recombinant proteins encompassing two intracellular domains of ethylene-recognizing sensor histidine kinase ETHYLENE RESPONSE1 (ETR1) from *Arabidopsis thaliana*. We demonstrate that conditions for expression and cell lysis found in our small-scale screen allow successful large-scale production of pure and functional domains of sensor histidine kinase, providing a strategy potentially transferable to other similar catalytic domains.

Key words Protein, Expression screen, Solubility screen, Growth conditions, Histidine kinase, Sample preparation

1 Introduction

Protein production, purification, and identification of protein stabilizing conditions is an indispensable step in generating proteins suitable for in vitro structural and functional studies. Creating the expression clone using a proper host strain and selecting suitable expression and purification conditions varies from protein to protein [1]. Even though protein properties can be predicted based on sequence [2], the most efficient, and in many cases the only, strategy for optimization of conditions for protein expression and purification is still trial and error. Determining optimal conditions for both high yield and soluble protein production on a small-scale facilitates the process and provides the necessary basis for large-scale protein production. Large crystallography laboratories often develop protocols for high-throughput protein expression and

Claire E. Eyers (ed.), *Histidine Phosphorylation: Methods and Protocols*, Methods in Molecular Biology, vol. 2077, https://doi.org/10.1007/978-1-4939-9884-5_2,

solubility screening in 96-well plates, relying on high levels of automation [3–6], vector suites [7, 8], and sophisticated detection systems [9]. However, this kind of "heavy artillery" approach is not always possible, particularly in smaller laboratories with limited resources. Thus, there is a strong need for simple, small-scale, high-throughput screening tests, which do not require expensive instrumentation, but still save time, effort, and costs. To achieve this, we have designed a simple protocol employing a TissueLyser II system from Qiagen, routinely used for DNA extraction from plant tissues. The TissueLyser II system is a bead mill that effectively disrupts up to 48 or 192 samples (depending on the adapter used) in parallel without danger of cross-contamination. The efficient extraction of intracellular proteins with reproducible results allows for simultaneous optimization of several soluble protein expression parameters.

Having optimized the protocol for recombinant proteins expressed in *Escherichia coli*, we use it routinely as part of our protein production workflow. In the screening protocol presented in this chapter, we detail how four parameters can be evaluated for optimal production of soluble proteins: the bacterial expression strain; pH of the culture media; temperature and duration of induction; composition and pH of lysis buffer. Other variables including the use of different expression vectors, expression hosts, or inducer concentration can be tested in a similar manner. Expression and prepurification conditions identified in this type of small-scale screen are easily transferable to large-scale preparatory experiments, as we demonstrate with two recombinant proteins comprising the intracellular domains of ethylene receptor ETHYLENE RESPONSE1 (ETR1) from *Arabidopsis thaliana*. The purification protocol and histidine kinase autophosphorylation assay confirming their functionality is briefly described.

2 Materials

Prepare all the media using distilled water. For buffers and solutions, use Milli-Q ultrapure water and analytical grade reagents.

2.1 Cultivation Media and Components for Bacterial Growth

1. Liquid Luria–Bertani (LB) medium: Weigh 25 g of LB-medium salts (10 g Bacto tryptone, 10 g NaCl, 5 g yeast extract) and add water to a volume of 1 L. Mix to dissolve and sterilize by autoclaving at 121 °C for 20 min.
2. Terrific Broth (TB) medium: Weigh 12 g tryptone, 24 g yeast extract, and transfer to a 1 L graduated cylinder. Add 4 mL of glycerol and water to a volume of 875 mL (*see* **Note 1**). Sterilize by autoclaving at 121 °C for 20 min. Immediately before use add 100 mL of sterile 1 M K-phosphate buffer to

Table 1
K-phosphate buffer preparation table

pH	Volume of 1 M K_2HPO_4 (mL)	Volume of 1 M KH_2PO_4 (mL)
5.8	8.5	91.5
6.0	13.2	86.8
6.2	19.2	80.8
6.4	27.8	72.2
6.6	38.1	61.9
6.8	49.7	50.3
7.0	61.5	38.5
7.2	71.7	28.3
7.4	80.2	19.8
7.6	86.6	13.4
7.8	90.8	9.2
8.0	94.0	6.0

Adapted from [10]

adjust the pH, selection antibiotics, and 25 mL of sterile 40% glucose (final 1%) to control basal expression.

3. K-phosphate buffer: Mix 1 M K_2HPO_4 and 1 M KH_2PO_4 salt solutions according to Table 1 to obtain the required pH (we usually use pH values of 6.0, 7.0, and 8.0) [10]. Sterilize by autoclaving at 121 °C for 20 min (*see* **Note 2**).
4. 1 M solution of isopropyl β-D-1-thiogalactopyranoside (IPTG): Dissolve 2.38 g of IPTG in 10 mL of water. Sterilize via filtration with a 0.22 μm pore diameter syringe filter in a laminar-flow chamber. Store in sterile tubes as 1 mL aliquots at −20 °C.
5. 50 mL sterile, glass conical flasks.
6. Sterile bacterial tubes with caps.
7. Incubation shaker with adjustable temperature control.
8. Glycerol stocks of expression strains of *E. coli* carrying a vector with the gene of interest (*see* **Note 3**). Here we used cDNA of *Arabidopsis thaliana* ETR1 histidine kinase [intracellular portion including histidine kinase (HK) domain only or the HK domain together with C-terminal receiver domain (RD)] fused with N-terminal ubiquitin and hexahistidine (His6) tags in pETM-60 vector. Store expression strains in 10% glycerol at −80 °C.

9. 1000× concentrated stocks of the required antibiotics, defined by the resistance markers of the *E. coli* strains and the vector carrying the gene of interest. Calculate final concentrations of antibiotics in the growth media according to the protocols of *E. coli* strains providers. Store the aliquot stocks at −20 °C.
10. Laminar-flow chamber (*see* **Note 4**).
11. UV-Vis spectrophotometer and cuvettes for optical density (OD) measurement.
12. 1.5 mL microcentrifuge tubes.
13. Multichannel and automatic pipettes.

2.2 Protein Extraction

Prepare a sufficient volume of lysis buffer (250 μL per condition, *see* also Subheading 3.2) to permit testing of the required number of conditions.

1. MES lysis buffer: 50 mM MES, 0.5 M NaCl, adjusted to pH 6.0 with 1 M NaOH. Immediately prior to cell lysis, add (a) dithiothreitol (DTT) to a final concentration of 10 mM, (b) 0.1% (v/v) Triton X-100, and (c) 200 μg/mL of lysozyme (*see* **Notes 5** and **6**).
2. Tris lysis buffer: 50 mM Tris, 0.5 M NaCl, adjusted to pH 7.5 with 6 M HCl. Immediately prior to cell lysis, add (a) DTT to a final concentration 10 mM, (b) 0.1% (v/v) Triton X-100, and (c) 200 μg/mL of lysozyme (*see* **Note 7**).
3. Glycine lysis buffer: 50 mM glycine, 0.5 M NaCl, adjusted to pH 10.0 with 1 M NaOH. Immediately prior to cell lysis, add (a) DTT to a final concentration 10 mM, (b) 0.1% (v/v) Triton X-100, and (c) 200 μg/mL of lysozyme.
4. Bead mill homogenizer: TissueLyser II from Qiagen with TissueLyser Adapter Set (2 × 96) containing collection microtubes (racked), sealed with collection microtube caps.
5. Glass beads, 3 mm in diameter (we use two per sample).
6. Centrifuge with cooling system and holders for TissueLyser adapters, for instance Eppendorf 5810/5810 R.
7. Water or vacuum aspirator (*see* **Note 8**).
8. Vortex.
9. Multichannel and automatic pipettes.

2.3 SDS-PAGE

1. 4× SDS reducing loading sample buffer: Mix 3.5 mL of glycerol (*see* **Note 9**), 2.5 mL of 1 M Tris–HCl (pH 6.8), 0.8 g SDS, 4 mg bromophenol blue, 0.62 g DTT, and 4 mL of water. Prepare 1 mL aliquots in 1.5 mL microcentrifuge tubes and store at −20 °C.

2. Molecular weight marker suitable for the size of the protein of interest (*see* **Note 10**).
3. SDS-PAGE running buffer: 25 mM Tris–HCl, 192 mM glycine, 0.1% SDS, pH approx. 8.3 (do not adjust). To prepare a 1 L stock of 10× concentrated buffer, mix 144.2 g glycine, 30.3 g Tris, 10 g SDS (*see* **Note 11**), fill to 1 L with water and store at room temperature. Dilute to 1× concentrated buffer with distilled water before use.
   (a) Resolving gel buffer: 1.5 M Tris–HCl adjusted to pH 8.8 with 6 M HCl. Store at 4 °C.
   (b) Stacking gel buffer: 0.5 M Tris–HCl adjusted to pH 6.8 with 6 M HCl. Store at 4 °C.
   (c) Acrylamide–Bis 30%. Store at 4 °C.
   (d) 10% SDS (*see* **Note 12**).
   (e) 10% Ammonium persulfate (APS). Store aliquots of 1 mL at −20 °C in 1.5 mL microcentrifuge tubes and thaw before use.
   (f) *N*,*N*,*N*′,*N*′-Tetramethylethane-1, 2-diamine (TEMED). Store at 4 °C.
4. 96-multiwell PCR plate with reusable caps.
5. Centrifuge with holders for 96-well plates.
6. Thermocycler suitable for 96-well plates.
7. Coomassie Brilliant Blue Dye R250 or G250 (*see* **Note 13**).
8. Mini-PROTEAN 3 cell system (Bio-Rad).
9. Plastic boxes for gel staining.
10. Rocking platform.
11. 1.5 mL microcentrifuge tubes.
12. Multichannel pipettes.
13. Fume hood.

2.4 Western Blotting

Store all solutions at room temperature, unless stated otherwise.

1. Methanol 100%.
2. Towbin's transfer buffer: 25 mM Tris–HCl, 150 mM glycine, 10% methanol, pH approx. 8.3 (do not adjust). Mix 3.03 g Tris–HCl and 11.72 g glycine and fill up to 900 mL with water. Add 100 mL of methanol to obtain 1 L. Store at 4 °C.
3. Ponceau-S red 0.5% (w/v) in 1% acetic acid solution. Store in the dark.
4. Tris-buffered saline (TBS; 20×): weigh 60.55 g Tris–HCl and 87.66 g NaCl. Make up to 450 mL with water, adjust to pH 8.0 with 6 M HCl and fill up to final 500 mL.

5. TBS containing 0.1% Tween 20 (TBST): 50 mM Tris–HCl, 150 mM NaCl, 0.1% Tween 20. Add 50 mL of 20× TBS to graduate 1 L cylinder, add 1 mL of Tween 20 (*see* **Note 9**) and make up to 1 L with water.
6. Blocking buffer: 5% (w/w) nonfat powdered milk in TBST (*see* **Note 14**).
7. Detection buffer: 100 mM Tris–HCl, 100 mM NaCl, 5 mM $MgCl_2$, adjust to pH 9.5 with 6 M HCl. 200 mL contains 2.42 g Tris–HCl, 0.2 g $MgCl_2$, and 1.17 g NaCl.
8. 4-Nitro-Blue Tetrazolium (NBT) Chloride solution: 100 mg/mL in 70% *N,N*-dimethylformamide (DMF). Prepare 1 mL aliquots and store at −20 °C in the dark.
9. 5-Bromo-4-chloro-3′-indolyl phosphate p-toluidine (BCIP) Salt solution: 50 mg/mL in 100% DMF. Prepare 1 mL aliquots and store at −20 °C in the dark.
10. Chromogenic substrate for alkaline phosphatase: Mix 15 mL of fresh detection buffer with 75 μL of NBT solution and 50 μL of BCIP for detection of proteins. Keep in the dark (*see* **Note 15**).
11. Antibodies specifically recognizing protein of interest or fusion tag. Here we use mouse monoclonal anti-polyhistidine (Sigma) primary antibodies diluted 1:10,000 in blocking buffer, and secondary anti-mouse IgG-alkaline phosphatase antibodies (Sigma) diluted 1: 20,000 in blocking buffer.
12. Polyvinylidene difluoride membrane, pore size 0.45 μm.
13. Whatman filter papers cut to the size of membrane (four per blot).
14. Mini-PROTEAN Tetra Cell and Tank Blotting System with cooling unit (Bio-Rad).
15. Scissors, pencil, and tweezers (*see* **Note 16**).
16. Roller for removing air bubbles.
17. Plastic boxes for membrane washing.
18. 1.5 mL microcentrifuge tubes.
19. 50 mL disposable tubes.

2.5 Protein Purification

The purification protocol is based on GE Healthcare's ÄKTA FPLC System official manual [11]. Filter all buffers through 0.2 μm nylon membrane filter before use. Degas all solutions in a sonication bath before use in FPLC system.

1. Protein lysate from large scale expression.
2. Sonicator or other cell disrupting equipment available in the laboratory.

3. High-speed centrifuge with fixed-angle rotor for 50 mL tubes (e.g., Beckman Coulter from Avanti series).
4. 50 mL high-speed centrifuge tubes.
5. GE Healthcare's ÄKTA FPLC System or similar.
6. Chromatography column corresponding to the amount of bacterial lysate, concentration of the protein and type of protein fusion tag used. We use a 1 mL HisTrap HP nickel-charged IMAC column.
7. Equilibration buffer optimal for the protein of interest. Choose the buffer according to purification method and results of solubility screening. We use 50 mM Tris–HCl, 0.5 M NaCl, 10 mM imidazole, 10% glycerol, 2.5 mM 2-mercaptoethanol, adjusted to pH 7.9 with 6 M HCl.
8. Elution buffer optimal for the protein of interest. Choose the buffer according to purification method and results of solubility screening. We use to the equilibration buffer supplemented with 500 mM imidazole.
9. Milli-Q water for ÄKTA system washing.
10. 20% ethanol.
11. Sonication bath.
12. Fraction collection box.
13. Syringe filter of pore diameter 0.22 μm.
14. 50 mL disposable tubes.
15. SDS-PAGE materials described in Subheading 2.3.
16. Protein concentrator spin columns suitable for protein size.
17. Centrifuge with cooling system suitable for protein concentrator spin columns.

2.6 Auto-phosphorylation Assay

Perform the assay with recombinant protein fused with the purification tag only when the tag does not influence protein function (must be tested beforehand). Otherwise, remove the tag with suitable protease and perform secondary purification step prior to the assay.

1. Histidine kinase protein (*see* **Note 17**).
2. 2 M $MgCl_2$: Dilute 0.4066 g of $MgCl_2$.6 H_2O in 1 mL of water and store at 4 °C.
3. 2 M $MnCl_2$: Dilute 0.395 g of $MnCl_2$.4 H_2O in 1 mL of water and store at 4 °C.
4. 1 M DTT: Dilute 1.542 g of DTT in 10 mL of water and prepare 0.5 mL aliquots in 1.5 mL microcentrifuge tubes. Store at −20 °C (*see* **Note 18**).

5. Kinase buffer stock: 50 mM Tris–HCl, 10% glycerol, adjust to pH 7.5 with 6 M HCl. Before assay, add the divalent cations of interest (*see* **items 2** and **3**) to a final concentration of 5 mM, and DTT to final concentration of 2 mM.
6. Stop buffer: same as 4× SDS reducing loading sample buffer (Subheading 2.3, **item 1**), 80 mM EDTA. Add 160 μL of 0.5 M EDTA to 840 μL of the 4× SDS reducing loading sample buffer.
7. 1 L of 10% citric acid for decontamination.
8. γ-^{32}ATP isotope.
9. SDS-PAGE materials described in Subheading 2.3 (*see* **Note 19**).
10. Imaging plate and cassette suitable for Typhoon FLA laser scanner.
11. Typhoon FLA laser scanner.
12. Wrapping foil for gel wrapping.
13. PCR-tube strips and caps.
14. 1.5 mL microcentrifuge tubes.
15. Filter tips.
16. Plexiglass shields.
17. Geiger counter.

3 Methods

3.1 Expression and Protein Solubility Test

Always manipulate growing cells under sterile conditions (*see* **Note 4**). After expression is finished, work cleanly but a sterile environment is no longer required. The optimization screen is demonstrated using a single bacterial strain carrying the gene of interest, grown in media with three different pH values, four temperatures and three lysis buffers (*see* **Note 20**). Use multichannel and automatic pipettes to speed up the pipetting process.

1. In a sterile bacterial tube inoculate 3 mL of LB medium containing the appropriate selection antibiotics with the glycerol stock of the expression *E. coli* strain carrying gene of interest to be tested. Incubate overnight at 37 °C, continuous shaking at 200 rpm. For graphical representation of the experimental setup, *see* Fig. 1.
2. Sterilize three 50 mL glass flasks each containing 20 mL of TB medium (one for pH 6.0, 7.0, and 8.0, respectively) with the appropriate selection antibiotics. Inoculate each flask with 800 μL of overnight culture from **step 1**.

Overnight culture in LB medium, 3 mL

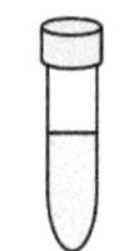

Inoculate 800 μL into each flask containing 20 mL of TB medium.

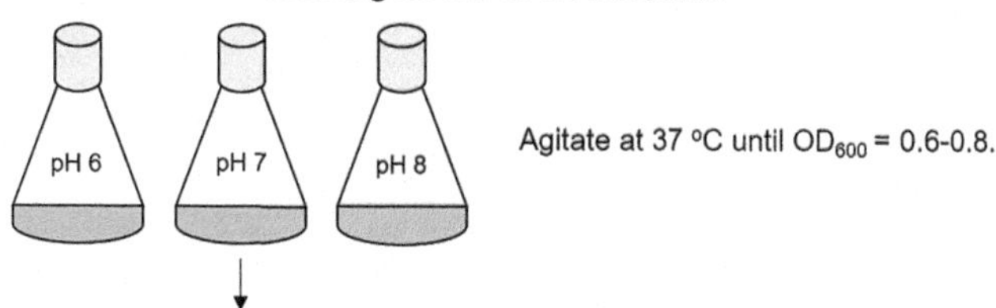

Sample 1 mL of non-induced culture from each flask,
spin down, remove the supernatant and store the pellet at -20 °C.
Induce protein expression with IPTG (20 μL of 1 M IPTG).

Split the content of each flask into 4 bacterial tubes (4 mL per tube) and grow them at different temperatures/times.

pH 6 7 8 — 18 °C overnight

pH 6 7 8 — 22 °C overnight

pH 6 7 8 — 28 °C overnight

pH 6 7 8 — 37 °C 3 h

Sample 1 mL from each tube and place separately in TissueLyser adapter according to the scheme in Table 2.
Centrifuge at 3220 x *g* at 4 °C for 2 minutes, remove supernatant with aspirator.

Fig. 1 Experimental setup of protein expression and solubility trials. The overnight bacterial culture from 3 mL of LB medium is used to inoculate three Erlenmeyer flasks each containing 20 mL of TB medium at three different pH values (6.0, 7.0, and 8.0). The cultures are grown at 37 °C shaking at 200 rpm until OD_{600} is 0.6–0.8. After sampling 1 mL of noninduced cells from each flask, protein expression is induced by addition of IPTG. From each flask, 4 mL is transferred into each of 4 bacterial tubes and grown under different temperature conditions (18 °C, 22 °C, and 28 °C overnight and 37 °C for 3 h). From each tube, three samples of 1 mL are used for solubility test applying different lysis buffers

3. Grow the cells at 37 °C, continuous shaking at 200 rpm until the optical density OD_{600} of the culture has reached 0.6–0.8. Measure the optical density with UV-Vis spectrophotometer.
4. Prepare nonsterile TissueLyser adapter by adding two glass beads in each adapter tube and describing each tube with numbers and colors. Example of sample numbering in the adapter is in Table 2 (*see* **Note 21**).
5. Sample 1 mL of cells before induction (non-induced control) from each flask and place them into TissueLyser adapter. Centrifuge at 3220 × *g* (4,000 rpm on rotor A-S-81 for Eppendorf 5810/5810 R centrifuge) for 2 min. Remove the supernatant by aspiration and store the pellets at −20 °C.

Table 2
An example of sample distribution and numbering in TissueLyser 96-well adapter

Temperature and growth time		TB medium pH 6.0			TB medium pH 7.0			TB medium pH 8.0			Non-induced samples			
		1	2	3	4	5	6	7	8	9	10	11	12	Cell type
18 °C o/n	A	**1**	*2*	3	**4**	*5*	6	**7**	*8*	9	*pH 6*			Strain 1
22 °C o/n	B	**10**	*11*	12	**13**	*14*	15	**16**	*17*	18	*pH 7*			
28 °C o/n	C	**19**	*20*	21	**22**	*23*	24	**25**	*26*	27	*pH 8*			
37 °C 3 h	D	**28**	*29*	30	**31**	*32*	33	**34**	*35*	36				
18 °C o/n	E	**1**	*2*	3	**4**	*5*	6	**7**	*8*	9	*pH 6*			Strain 2
22 °C o/n	F	**10**	*11*	12	**13**	*14*	15	**16**	*17*	18	*pH 7*			
28 °C o/n	G	**19**	*20*	21	**22**	*23*	24	**25**	*26*	27	*pH 8*			
37 °C 3 h	H	**28**	*29*	30	**31**	*32*	33	**34**	*35*	36				

Each condition is designed with a separate number. Samples in MES lysis buffer are in bold, samples in Tris–HCl lysis buffer are in italics, samples in glycine buffer are underlined. Noninduced cells were disrupted in Tris–HCl buffer

6. Induce each culture by addition of 20 μL of 1 M IPTG. Divide the cultures into four bacterial tubes, 4 mL per tube. Incubate each tube at a different temperature: 18 °C, 22 °C, 28 °C overnight, 37 °C for 3 h.
7. After incubation, sample 3 × 1 mL from each tube and pipette each aliquot into a corresponding tube in TissueLyser adapter (*see* Table 2).
8. Spin the adapter at 3220 × *g* for 2 min and remove the supernatant by aspiration.

3.2 Protein Extraction

To prevent protein degradation, always manipulate the protein samples on ice. Prechill the lysis buffers. Before choosing the pH range of lysis buffers to be tested, check the pI of the protein of interest to reduce unnecessary sample precipitation. Use multichannel and automatic pipettes to speed up the pipetting process.

1. Add 250 μL of lysis buffer to the tubes containing the bacterial pellets as depicted in Table 2. Add 250 μL of Tris–HCl lysis buffer to the tubes with pellets of noninduced cells and seal the tubes with caps. Vortex the adapter to resuspend the cells and keep on ice or at 4 °C in the fridge for 20 min. Optionally, add freeze/thaw step to facilitate the lysis.
2. Put the adapters into holders in TissueLyser (*see* **Note 22**) and shake 2–3 times for 10 s. Take 10 μL of total protein from each sample lysed with Tris–HCl buffer and transfer it into the 96-multiwell PCR plate in the positions outlined in Table 3.
3. Spin the adapters at 3220 × *g* for 10 min. Sample 10 μL of supernatant from each position and transfer it into the PCR multiwell plate in the manner described in Table 3.

Table 3
An example of sample distribution and numbering in PCR 96-well plates

Temperature and growth time after induction		TB medium pH 6.0			TB medium pH 7.0			TB medium pH 8.0			Total protein sample			Cell type
		1	2	3	4	5	6	7	8	9	10	11	12	
18 °C o/n	A	**1**	*2*	3	**4**	*5*	6	**7**	*8*	9	*2*	*5*	*8*	Strain 1
22 °C o/n	B	**10**	*11*	12	**13**	*14*	15	**16**	*17*	18	*11*	*14*	*17*	
28 °C o/n	C	**19**	*20*	21	**22**	*23*	24	**25**	*26*	27	*20*	*23*	*26*	
37 °C 3 h	D	**28**	*29*	30	**31**	*32*	33	**34**	*35*	36	*29*	*32*	*35*	
18 °C o/n	E	**1**	*2*	3	**4**	*5*	6	**7**	*8*	9	*2*	*5*	*8*	Strain 2
22 °C o/n	F	**10**	*11*	12	**13**	*14*	15	**16**	*17*	18	*11*	*14*	*17*	
28 °C o/n	G	**19**	*20*	21	**22**	*23*	24	**25**	*26*	27	*20*	*23*	*26*	
37 °C 3 h	H	**28**	*29*	30	**31**	*32*	33	**34**	*35*	36	*29*	*32*	*35*	

Samples in MES lysis buffer are in bold, samples in Tris–HCl lysis buffer are in italics, samples in glycine buffer are underlined. Samples obtained by lysing the cells in Tris–HCl buffer and taken before removing the cell debris via centrifugation are designated as "Total protein sample"

4. Add 20 μL of water and 10 μL of 4× SDS reducing loading sample buffer to each sample in the PCR multiwell plate. Boil the samples in the thermocycler at 95 °C for 5 min and spin for 30 s at room temperature using holders for 96-well plates.
5. Prepare samples of noninduced cells in the same way in PCR strips or spare rows of another TissueLyser adapter.

3.3 SDS-PAGE and Primary Protein Detection

The described SDS-PAGE procedure is based on the Bio-Rad Protean Tetra system manual [12, 13].

1. Choose the optimal resolving gel composition according to the size of the protein of interest:
   (a) 15% of acrylamide monomer: 12–43 kDa.
   (b) 10% of acrylamide monomer: 16–68 kDa.
   (c) 7% of acrylamide monomer: 36–94 kDa.
2. Prepare stacking and resolving gel mixtures in glass beakers or small bottles in the fume hood according to the Table 4. Assemble the SDS-PAGE gels according to the manufacturer (Bio-Rad in our case) instructions [13] (*see* **Note 23**).
3. Load protein samples (18 μL each) on 15-well SDS-PAGE gel. Run the gel at 20–25 mA per gel (1–2 h, depending on the gel density).
4. After electrophoresis, stain the gel with Coomassie Brilliant Blue dye in the plastic boxes according to the manufacturer's instructions.

Table 4
Preparation of stacking and running gels for SDS-PAGE, adapted from [13]

Chemicals	Resolving gel (per one gel)				Chemicals	Stacking gel (per one gel)
Monomer percentage	15%	12.5%	10%	7%	Monomer percentage	5%
H_2O	1.20 mL	1.60 mL	2.00 mL	2.55 mL	H_2O	800 μL
Resolving buffer	1.25 mL	1.25 mL	1.25 mL	1.25 mL	Stacking buffer	400 μL
30% Acrylamide–Bis	2.50 mL	2.10 mL	1.70 mL	1.15 mL	30% Acrylamide–Bis	250 μL
10% SDS	50 μL	50 μL	50 μL	50 μL	10% SDS	16 μL
10% APS	50 μL	50 μL	50 μL	50 μL	10% APS	25 μL
TEMED	3 μL	3 μL	3 μL	3 μL	TEMED	1 μL

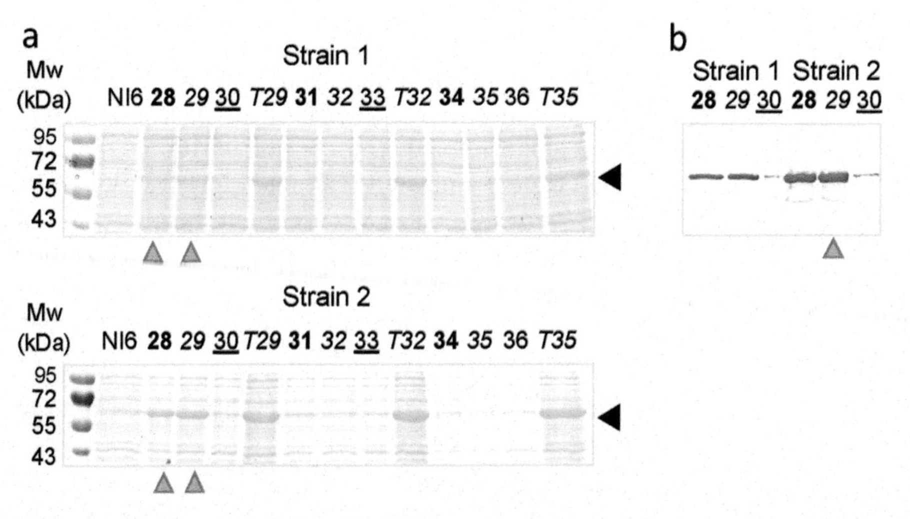

Fig. 2 Example results from expression and solubility test. (**a**) SDS-PAGE gels of lysates from two different *E. coli* strains (Strain 1: Rosetta 2; Strain 2: ER2566), both overexpressing truncated ETR1 protein [histidine kinase and C-terminal receiver domain ($ETR1_{HK-RD}$)] under different conditions as described in Table 3. The recombinant protein contains $ETR1_{HK-RD}$ fused with N-terminal ubiquitin and hexahistidine (His6) tag (total size 56.6 kDa). T—total bacterial protein lysates (obtained by lysing the cells in Tris–HCl buffer and taken before removing the cell debris via centrifugation); NI—noninduced samples. The gels were stained with Bio-Safe Coomassie G-250. Black arrow indicates position of the recombinant protein. Green arrows indicate the conditions that were evaluated to be the best in terms of protein solubility. PageRuler Prestained Protein Ladder (ThermoFisher Scientific) was used as a molecular weight marker. (**b**) Protein identification via immunoblotting. Only the best soluble protein samples were selected for analysis. The conditions evaluated as being optimal for further large-scale protein expression are 28 and 29: *E. coli* ER2566, TB medium pH 6.0, 3 h induction at 37 °C, lysis buffer: MES pH 6.0 (28) or Tris–HCl pH 7.5 (29)

5. To quantify the condition-specific protein yield on SDS-PAGE gels, use GelAnalyzer software or similar. Based on the results of the first run, run the most soluble (showing the highest supernatant-to-total protein band intensity ratio) samples again on SDS-PAGE gels and confirm protein identity via immunoblotting (Fig. 2a, b).

3.4 Transfer of Proteins on Membrane

1. Run SDS-PAGE gels with the samples selected as described in Subheading 3.3.
2. Cut the membrane and four pieces of Whatman filter paper according to the size of the gel (*see* **Notes 24** and **25**). Soak the membrane in 100% methanol for 1 min to decrease hydrophobicity and then rinse in distilled water for 2 min.
3. Equilibrate the SDS-PAGE gel, membrane and filter papers in Towbin's transfer buffer for at least 10 min.
4. Prepare the gel sandwich and proceed with the wet blotting according to the manufacturer's instructions [13, 14] (*see* **Note 26**).
5. Check the effectiveness of protein transfer from the gel to the membrane by staining the membrane with Ponceau-S red (3 min) and then rinse in water (*see* **Note 27**).
6. If prestained molecular weight marker is not used, mark the individual positions of the protein markers on the membrane with a pencil.
7. If necessary, determine protein elution efficiency from the gel, stain the blotted gel with Coomassie Brilliant Blue dye.

3.5 Western Blotting

1. Incubate the membranes in blocking buffer for 45 min with gentle agitation or overnight at 4 °C.
2. Add primary antibody diluted in blocking buffer and incubate for 1 h with gentle agitation at room temperature or overnight at 4 °C (*see* **Note 28**).
3. Wash the membrane 3× in TBST buffer, 5 min each time.
4. Add secondary antibody conjugated with alkaline phosphatase diluted in blocking buffer and incubate for 45 min with gentle agitation.
5. Wash the membrane 3× in TBST buffer, 5 min each time.
6. Add detection buffer and incubate for 10 min.
7. Incubate the membrane with a chromogenic substrate for alkaline phosphatase in the dark until the blue-purple signal appears (Fig. 2b).

3.6 Protein Purification

Perform large-scale expression of the protein, applying the results from the small-scale expression and solubility tests as described in

Subheading 3.1. Perform purification at room temperature only when the protein of interest is stable under these conditions, otherwise (or if unknown), protein purification should be performed at 4 °C. Before purification, let the buffers temperate for 1 h at correct temperature according to purification conditions and degas them via sonication (*see* **Note 29**).

1. Prepare bacterial lysate from 1 L of bacterial culture using the optimal expression conditions defined Subheadings 3.1–3.3. Use the lysis buffer giving the best results in the small-scale expression and solubility test. For 1 g of bacterial cells we typically use 5 mL of lysis buffer. Resuspend the cells on ice by pipetting up and down (*see* **Note 30**).
2. Leave bacterial cells in lysis buffer on ice for 20 min, then disrupt the cells (in our case by sonication). Spin the lysate for 40 min in 4 °C 48298 × *g* (20,000 rpm on JA-25.50 Fixed-Angle Rotor by Beckman Coulter for Avanti J Centrifuge Series).
3. Filter the supernatant using syringe filter of pore diameter 0.22 μm to remove any debris.
4. Perform the purification according to the manufacturer's instructions of the column and the FPLC system (GE Healthcare's ÄKTA FPLC System in our case) [11].
5. Using SDS-PAGE, identify and collect the fractions containing the protein of interest.
6. Concentrate the sample using protein concentrator spin columns.
7. Flash-freeze the protein sample in liquid nitrogen and store at −80 °C, or use directly for the autophosphorylation assay.

3.7 Autophosphorylation Assay

Always work safely while handling radioactive solutions. Use gloves, lab coat, and safety glasses and work behind plexiglass shields. Use a Geiger counter to monitor possible contamination, and filter tips for pipetting. In our example histidine kinase assay, we used protein with proteolytically cleaved fusion tag, based on a procedure described in [15].

1. Prechill SDS-PAGE electrophoresis buffer (*see* **Note 19**).
2. Just before assay, freshly prepare the kinase buffer by adding the necessary cations to the kinase buffer stock solution: 5 mM final concentration of each ion. Add DTT to a final concentration of 2 mM (*see* **Note 31**). In our assays we prepared three different buffers containing:
   (a) 5 mM $MgCl_2$.
   (b) 5 mM $MnCl_2$.
   (c) 5 mM $MgCl_2$ + 5 mM $MnCl_2$.

3. Centrifuge protein samples before assay for 10 min at 4 °C, 20,800 × *g* (14,000 rpm on FA-34-30-11 rotor for Eppendorf 5810/5810 R centrifuge) and determine the total protein concentration via Bradford assay [16].
4. Use the protein histidine kinase at a final concentration of 3 μM in the assay (this may require optimization dependent on the specific activity of the enzyme preparation). Prepare samples of proteins separately in each buffer to be tested (10 μL each).
5. Add 2 μCi/mmol γ-^{32}ATP (0.2 μL) to each sample.
6. Incubate the samples at room temperature and stop the reaction after 60 min by adding 4 μL of stop buffer.
7. Load the samples on SDS-PAGE gel without boiling (*see* **Note 32**).
8. After electrophoretic separation, cut the bottom 3–4 mm of the gel to remove free γ-^{32}ATP and the upper (stacking) layer with plastic Mini-Protean gel releaser. Wrap the gel with foil and place it into the imaging cassette.
9. Keep the gels in the imaging cassette overnight and scan the next day using the Typhoon FLA scanner (Fig. 3).
10. Stain the gels with Coomassie Brilliant Blue dye (*see* **Note 33**).

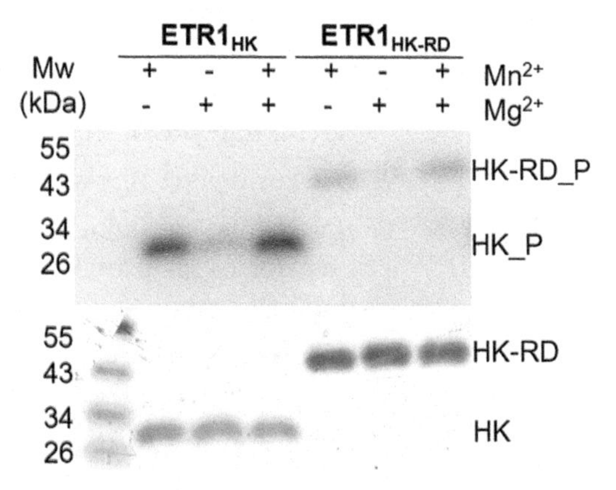

Fig. 3 Radiogram showing incorporation of ^{32}P into histidine kinase domain of ETR1 (ETR1$_{HK}$) and ETR1$_{HK\text{-}RD}$ (30.8 kDa and 45.9 kDa, respectively) via its autokinase activity. The purification tag was proteolytically removed from both proteins. Autophosphorylation of the protein is visible as dark bands in positions of proteins of interest, demonstrating in vitro protein activity due to ^{32}P incorporation (_P). The corresponding Coomassie-stained SDS-PAGE gel as a protein loading control is shown below the radiogram. We observed higher autophosphorylation activity of both ETR1$_{HK}$ and ETR1$_{HK\text{-}RD}$ in the presence of manganese ions, confirming previously published results [18, 19]. The presence of the C-terminal receiver domain in ETR1$_{HK\text{-}RD}$ appears to downregulate the level of (auto)phosphorylation when compared with ETR1$_{HK}$. The prestained protein ladder (PageRuler, ThermoFisher Scientific) was used as a molecular weight marker

4 Notes

1. The chemicals do not need to be dissolved fully in the water at room temperature, but will dissolve during autoclaving.
2. K-phosphate buffer at the required pH can be prepared in larger quantities and stored after sterilization at room temperature. pH does not need to be adjusted.
3. Bacterial strains should be selected according to the specific needs of the protein to be produced (*see* ref. 17).
4. If a laminar-flow chamber is not available, the bacterial work can be performed on the lab bench using a Bunsen burner (or similar). Under these conditions, it is important to work with caution to avoid contamination of media and reagents.
5. MES buffer deteriorates relatively quickly (approx. 10 days). To extend its shelf life, sterilize with membrane nylon filters of pore diameter 0.22 μM. Store at 4 °C.
6. DTT can be stored in a form of 1 M aliquots in 1.5 mL microcentrifuge tubes at −20 °C. The solution needs to be completely thawed before adding it into buffers.
7. Tris–HCl buffer should be prepared in double the quantity compared to other buffers due to higher consumption (cell lysis in control samples of noninduced cells for each growth condition).
8. Put a clean pipette tip on the top of aspirator hose to allow careful removal of the supernatant.
9. Use cut pipette tips while working with glycerol or Tween 20.
10. If possible, prestained molecular weight markers should be used for easier protein identification, following immunoblotting and autophosphorylation assay.
11. Weigh SDS in fume hood to prevent inhalation.
12. Store at room temperature. In rooms that are air conditioned below approx. 20 °C, SDS might start to precipitate.
13. We prefer Bio-Safe Coomassie G-250 dye as it allows destaining of gels with water, not methanol/acetic acid.
14. Blocking buffer can be stored in the fridge up to 1 week without preservatives.
15. Chromogenic substrate for alkaline phosphatase is light- and heat-sensitive.
16. Always use tweezers for manipulation of the membrane to prevent smears and contamination.
17. If the protein concentration is low, concentrate it before assay or exchange protein storage buffer for stock kinase buffer.

Measure the protein concentration before the kinase assay using a Bradford assay or similar.

18. Always thaw the aliquot of DTT completely before use. Freeze it again immediately after use to prevent DTT from degradation.
19. 1× SDS-PAGE electrophoresis buffer should be prechilled to prevent distorted ("smiling") bands.
20. Always calculate the amount of required medium, flasks and buffers in advance according to the number of conditions and/or cell strains assayed.
21. Each condition has its own following number. Each lysis buffer is marked with different color to prevent pipetting errors. Every tested strain has same pattern for numbering conditions. To prevent mismatching the samples from different strains always use separate rack of different color for SDS-PAGE samples from each strain.
22. Check if the caps are tight before cell lysis and make sure that the adapters are tightened well in the holders.
23. Use a 5 mL pipette to pour the acrylamide mixture between the glass plates.
24. Cut one corner of the membrane to mark the orientation.
25. Always wear gloves when handling membranes to prevent contamination and smears.
26. In case of small proteins, Semi-Dry blot transfer can be performed using, for example, Trans-Blot® Turbo™ Transfer System (Bio-Rad).
27. Ponceau S-Red can be reused repeatedly.
28. Antibody-containing solutions can be stored for up to 1 week at 4 °C without preservatives and reused once.
29. Untighten the caps of the solutions to prevent overpressure inside the bottles.
30. For bacterial pellet resuspension use preferably 10 mL pipette with cut tips.
31. The amount of assay buffer should be calculated based on the number of samples that will be run and if the desalting step will be needed.
32. Boiling of protein samples decreases the signal intensity of radioactive phosphate (phosphorylated His is thermolabile).
33. If the gels are stained using Bio-Safe Coomassie Blue, the dye can be reused once, but care should be taken to store appropriately given the possible ^{32}P contamination. If necessary, gels can be decontaminated before scanning by submerging them in water with 3% citric acid for several days prior to scanning.

Acknowledgments

The work was supported by the Ministry of Education, Youth and Sports of the Czech Republic under the projects CEITEC 2020 (LQ1601).

References

1. Structural Genomics C, China Structural Genomics C, Northeast Structural Genomics C et al (2008) Protein production and purification. Nat Methods 5:135–146
2. Mizianty MJ, Kurgan L (2011) Sequence-based prediction of protein crystallization, purification and production propensity. Bioinformatics 27:i24–i33
3. Vincentelli R, Canaan S, Offant J et al (2005) Automated expression and solubility screening of His-tagged proteins in 96-well format. Anal Biochem 346:77–84
4. Braun P, Labaer J (2003) High throughput protein production for functional proteomics. Trends Biotechnol 21:383–388
5. Berrow NS, Bussow K, Coutard B et al (2006) Recombinant protein expression and solubility screening in Escherichia coli: a comparative study. Acta Crystallogr D Biol Crystallogr 62:1218–1226
6. Vincentelli R, Cimino A, Geerlof A et al (2011) High-throughput protein expression screening and purification in Escherichia coli. Methods 55:65–72
7. Correa A, Ortega C, Obal G et al (2014) Generation of a vector suite for protein solubility screening. Front Microbiol 5:67
8. Bjerga GE, Arsin H, Larsen O et al (2016) A rapid solubility-optimized screening procedure for recombinant subtilisins in E. coli. J Biotechnol 222:38–46
9. Listwan P, Terwilliger TC, Waldo GS (2009) Automated, high-throughput platform for protein solubility screening using a split-GFP system. J Struct Funct Genom 10:47–55
10. Cold Spring Harb Protoc. 2006. doi:https://doi.org/10.1101/pdb.tab19
11. https://www.auburn.edu/~duinedu/manuals/HisTrapHP.pdf
12. Laemmli UK (1970) Cleavage of structural proteins during the assembly of the head of bacteriophage T4. Nature 227:680–685
13. http://www.bio-rad.com/webroot/web/pdf/lsr/literature/10007296D.pdf
14. Towbin H, Staehelin T, Gordon J (1979) Electrophoretic transfer of proteins from polyacrylamide gels to nitrocellulose sheets: procedure and some applications. Proc Natl Acad Sci U S A 76:4350–4354
15. Ueno TB, Johnson RA, Boon EM (2015) Optimized assay for the quantification of histidine kinase autophosphorylation. Biochem Biophys Res Commun 465:331–337
16. Bradford MM (1976) A rapid and sensitive method for the quantitation of microgram quantities of protein utilizing the principle of protein-dye binding. Anal Biochem 72:248–254
17. Rosano GL, Ceccarelli EA (2014) Recombinant protein expression in Escherichia coli: advances and challenges. Front Microbiol 5:172
18. Gamble RL, Coonfield ML, Schaller GE (1998) Histidine kinase activity of the ETR1 ethylene receptor from Arabidopsis. Proc Natl Acad Sci U S A 95:7825–7829
19. Moussatche P, Klee HJ (2004) Autophosphorylation activity of the Arabidopsis ethylene receptor multigene family. J Biol Chem 279:48734–48741

Chapter 3

SDS-PAGE and Dot Blot Autoradiography: Tools for Quantifying Histidine Kinase Autophosphorylation

Jonathan T. Fischer, Ilana Heckler, and Elizabeth M. Boon

Abstract

Histidine kinases play a vital role in bacterial signal transduction. However, methods for studying the activity of histidine kinases in vitro are limited in comparison to those for investigating serine, threonine, and tyrosine kinases, largely due to the lability of the phosphoramidate (P-N) bond. Here, we describe two useful methods for quantifying histidine kinase autophosphorylation: SDS-PAGE autoradiography and dot blot autoradiography/scintillation counting.

Key words Quantification, Histidine kinase, Autophosphorylation, Autoradiography, SDS-PAGE, Dot blot, Two-component signaling

1 Introduction

Bacteria utilize two-component signal transduction pathways to respond to stimuli and changing conditions. Two-component signal transduction pathways feature histidine kinases, which bind 5'-ATP to autophosphorylate on a conserved histidine residue. The phosphate is subsequently transferred from the histidine residue to a conserved aspartate on a receiver domain [1–3]. Histidine kinase autophosphorylation is sensitive to sensory input. In many cases, the autophosphorylation of a histidine kinase is controlled by a sensory domain, which exists either as a stand-alone accessory protein, or on the same polypeptide as the kinase [4–7]. Receiver domains are typically found in response regulator proteins. Phosphorylation of the receiver domain of a response regulator usually elicits a change in its activity [8, 9]. In other, atypical, two-component signal transduction systems, a receiver domain is found on the same polypeptide as the histidine kinase. In these instances, phosphotransfer occurs from the kinase domain to its internal receiver

Jonathan T. Fischer and Ilana Heckler contributed equally to this work.

Claire E. Eyers (ed.), *Histidine Phosphorylation: Methods and Protocols*, Methods in Molecular Biology, vol. 2077,
https://doi.org/10.1007/978-1-4939-9884-5_3,

domain, followed by an additional transfer step to a histidine phosphotransfer protein (Hpt). The phosphate can then finally be transferred from the Hpt protein to the receiver domain of a response regulator [10]. Thus, two-component signaling pathways use histidine kinases to translate sensory information into a response.

Methods available for studying histidine kinase autophosphorylation are quite limited. Studies of phosphorylated hydroxyamino acids such as phosphoserine, phosphothreonine, and phosphotyrosine benefit greatly from their acid stability [11–13]. However, the acid lability of phosphohistidine restricts studies of histidine kinase activity to a select few methods [14–16]. Despite this challenge, quantification of histidine kinase autophosphorylation, as well as phosphotransfer profiling to a response regulator, can be accomplished using currently available methodology. In this chapter, two methods for quantifying histidine kinase activity, using (1) SDS-PAGE (Subheading 3.1) and (2) dot blot autoradiography/scintillation counting (Subheading 3.2) assays, are described in detail and representative data from each method is shown. The two methods can be utilized in tandem to complement one another when characterizing and quantifying histidine kinase activity. Alternatively, one method may be found to be more suitable than the other for the desired experiment. The SDS-PAGE method allows for 2D resolution defining the mass of the phosphorylated protein, which is useful when characterizing more complex reactions, while the dot blot method has the advantage of being higher throughput.

2 Materials

Prepare all solutions in double distilled ultrapure water (ddH_2O). Store and prepare all solutions at room temperature, unless stated otherwise.

2.1 SDS-PAGE Autoradiography

2.1.1 Histidine Kinase Autophosphorylation Reaction

1. 1 M Tris–HCl, pH 8: Weigh 12.1 g Tris and transfer to a glass graduated cylinder. Dissolve the buffer completely in ~50 mL of ddH_2O. Adjust the pH to 8.0 using HCl. Adjust volume to 100 mL.
2. 2.5 M KCl: Weigh 3.7 g KCl and transfer to a conical tube. Dissolve in ~10 mL of ddH_2O, then bring the volume to 20 mL.
3. 1 M $MgCl_2$: Weigh 4 g $MgCl_2 \cdot 6H_2O$ and prepare a 20 mL solution as described in **step 2**.
4. 100 mM Dithiothreitol (DTT): Weigh 15.4 mg DTT and transfer to an Eppendorf tube. Dissolve in 1 mL of ddH_2O (*see* **Note 1**).
5. Kinase assay buffer: 100 mM Tris–HCl pH 8, 50 mM KCl. To a 100 mL graduated cylinder, add 10 mL of the 1 M Tris–HCl stock and 2 mL of the 2.5 M KCl stock. Bring the volume to 100 mL with ddH_2O.

6. Purified histidine kinase of interest (*see* **Note 2**).
7. 100 mM 5′-ATP.
8. [γ-^{32}P]-ATP.
9. 2× reaction mix (*see* **Note 3**): To an Eppendorf tube, add the kinase assay buffer, the 1 M $MgCl_2$ stock to a final concentration of 10 mM, the 100 mM DTT to a final concentration of 2 mM, 5′-ATP to a final concentration of 5 mM and 50 μCi/μ mol [γ^{32}P]-ATP (*see* **Note 4**).
10. Geiger counter.
11. Leucite, acrylic, or Plexiglass shield.
12. SDS loading buffer (5×): 0.3 M Tris–HCl (pH 6.8), 10% (w/v) SDS, 0.1% (v/v) bromophenol blue. Immediately before use, add 25% (v/v) β-mercaptoethanol.

2.1.2 SDS-PAGE

1. Gel assembly equipment: Mini-PROTEAN® casting stand, Mini-PROTEAN® glass spacer plates with 0.75 mm integrated spacers, Mini-PROTEAN® glass short plates, 10-well comb with 0.75 mm thickness, buffer tank and lid, PowerPac basic power supply, electrode assembly.
2. Resolving gel buffer: 1.5 M Tris–HCl, pH 8.8. Weigh 181.7 g Tris and transfer to a glass graduated cylinder. Dissolve Tris completely in ~800 mL of ddH_2O. Adjust pH to 8.8 using HCl. Bring volume up to 1 L.
3. Stacking gel buffer: 1 M Tris–HCl, pH 6.8. Weigh 121.1 g Tris and prepare a 1 L solution as stated in **step 2**.
4. 40% (w/v) Acrylamide–Bis solution (29:1): Store at 4 °C.
5. Ammonium persulfate: 10% (w/v) solution in water (*see* **Note 5**).
6. Sodium dodecyl sulfate (SDS): 10% (w/v) solution in water (*see* **Note 6**).
7. TEMED: Store at 4 °C.
8. SDS-PAGE gel running buffer (10×): Weigh 140 g glycine, 30 g Tris, 10 g SDS and transfer to a graduated cylinder. Dissolve in about 500 mL of ddH_2O. Bring volume up to 1 L. Prepare 1× running buffer by adding 100 mL 10× SDS-PAGE gel running buffer to a graduated cylinder and bringing the volume up to 1 L with ddH_2O.
9. Small Tupperware container or similar plastic container with lid.
10. Gel releaser or spatula (to remove gel from glass plates).
11. Rocker.

12. Staining solution: Dissolve 1 g of Coomassie Brilliant Blue in a 1 L solution of 500 mL methanol, 100 mL acetic acid, 400 mL ddH_2O.
13. Destaining solution: Prepare a 1 L solution of 300 mL methanol, 100 mL acetic acid, 600 mL ddH_2O.

2.2 Dot Blot Autoradiography/Scintillation Counting

2.2.1 Histidine Kinase Autophosphorylation Reaction

1. Prepare materials 1–9 from Subheading 2.1.1.
2. 250 mM H_3PO_4: To 983 mL ddH_2O, pipet 17 μL concentrated H_3PO_4 (85% w/w). Aliquot 3 μL into reaction tubes (*see* **Note 7**). Place on ice.

2.2.2 Dot Blot Apparatus and Reagents

1. 25 mM H_3PO_4: To 1 L ddH_2O, pipet 1.7 mL concentrated H_3PO_4 (85% w/w) to a final concentration of 25 mM. Place on ice.
2. 96-well dot blot microfiltration apparatus.
3. Nitrocellulose membrane 8 × 12 cm, 0.2 μm pore size.
4. Vacuum source (e.g., Bio-Rad Hydrotech vacuum pump).
5. Vacuum filter flask.
6. Vacuum tubing (*see* **Note 8**).
7. Rocker.

2.2.3 Scintillation Counting (See Note 9)

1. Ponceau S staining solution: 0.5% (w/v) Ponceau S, 1% (v/v) acetic acid. Pipet 200 μL concentrated acetic acid (17.4 M) into 19.8 mL ddH_2O. Weigh 100 mg Ponceau S and add to 1% (v/v) acetic acid solution.
2. Scintillation counter (PerkinElmer).
3. Scintillation vials.
4. Scissors or cork borer (to cut out spots).
5. Forceps.

2.3 Autoradiography (Same for Both SDS-PAGE and Dot Blot Assays)

1. Phosphor screen.
2. Phosphor screen scanner (Typhoon imaging systems).
3. Image eraser (white light source).

3 Methods

CAUTION! These protocols are to be carried out behind suitable beta radiation shielding such as Leucite, acrylic, or Plexiglas. Ensure that all radioactive waste generated is transferred to appropriate labelled waste containers and disposed of according to on-site requirements. Keep a Geiger counter near the workspace to monitor for potential contamination.

3.1 SDS-PAGE Autoradiography

3.1.1 Histidine Kinase Autophosphorylation Reaction for SDS-PAGE Autoradiography

1. Pipet the histidine kinase of interest into a reaction tube. The concentration of the histidine kinase should be 2× more concentrated than the desired final concentration in the reaction (*see* **Note 10**). Dilute with kinase assay buffer if necessary.
2. Initiate the histidine kinase reaction by adding 1 volume of the 2× reaction mix to 1 volume of the histidine kinase. Gently pipet up and down to mix the reaction components. Allow the autophosphorylation reaction to proceed for the desired amount of time (*see* **Note 11**). Monitor with a timer.
3. Quench the reaction(s) by pipetting 20 μL of the reaction into 5 μL 5× SDS loading dye (*see* **Note 12**). Pipet up and down to mix. Keep the reaction tube(s) on ice until ready to load the gel. Repeat until all time points/reactions have been quenched.

3.1.2 12.5% SDS-PAGE Gel Electrophoresis

1. Cast the resolving gel layer by mixing 2.1 mL ddH_2O, 1.5 mL acrylamide solution, 1.3 mL resolving buffer, 50 μL 10% (w/v) SDS, 50 μL 10% (w/v) APS, and 2 μL TEMED in a conical tube. After addition of the TEMED, thoroughly mix components by pipetting up and down, then immediately cast gel within the glass gel cassette sandwich, leaving room for the stacking layer (*see* **Note 13**). Overlay the resolving layer with ethanol. Wait 30 min to allow the resolving layer to polymerize before pouring off the ethanol and moving on to the next step (*see* **Note 14**).
2. Cast the stacking gel layer by mixing 1.48 mL ddH_2O, 250 μL acrylamide solution, 250 μL stacking buffer, 20 μL 10% (w/v) SDS, 20 μL 10% (w/v) APS, and 2 μL TEMED in a conical tube. After addition of the TEMED, pipet up and down, then immediately pour over the top of the resolving layer, having removed the ethanol. Promptly insert a 10-well comb, ensuring that air bubbles are not introduced and that the comb is level. Wait at least 30 min to allow the stacking layer to polymerize before proceeding.
3. Carefully remove the 10-well comb by sliding it up vertically.
4. Assemble the gel cassette sandwich within the electrode assembly and place in the buffer tank as per the manufacturer's recommendations.
5. Add the 1× SDS-PAGE gel running buffer to the inside chamber so that it completely covers the tops of the wells (*see* **Note 15**). Add ~200 mL 1× running buffer to the outside of the chamber so that the electrode is covered.

6. Load a protein ladder standard into the first well, followed by the quenched reaction(s) from Subheading 3.1.1 (*see* **Note 16**).
7. Electrophorese at a constant voltage (120 V) for 1 h, or until the dye front runs 1 cm from the bottom of the gel.
8. Following electrophoresis, pry open the gel cassette sandwich with a gel releaser or spatula. Remove the stacking layer and the dye front using a spatula.
9. Transfer the gel directly to a small Tupperware container or similar (*see* **Note 17**). Add enough ddH_2O to cover the gel completely. Cover the container and place on a rocker for 1 min (*see* **Note 18**).
10. Carefully pour out the water from the container into a radioactive waste bucket, ensuring that the gel does not slide out (*see* **Note 19**). Add enough staining solution to cover the gel. Cover the container and place on a rocker for 30 min.
11. Pour out the staining solution into an appropriate waste container. Immediately add enough destaining solution to cover the gel. Cover the container and place on a rocker for 10 min.
12. Pour out the used destaining solution into a beaker. Use a Geiger counter to check the radioactivity of the used destaining solution before disposing of it into an appropriate waste container. Add fresh destaining solution to the gel. Repeat the destaining **step 5** more times, at 10 min intervals, or until the used destaining solution does not appear to be radioactive as determined by the Geiger counter.
13. Place the destained gel in a gel dryer. Dry the gel at 70 °C for 1 h (*see* **Note 20**).
14. Once the gel is dry, obtain an image of the gel with either a camera or a scanner, so that the relative intensities of the protein bands can be used for normalization of ^{32}P incorporation in the next section. This is needed to control for protein loading discrepancies (Fig. 1a).

3.1.3 Autoradiography and Band Quantification Following SDS-PAGE

1. Prior to exposing the dried gel to the storage phosphor screen, expose the phosphor screen to white light for at least 5 min. This ensures that any residual image that may be held by the screen is erased.
2. Make sure the phosphor screen is clean and dry. If necessary, gently clean the screen with a phosphor screen cleaning solution approved by the screen manufacturer and wipe dry. Place the dried gel in the storage phosphor screen cassette, lay the phosphor screen face-down over the gel, and close the cassette. Do not open the cassette or move the gel until exposure is

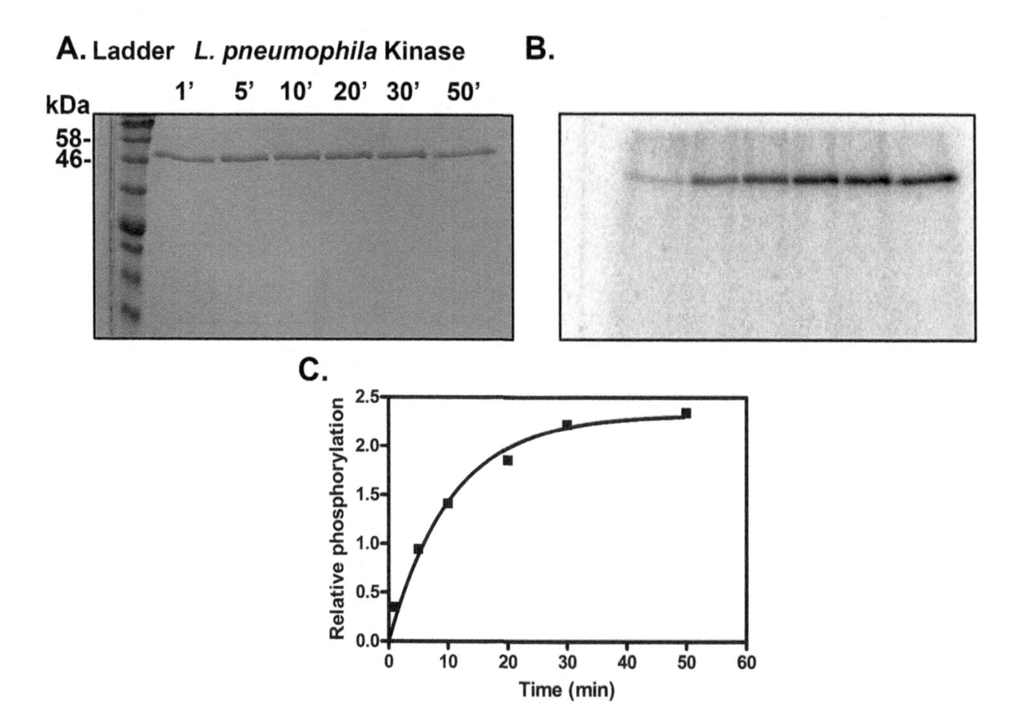

Fig. 1 Autophosphorylation of a *Legionella pneumophila* histidine kinase (lpg0278) as quantified by SDS-PAGE autoradiography. (**a**) Photograph of the SDS-PAGE gel after electrophoresis and staining/destaining. Lpg0278 was incubated with ^{32}P-labeled ATP for a total of 50 min. Aliquots of the reaction were removed at certain time points and quenched. The first lane contains a protein ladder, followed by the quenched autophosphorylation reactions from the 1-, 5-, 10-, 20-, 30-, and 50-min time points. The band visible at ~53 kDa corresponds to the molecular weight of lpg0278 (53.5 kDa). (**b**) Autoradiography image of the SDS-PAGE gel from (**a**). An increase in the kinase band intensity, with increasing time, is apparent. (**c**) Relative autophosphorylation of lpg0278 over time. ImageJ was used to quantify the band intensities from (**a**) and (**b**). Values of kinase autophosphorylation were normalized to protein loading band intensities

complete, as gel shifting during exposure will cause a double or smeared image to be captured.

3. Expose the gel to the phosphor screen for at least 4 h, or as long as the phosphor screen manufacturer suggests.
4. Once exposure is completed, remove the phosphor screen from the cassette and scan with a phosphor scanner (Fig. 1b). If the image is not satisfactory, repeat **steps 1–3** to re-expose the screen to the gel, adjusting the exposure time if the intensity is too low or high. Once happy with the image, dispose of the radioactive gel in the appropriate waste container.
5. Use ImageJ to quantify band intensities (Fig. 1c). Normalize the intensity of ^{32}P incorporation arising from autophosphorylation using the autoradiography image according to the Coomassie stained kinase protein band intensity from the image from **step 15** of Subheading 3.1.2.

3.2 Dot Blot Assay

3.2.1 Histidine Kinase Autophosphorylation Reaction for Dot Blot Assay

1. Pipet the histidine kinase of interest into a reaction tube. The concentration of the histidine kinase should be 2× more concentrated than the desired final concentration in the reaction (*see* **Note 10**). Dilute with kinase assay buffer if necessary.
2. Initiate the histidine kinase reaction by adding 1 volume of the 2× reaction mix to 1 volume of the histidine kinase. Gently pipet up and down to mix the reaction components. Allow the autophosphorylation reaction to proceed for the desired amount of time. Monitor with a timer. For time course assays, it is typical to take several (5–10) time points between 0 and 90 min (*see* **Note 21**). Assays are typically done either at 25 or 37 °C, depending on the protein.
3. Quench time points by pipetting 27 μL of the reaction into 3 μL ice cold 250 mM H_3PO_4. Mix thoroughly by pipetting up and down. Immediately place quenched reaction on ice. Repeat until all reactions have been quenched.
4. Quench the "no-kinase" control reaction in the same manner as **step 3**. Multiple time points of the control are not necessary, but triplicate repeats are recommended for improved accuracy.

3.2.2 Spotting Reactions on Nitrocellulose

1. Assemble the 96-well dot blot apparatus with the nitrocellulose positioned such that the apparatus is sealed when vacuum is applied. Connect the apparatus to the side-arm vacuum flask to collect filtrate (*see* **Note 22**).
2. Connect the side-arm flask to a vacuum source, such as a Bio-Rad HydroTech vacuum pump. Apply vacuum. Check that the apparatus is sealed, and a vacuum is present in all the wells (*see* **Note 23**).
3. Position beta shielding (e.g., leucite shield), such that the dot blot apparatus and vacuum flask are behind the shield.
4. Carefully pipet 25 μL of quenched reaction directly onto the membrane in the well. Repeat until all samples have been loaded onto the membrane, and all liquid has passed through the membrane.
5. Wash each reaction well with 500 μL 25 mM H_3PO_4. Repeat until all reaction wells have been washed, and all liquid has passed through the membrane.

3.2.3 Apparatus Disassembly and Washing of the Nitrocellulose Membrane

1. Turn off vacuum source. Carefully disassemble apparatus and, with forceps, remove the now radioactive nitrocellulose membrane.
2. Transfer the membrane into a Tupperware container (or similar). Carefully pour 100 mL ice cold 25 mM H_3PO_4 wash solution onto the membrane.

3. Place the membrane on a rocker to wash away unbound ATP. Wash the membrane with 100 mL 25 mM H_3PO_4 3–5 times for 10 min. When decanting, check the used wash solution for radioactivity with a Geiger counter. Continue until no radioactivity is present in the used wash solution.
4. Allow the membrane to air dry until the membrane is completely dry, usually about 20 min.

3.2.4 Autoradiography

1. Once the membrane has dried, carefully wrap the nitrocellulose membrane in plastic wrap.
2. Follow the procedure outlined in Subheading 3.1.3 to quantify the intensity of ^{32}P-labeled histidine kinase spots (Fig. 2a).

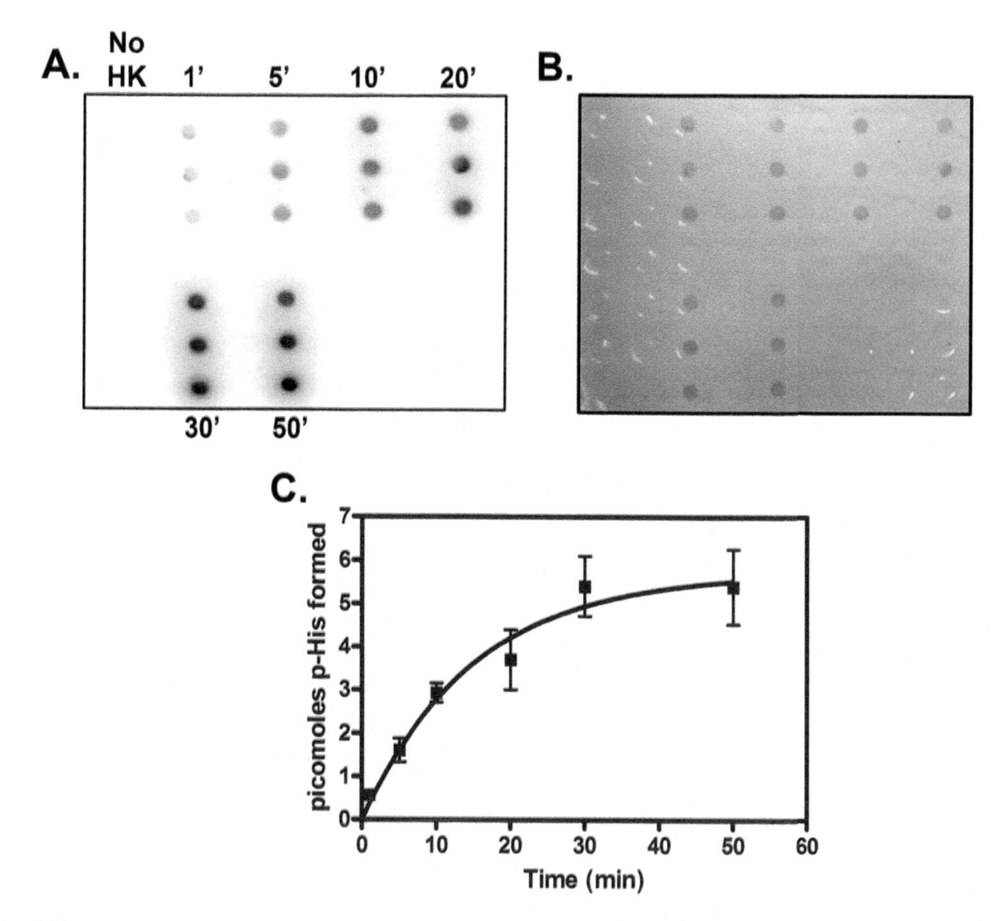

Fig. 2 Representative results of the autophosphorylation of a *Legionella pneumophila* histidine kinase as quantified by dot blot scintillation counting. (**a**) Autoradiograph of the spotted nitrocellulose membrane obtained using a Typhoon Scanner. Increasing intensity is apparent as reaction time increases. Triplicate spots of each time point appear uniform in intensity. (**b**) Ponceau stain of the dot blot. Protein loading is even throughout all reaction spots. No protein is present in the no kinase control spots. (**c**) Quantification of histidine kinase autophosphorylation over time using scintillation counting. The results are very similar to those obtained using SDS-PAGE autoradiography (Fig. 1c)

3.2.5 Scintillation Counting of Spots

1. Place the nitrocellulose membrane in a Tupperware (or similar) container. Pour 20 mL Ponceau S staining solution onto the nitrocellulose membrane. Place on a rocker for 2 min, ensuring the entire membrane is exposed to the stain.
2. Decant Ponceau S solution. Wash membrane with ddH_2O until only stained dots containing protein remain (Fig. 2b).
3. Cut out each stained dot using scissors or a cork borer (*see* **Note 24**). Using forceps, carefully transfer the cut-out dots directly to scintillation vials. Repeat until all dots have been cut out (*see* **Note 25**).
4. Carefully pipet serial dilutions of the 2× [γ^{32}P]-ATP reaction mix onto small (2 × 2 cm) squares of nitrocellulose or filter paper (*see* **Note 26**). Calculate and record the number of picomoles of ATP in each spot. Using forceps, carefully transfer the spotted nitrocellulose squares to scintillation vials. These will be used to generate a standard curve of counts per minute (CPM) per picomole of ATP, that is, determine the specific activity of the γ^{32}P-ATP.
5. Use a scintillation counter (PerkinElmer) to quantify CPM for each sample.
6. Create a standard curve of the serial dilutions of the γ^{32}P-ATP standards plotting CPM versus picomoles of ATP to determine the specific activity (*see* **Note 27**).
7. Subtract the CPM of the no-kinase control dots from the rest of the samples. Using the slope of the standard curve generated in **step 6**, calculate the picomoles of phosphorylated histidine kinase for each dot. Plot the results as a function of time (Fig. 2c).

4 Notes

1. Prepare the DTT solution fresh before each experiment. Vortexing the tube will help the DTT dissolve.
2. We purified a histidine kinase (gene ID 19831845) from *Legionella pneumophila* to generate the representative results shown. The kinase was cloned into the C-terminal His-tag expression vector pET-20b using NdeI and XhoI restriction sites. The plasmid was transformed into *E. coli* BL21 (DE3) pLysS. Cells were grown in TB media with ampicillin (100 μg/ml) and chloramphenicol (34 μg/ml) with agitation (250 rpm) at 37 °C to an OD_{600} of 0.6. Protein expression was induced with IPTG at a final concentration of 25 μM, and cultures were grown overnight at 16 °C. Cells were harvested by centrifugation (4424 rcf for 15 min), lysed by sonication, centrifuged to clear lysate (41399 rcf for 60 min), and the His-tagged kinase

was purified using Ni-NTA agarose. The kinase was then desalted into the kinase assay buffer (100 mM Tris–HCl, 50 mM KCl) on a PD-10 desalting column. The protein was stored in the kinase assay buffer at −80 °C until the day of the experiment. Concentration of the protein was measured by UV-Vis.

3. The total reaction volume is dependent on the type of experiment. A larger reaction volume is needed to perform a time-dependent reaction, since multiple aliquots are taken out from the ongoing reaction over time and quenched. The total reaction volume will depend on the number of time points desired.
4. The specific activity of [γ^{32}P]-ATP necessary for adequate detection of phosphorylated histidine kinase will vary depending on the catalytic activity of the kinase, as well as the total amount of kinase in the sample. Adjust accordingly based on initial results.
5. Prepare the 10% (w/v) ammonium persulfate solution fresh before each experiment for best results.
6. A magnetic stir bar may be necessary to assist with dissolving.
7. For the dot blot assay, we typically process 15–30 samples (5–10 time points in triplicate), although it is easy to do more if desired.
8. Tubing will be needed to connect the vacuum to the vacuum filter flask, and to connect the filter flask to the dot blot microfiltration apparatus. Be sure to use enough tubing so that the dot blot apparatus and the vacuum filter flask can be kept behind the leucite shielding without danger of tipping over.
9. Scintillation counting is recommended for precise quantification of kinase activity, as a standard curve can be easily generated using this method. Quantification of Dot Blots using autoradiography is optional, as scintillation counting will be performed to quantify phosphohistidine formation.
10. To obtain adequate signal, we recommend that the final kinase concentration in the reaction is at least 1 μM.
11. For time course assays, it is typical to take several (5–10) time points between 0 and 90 min. Assays are typically done either at 25 or 37 °C, depending on the protein.
12. If monitoring the reaction progression over time, remove aliquots during the reaction and transfer to Eppendorf tubes containing the 5× SDS loading dye (25% of the volume of the aliquot). Pipet up and down to mix.
13. Pour the resolving layer so that after the stacking layer is poured, there will be a 1 cm gap between the bottom of the well comb teeth and the top of the resolving layer.

14. For best results, gels should be prepared on the day of the experiment prior to initiating reactions.
15. Check for leaks and remedy if needed. If the level of running buffer falls below the tops of the wells, the gel will not run properly.
16. We recommend loading a maximum of 25 μL per lane for optimal band quality.
17. If the gel has dried at all between the previous step and this transfer step, wet the gel with ddH_2O using a squirt bottle before trying to peel it from the glass plates. This will reduce the chance of ripping the gel.
18. This step removes excess running buffer, which can interfere with the staining of the gel.
19. We have found that lightly placing a spatula on the gel can keep the gel from sliding.
20. If a gel dryer is not available, an alternative gel drying method is to use gel drying film (25 × 28 cm). Soak 2 pieces of cellulose gel drying film in gel drying solution (per 1 L: 400 mL methanol, 100 mL glycerol, 75 mL acetic acid, 425 mL ddH_2O) for 5 min. Lay 1 piece of film onto a flat wet surface and carefully place the gel on top of the film. Cover with the second piece of film, smoothing out any air bubbles. Position in a gel drying frame. Secure the frame with binder clips and allow the gel to dry overnight.
21. To improve precision, it is recommended to set up reactions in triplicate, as the dot blot assay allows for higher throughput of samples than SDS-PAGE. Each time point can be done in triplicate for a total of 15–30 samples.
22. The tubing connecting the side-arm flask to the apparatus should be fitted in a holed rubber stopper to create a seal at the mouth of the flask.
23. Leaky connections in the system can be fortified with parafilm or plastic wrap to improve vacuum strength.
24. It is not necessary to cut perfectly along the edge of the circle. Cut pieces big enough so that they can be handled easily with forceps.
25. Scintillation cocktail is not necessary as ^{32}P is readily detectable via Cherenkov radiation.
26. Do not pipet more than 2 μL of each serial dilution onto the 2 × 2 cm nitrocellulose square. Use a 1:100 dilution of the reaction mix as a starting point for serial dilutions, that is, 1:100, 1:200, 1:400, 1:800 serial dilutions for the standard curve.

27. The slope of this line should be fit with linear regression with $R^2 > 0.95$. Set the y-intercept to 0. The slope can be used to convert the CPM of each dot to picomoles of phosphorylated histidine kinase present.

Acknowledgments

This work was supported by NIH GM118894 and NSF CHE-1607532.

References

1. Parkinson JS, Kofoid EC (1992) Communication modules in bacterial signaling proteins. Annu Rev Genet 26:71–112
2. Stock AM, Robinson VL, Goudreau PN (2000) Two-component signal transduction. Annu Rev Biochem 69:183–215
3. Jung K, Fried L, Behr S, Heermann R (2012) Histidine kinases and response regulators in networks. Curr Opin Microbiol 15:118–124
4. Szurmant H, Bu L, Brooks CL, Hoch JA (2008) An essential sensor histidine kinase controlled by transmembrane helix interactions with its auxiliary proteins. Proc Natl Acad Sci U S A 105:5891–5896
5. Hossain S, Boon EM (2017) Discovery of a novel nitric oxide binding protein and nitric oxide-responsive signaling pathway in Pseudomonas aeruginosa. ACS Infect Dis 3:454–461
6. Arora DP, Boon EM (2012) Nitric oxide regulated two-component signaling in *Pseudoalteromonas atlantica*. Biochem Biophys Res Commun 421:521–526
7. Price MS, Chao LY, Marletta MA (2007) *Shewanella oneidensis* MR-1 H-NOX regulation of a histidine kinase by nitric oxide. Biochemistry 46:13677–13683
8. Nixon BT, Ronson CW, Ausubel FM (1986) Two-component regulatory systems responsive to environmental stimuli share strongly conserved domains with the nitrogen assimilation regulatory genes ntrB and ntrC. Proc Natl Acad Sci U S A 83:7850–7854
9. Ronson CW, Nixon BT, Ausubel FM (1987) Conserved domains in bacterial regulatory proteins that respond to environmental stimuli. Cell 49:579–581
10. Uhl MA, Miller JF (1996) Integration of multiple domains in a two-component sensor protein: the Bordetella pertussis BvgAS phosphorelay. EMBO J 15:1028–1036
11. Gallo R, Provenzano C, Carbone R, Di Fiore PP, Castellani L, Falcone G, Alemà S (1997) Regulation of the tyrosine kinase substrate Eps8 expression by growth factors, v-Src and terminal differentiation. Oncogene 15:1929–1936
12. Wei Y, Yu L, Bowen J, Gorovsky MA, Allis CD (1999) Phosphorylation of histone H3 is required for proper chromosome condensation and segregation. Cell 97:99–109
13. Sarg B, Helliger W, Talasz H, Förg B, Lindner HH (2006) Histone H1 phosphorylation occurs site-specifically during interphase and mitosis: identification of a novel phosphorylation site on histone H1. J Biol Chem 281:6573–6580
14. Stock JB, Stock AM, Mottonen JM (1990) Signal transduction in bacteria. Nature 344:395–400
15. Kee JM, Muir TW (2012) Chasing phosphohistidine, an elusive sibling in the phosphoamino acid family. ACS Chem Biol 7:44–51
16. Ueno TB, Johnson RA, Boon EM (2015) Optimized assay for the quantification of histidine kinase autophosphorylation. Biochem Biophys Res Commun 465:331–337

Chapter 4

A Quantitative Method for the Measurement of Protein Histidine Phosphorylation

Paul V. Attwood

Abstract

The method described in this chapter provides a quantitative means of assaying for protein histidine phosphorylation and thus protein histidine kinase activity, even in the presence of other protein kinases, for example, serine/threonine or tyrosine kinases. The method involves the measurement of ^{32}P, derived from [γ^{32}P]ATP, incorporation into phosphohistidine in a protein substrate. The method makes use of the differential stabilities of phosphohistidine and the common phosphohydroxyamino acids to alkali and acid treatments to measure phosphohistidine incorporation. Phosphoserine and phosphothreonine are depleted by alkali treatment, while phosphohistidine, which is alkali-stable, is removed by acid treatment. Phosphotyrosine is stable to both alkali and acid treatments. The method is filter-based and allows for rapid assay of multiple protein histidine kinase samples, for example, screening for histidine kinase activity, allowing for the calculation of specific activity. In addition, quantitative time-course assays can also be performed to allow for kinetic analysis of histidine kinase activity.

Key words Phosphohistidine, Histidine kinase, Quantitative assay, Phosphotyrosine, Cherenkov, Radioactive phosphate

1 Introduction

In recent years there has been an explosion of interest in protein histidine phosphorylation and the associated histidine kinases and phosphatases, especially in vertebrates (for reviews *see* [1–3]). This has led to the development of a number of techniques for the detection of protein histidine phosphorylation including phosphohistidine-specific antibodies [4, 5] and a number of mass spectrometric techniques [6–8]. These methodologies are aimed at specifically detecting the presence of phosphohistidine in proteins, often in cellular extracts, and are at best, semiquantitative. The focus of this chapter is the description of a protocol for the quantitative measurement of phosphorylation of histidine in a protein substrate by a histidine kinase, allowing for the assay of histidine kinase activity, even in the presence of the more common serine/

Claire E. Eyers (ed.), *Histidine Phosphorylation: Methods and Protocols*, Methods in Molecular Biology, vol. 2077, https://doi.org/10.1007/978-1-4939-9884-5_4, © Springer Science+Business Media, LLC, part of Springer Nature 2020

threonine and tyrosine kinases. This particular approach is based on a less specific assay developed by Wei and Matthews [9]. The method is a filter-based, fixed time point assay that involves measurement of the incorporation of ^{32}P from [γ^{32}P]ATP into histidine residues in a protein substrate (histone H4 in this example) by incubation with a histidine kinase preparation. The specificity of the assay for histidine phosphorylation comes from employing the differences in stability to acid or alkali treatment between phosphohistidine and the more commonly occurring phosphoamino acids (viz., phosphoserine, phosphothreonine, and phosphotyrosine), with the reaction mixture being split and differentially treated prior to evaluating the remaining ^{32}P incorporation. Phosphohistidine is alkali-stable and acid-labile, while phosphoserine and phosphothreonine are alkali-labile and acid stable, phosphotyrosine is alkali-stable and also acid-stable [10]. Thus, the difference in incorporated radioactivity between base-treated and acid-treated fractions can be used to deduce the acid-labile phosphorylation arising due to ^{32}P-phosphohistidine formation. By measuring the specific radioactivity of the [γ^{32}P]ATP and the concentration of protein in the assay, it is possible to determine the stoichiometry of formation of phosphohistidine in the substrate protein at any time. A diagrammatic overview of the procedure is given in Fig. 1. The presence of specific phosphoamino acids in the substrate protein can be confirmed by phosphoamino acid analysis of the phosphorylated substrate protein as described by Tan et al. [11].

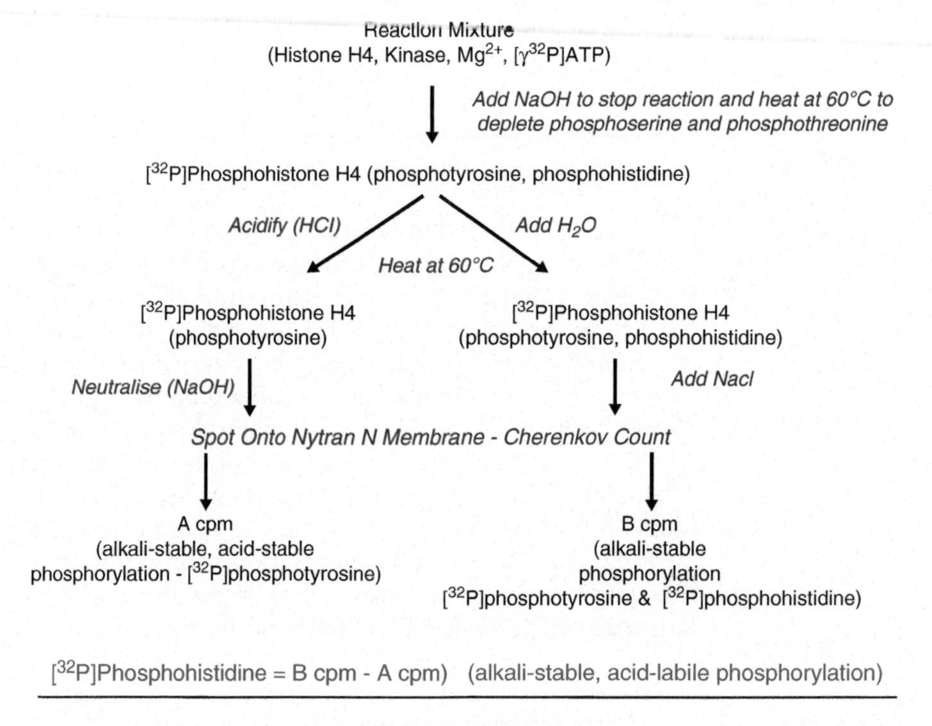

Fig. 1 Schematic flow diagram giving an overview of the assay procedure

2 Materials

1. Nytran N membrane (Schleicher and Schuell).
2. [γ^{32}P]ATP (6000 Ci/mmol) from NEN Radiochemicals.
3. Tris–HCl/$MgCl_2$: 150 mM Tris–HCl, pH 7.5 and 150 mM $MgCl_2$, kept at 4 °C.
4. Substrate protein, example: 5 mg/ml histone H4 purified from Type II-S histone (Sigma-Aldrich) as described by Tan et al. [11] in water, aliquots kept frozen at −20 °C.
5. ATP: 1 mM ATP in water, pH adjusted to 7.0 with NaOH, kept frozen in small aliquots at −20 °C, any aliquot remaining once thawed is discarded.
6. Phosphorylation reaction mixture: 15 mM Tris–HCl (pH 7.5), 15 mM $MgCl_2$, 100 μM ATP, 50 μCi/ml [γ^{32}P]ATP, 1.25 mg/ml substrate protein, (50 mM NaF and 0.2 μg/ml okadaic acid—*see* **Note 1**).
7. Control phosphorylation reaction mixture: as in 6, except the substrate protein is omitted and replaced with an equal volume of water.
8. Alkali stop solution: 3 M NaOH, 2 mM ATP.
9. Acid-treatment solution: 3.6 M HCl.
10. Neutralizing solution: 3.6 M NaOH.
11. Salt solution: 3.6 M NaCl.
12. Protein histidine kinase in 15 mM Tris–HCl, pH 7.5 (containing 20% glycerol if stored at −80 °C), examples: regenerating rat liver nuclear protein histidine kinase in nuclear protein extract as described by Tan et al. [11] (0.3 mg/ml) and yeast protein histidine kinase purified as described by Huang et al. [12] (0.33 mg/ml).
13. Membrane preparation solution: 10 mM tetrasodium pyrophosphate, 1 mM ATP (pH should be 10.0 without adjustment).
14. Membrane washing solution: 10 mM tetrasodium pyrophosphate (pH should be 10.0 without adjustment).
15. Scintillation vials: Beckman Mini PolyQ Scintillation Vials.

3 Methods

3.1 Preparation of Nytran N Membrane

1. Cut a sheet of Nytran N membrane of a size to accommodate the total required number of assays and the controls. Each assay requires a 2 × 2 cm square of membrane, therefore 50 assays with controls will require a sheet 20 × 10 cm.

2. Using a soft pencil, mark squares of 2×2 cm, numbering each square in pencil in one corner.
3. Place the sheet in a lidded container with 100–200 ml (depending on the size of the sheet and the container) of membrane preparation solution. Incubate overnight at 4 °C on a rocker.
4. Remove the sheet from the solution and place on a piece of plastic gutter-guard (a strip of plastic mesh with 1×1 cm square holes separated by 1–2 mm strips of plastic).
5. Bend down the gutter-guard at each end to form a platform to raise the sheet above the bench. Leave at room temperature to dry. A number of sheets can be prepared at the same time and they can be kept in a sealed plastic bag at −80 °C for future use. The purpose of this preincubation is to reduce background radioactivity due to nonspecific adsorption of $^{32}P_i$ and [γ^{32}P] ATP.

3.2 Preparation of Reaction Solutions

Safety note: in all subsequent steps in the protocol, radioactive samples and waste should be handled behind Perspex screens. When handling radioactivity, it is imperative that you wear suitable protective clothing and a personal radiation monitor (as mandated by local institutional regulations). A suitable portable radiation detector should be on hand to monitor contamination. All radioactive waste should be disposed of as mandated by local regulations.

Prepare sufficient phosphorylation reaction mixture and control phosphorylation reaction mixture in disposable plastic tubes to allow 20 μl per sample for the required number of assays: three acid-treated samples and three non–acid-treated samples per assay, and an equal number of controls, plus an additional two or three. After preparation of the stock reaction (or control) solutions, the required volume can be aliquoted for the assays as appropriate (*see* **Note 2**).

1. Add stock Tris–HCl/$MgCl_2$ to the disposable plastic tube, allowing 2 μl per assay.
2. Add the substrate solution, allowing 25 μg per assay for the phosphorylation reaction mixture. For the control phosphorylation reactions add the equivalent amount of water.
3. Add the stock ATP solution (2 μl per assay). Any NaF/okadaic acid should then added (if required, *see* **Note 1**).
4. Add [γ^{32}P]ATP, to the phosphorylation and control phosphorylation reactions tubes, allowing 1 μCi per assay. If required, the volume can be made up to the equivalent of final 20 μl per assay by addition of water (*see* **Note 2** for pipetting radioactive solutions).
5. Aliquot the reaction mixtures into small lidded tubes (e.g., 1.5 ml Eppendorf tubes; 20 μl per assay) in a plastic rack in a

lidded Perspex box (to shield the radiation). If the experiment is to measure a reaction time course on a single protein histidine kinase sample, it is beneficial to include all assays in a single tube and remove aliquots at the required time points (*see* **Note 3**).

6. Dilute the protein histidine kinase with 15 mM Tris–HCl, pH 7.5 so that addition of 30 μl per assay will deliver the required amount of kinase and bring the total reaction volume to 50 μl per assay. If the reaction is required to be performed at a set temperature, above room temperature, both the reaction solutions and the protein histidine kinase solutions should be prewarmed (e.g., to 30 or 37 °C) in an incubator prior to initiation of the reactions and the reactions placed back in the incubator for the reaction period.

3.3 Protein Kinase Reaction

1. Initiate the protein kinase reaction by addition of 30 μl of protein histidine kinase solution per sample. At the same time also add of 30 μl of protein histidine kinase solution to each of the controls. Time the reaction from the point of addition of the kinase (*see* **Note 4** for zero-time measurements).
2. At the end of the incubation time, add 6 μl of alkali stop solution (to each assay (including the control sample) and mix by pipetting up and down. Incubate the tubes at 60 °C for 30 min.
3. To the acid-treatment samples only add 10 μl acid-treatment solution. To the non–acid-treated control samples, add 10 μl water. Incubate the samples at 60 °C for 30 min.
4. Neutralize the acid-treated samples by addition of 10 μl neutralizing solution (3.6 M NaOH). To the non–acid-treated samples add 10 μl salt solution (3.6 M NaCl).
5. Place the preprepared sheet of Nytran N membrane on the gutter-guard support. Carefully apply a 50 μl aliquot of each sample to each 2 × 2 cm square of pretreated Nytran N membrane, avoiding any adsorption to neighbouring squares and concentrating the sample in the centre of square. Note which sample is applied to which numbered square.
6. Place the membrane in a lidded container on a rocker behind a Perspex screen. Wash with 3 × 300 ml membrane washing solution for 15 min per wash at room temperature.
7. Dry the membrane using an infrared lamp positioned directly above the membrane at a height of 50 cm, for about 15–20 min (until dry). Alternatively, use a hair dryer on a low-heat setting, or leave to air-dry for 6 h to overnight.
8. Cut each square carefully from the membrane and place in a scintillation vial for Cherenkov counting.

3.4 Determination of Radioactivity in the Phosphorylated Substrate Protein Attributable to Likely Phosphohistidine Formation and That Attributable to Likely Phosphotyrosine Formation

1. Calculate the radioactivity attributable to likely phosphohistidine by subtracting the radioactivity in the acid-treated reaction sample (**A** cpm) from that of the non–acid-treated samples (**B** cpm), that is, (**B** − **A** = **X** cpm). This value accounts for likely phosphohistidine in both the substrate protein and proteins in the histidine kinase preparation.
2. Perform the same calculation for the control samples (**C** cpm and **D** cpm, respectively) to determine the radioactivity attributable to likely phosphohistidine formation in the proteins in the histidine kinase preparation only, that is, **D** − **C** = **Y** cpm.
3. Calculate the radioactivity attributable to the likely phosphohistidine formation in the substrate protein only by subtraction the total acid-labile phosphorylation in the histidine kinase preparation from the phosphohistidine in both the substrate protein and proteins in the histidine kinase preparation that is, **X** − **Y** = **Z** cpm.
4. The radioactivity attributable to likely phosphotyrosine formation in the substrate protein is given by **A** − **C** = **E** cpm. Phosphoamino acid analysis of the phosphorylated substrate can be performed to confirm attributions of radioactivity to specific phosphoamino acids (*see* **Note 5**).

3.5 Determination of the Specific Radioactivity of the ATP Used in the Reaction Mixture

To calculate stoichiometry of phosphorylation of the substrate protein, it is necessary to determine the specific activity of the [γ^{32}P]ATP used in the assay.

1. Dilute a 10 μl aliquot of one of the reaction samples and one of the control samples 1:4 with water. Apply 10 μl of these solutions to each of six squares of the membrane (*see* **Note 6**). Dry the membrane without washing. Cut out the squares and place in scintillation vials for Cherenkov counting.
2. Calculate the specific radioactivity of the ATP (cpm/pmol) in the samples on the basis of the known amount (pmol) of total ATP (ATP + [γ^{32}P]ATP) in the sample (the [γ^{32}P]ATP contributes a negligible amount of ATP). The amount of likely phosphohistidine (**Z** cpm) and phosphotyrosine (**E** cpm) in the substrate protein can then be converted into pmol phosphate incorporated into the substrate protein in the likely forms of phosphohistidine and phosphotyrosine respectively (*see* examples of time-courses of reactions from Tan et al. [11] in Figs. 2 and 3). Stoichiometries of phosphorylation of the substrate protein can then be estimated from the known amount of substrate protein present and the specific kinase activities in terms of pmol phosphate incorporated into the substrate protein per minute per mg protein in the histidine kinase preparation.

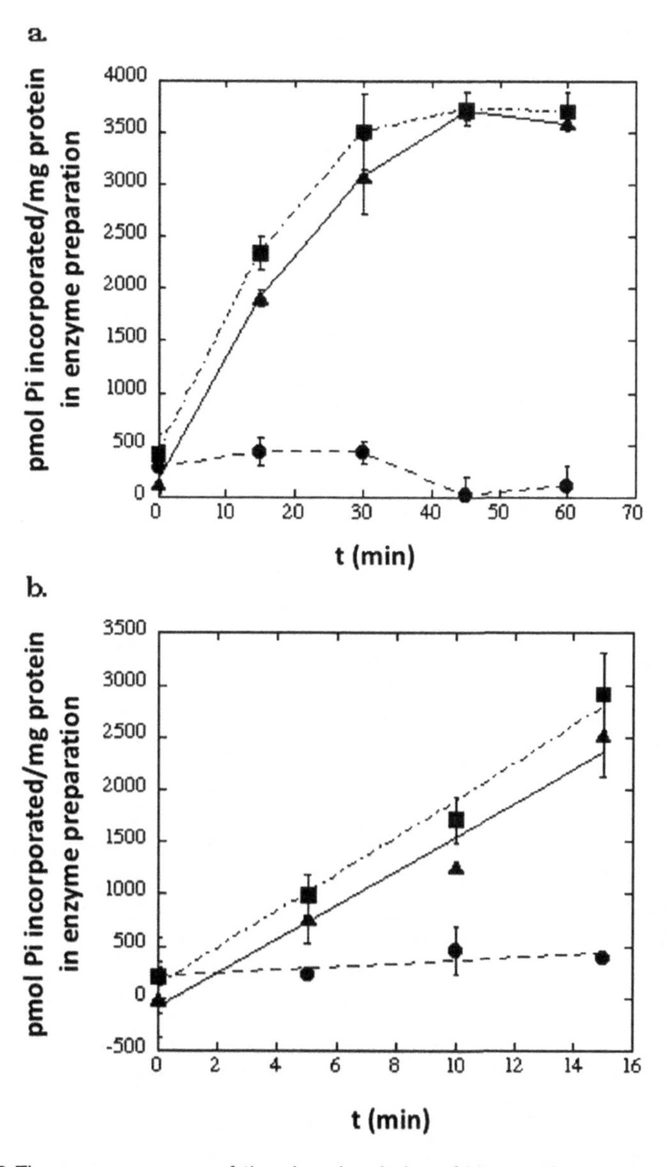

Fig. 2 Time-course assays of the phosphorylation of histone H4 by the nuclear protein extract from regenerating rat liver (Subheading 2, **item 12**). Assays were performed in triplicate and the data are means ± S.E. Time-courses of alkali-stable phosphorylation (filled square); alkali-stable/acid-stable phosphorylation (filled circle); and acid-labile phosphorylation (alkali-stable phosphorylation—alkali-stable/acid-stable phosphorylation) are shown (filled diamond). (**a**) is the time course of the reactions over 60 min, while (**b**) is the time-course of the same reactions over 15 min, lines in (**b**) are linear least squares regression fits of the data. From the linear time courses in (**b**) the specific activities of the phosphorylation reactions were calculated to be alkali-stable/acid-stable phosphorylation (tyrosine kinase activity) = 14 pmol/mg/min; acid-labile phosphorylation (histidine kinase activity) 162 pmol/mg/min. Reproduced from Tan et al. [11] with permission from Elsevier Press, Netherlands

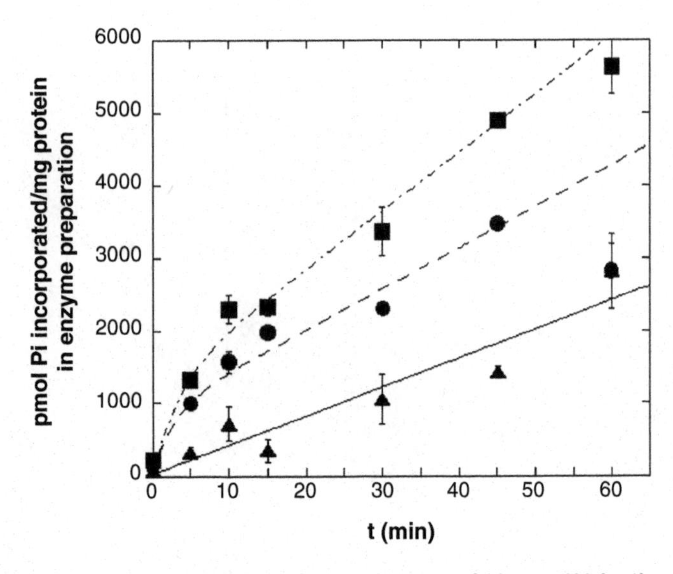

Fig. 3 Time-course assays of the phosphorylation of histone H4 by the yeast histidine kinase preparation (Subheading 2, **item 12**). Assays were performed in triplicate and the data are means ± S.E. Time courses of alkali-stable phosphorylation (filled square); alkali-stable/acid-stable phosphorylation (filled circle); and acid-labile phosphorylation (alkali-stable phosphorylation—alkali-stable/acid-stable phosphorylation) are shown (filled diamond). Lines are nonlinear least squares fits of the data (filled square and filled circle) to an equation describing a first-order exponential approach to steady state or linear least squares regression analysis (filled diamond). The acid-stable/alkali stable phosphorylation (tyrosine kinase activity) exhibits a burst phase in the approach to steady state, which has a specific activity of 57 pmol/mg/min. On the other hand, the acid-labile phosphorylation (histidine kinase activity) is linear throughout the time-course and its specific activity is 30 pmol/mg/min. Thus, there appears to be a somewhat greater tyrosine kinase activity than histidine kinase activity in the preparation. Reproduced from Tan et al. [11] with permission from Elsevier Press, Netherlands

4 Notes

1. Optional—depending on the purity of the kinase and substrate and their potential contamination with protein phosphatases.
2. Use of pipette tips containing a barrier filter (often used for pipetting bacterial suspensions) will prevent contamination (and tedious decontamination) of pipettes.
3. For measurement of a time-course reaction where measurements are made at various times (on the reaction and control), it is better to have single reaction mixture (and control reaction mixture) from which 50 μl aliquots can be removed at different times and added to the alkali stop solution. Use of a "master mix" in this manner removes the variability of having separate reaction mixtures for each time point.

4. For zero-time measurements the alkali stop solution is added to the reaction and control mixtures prior to addition of the protein histidine kinase preparation, and they are immediately heated at 60 °C for 30 min. The rest of the procedure is then as described in Subheading 3.

5. It is generally good practice to check that the inferred attribution of radioactivity incorporation into specific phosphoamino acids in the assays described above is appropriate. This is relatively easily done by phosphoamino acid analysis of the phosphorylated substrate protein as described by Tan et al. [11]. Briefly, the best approach is to isolate the phosphorylated substrate protein from the reaction mixture (prior to acid treatment, but after alkali treatment—followed by neutralization) by immunoprecipitation, assuming an antibody is available for the substrate protein. The immune precipitate is washed with 3 × 1 ml of 1 mM ATP in 10 mM tetrasodium pyrophosphate to remove any bound $^{32}P_i$ and $[\gamma^{32}P]$ATP. The phosphoprotein is then digested with 10 μg pronase E (*Streptomyces griseus* protease) in in 40 μl 20 mM NH_4HCO_3 overnight at room temperature and after vacuum drying of the digest it is resuspended in 10 μl water. Half of the digest is subjected to acid-treatment with 0.6 M HCl at 60 °C for 30 min. 1 μl aliquots of the acid-treated digest and non–acid-treated digest on then spotted on to a reverse-phase TLC plate (Merck RP-18 F_{254}s). Two microgram quantities of authentic phosphoamino acid standards are also spotted on to the plate. The samples are then resolved by running the plate with 100 ml of 72:10:9 mixture of ethanol, 28% ammonia solution, and water. The plate is dried and phosphor-imaged to detect the ^{32}P-phosphoamino acids, signals identified as corresponding to phosphohistidine, should have disappeared in the acid-treated samples while those identified as phosphotyrosine should not have disappeared. The plate can then be sprayed with ninhydrin to visualize the phosphoamino acid standards. An example of this type of analysis is taken from Tan et al. [11] and shown in Fig. 4.

   If no antibody is available against the substrate protein, phosphoamino acid analysis should be performed on both reaction samples and control samples. Some form of microdialysis should be performed prior to digestion, to remove as much unincorporated radioactivity as possible and to get the sample into 20 mM NH_4HCO_3. The control samples will have ^{32}P-phosphoamino acids derived solely from the proteins in the histidine kinase preparation while the reaction samples will have these ^{32}P-phosphoamino acids, plus those from the substrate protein. Differences in phosphoamino acids present or intensities of phosphoimages of phosphoamino acids will indicate which are present in the substrate protein.

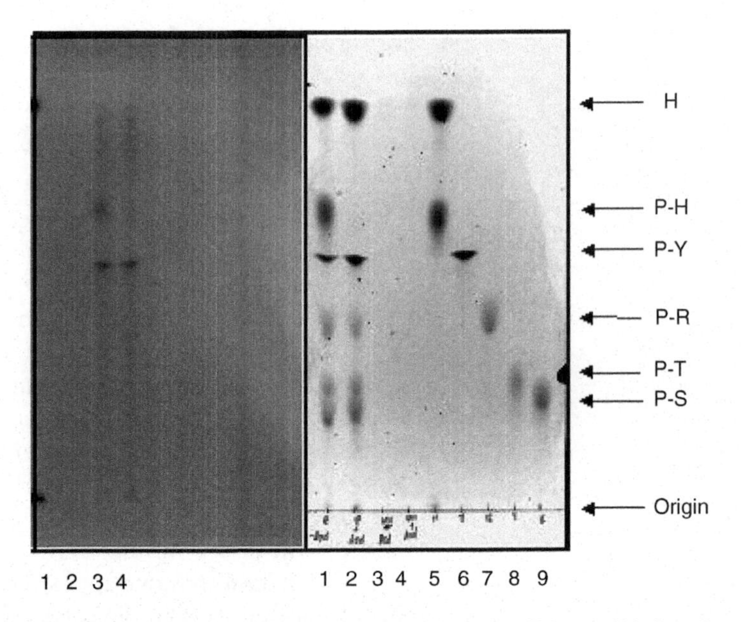

Fig. 4 Reverse-phase TLC phosphoamino acid analysis of histone H4 phosphorylated by the yeast histidine kinase preparation (Fig. 3). On the left is the phosphoimage of the TLC plate and on the right is the image of the ninhydrin-stained plate. A digest of immunoprecipitated phosphorylated histone H4 was prepared and the TLC was run as described in **Note 5**. The non–acid-treated digest was run in Lane 3 and the acid-treated digest was run in Lane 4. Lanes 1 and 2 contain mixtures of phosphoamino acid standards, with the non–acid-treated standards in Lane 1 and the acid-treated standards in Lane 2. Individual phosphoamino acid standards were run in Lanes 5–9 as indicated. The radioactivity seen in the phosphoimage, that migrates to the same point as the phosphohistidine standard (Lane 3), disappears in the acid-treated sample (Lane 4) as might be expected. This phenomenon also occurs in the ninhydrin-stained, acid-treated phosphoamino acid standard mixture (Lane 2), where the signal due to phosphohistidine disappears. The other major radioactive signal, which appears to be more intense than that corresponding to phosphohistidine, migrates similarly to the phosphotyrosine standard and behaves like phosphotyrosine in that it is not affected by acid treatment. Thus, the phosphoamino acid analysis supports the inferred attributions of phosphorylating activity of the kinase assays (Fig. 3). Reproduced from Tan et al. [11] with permission from Elsevier Press, Netherlands

Phospholysine is another alkali-stable, acid-labile phosphoamino acid [13], although it appears to be less common than phosphohistidine in biological samples. It comigrates with phosphoserine on reverse-phase TLC analysis [14] and is not commercially available. It must be prepared, usually by phosphorylation of polylysine with $POCl_3$ followed by alkali hydrolysis [15].

6. Ideally this should be done using each sample. However, the preparation of a "master" phosphorylation reaction mixture

and a control phosphorylation reaction mixture means that measurement of radioactivity in a reaction sample and a control sample is sufficient. (It does not seem to matter whether acid-treated samples or non–acid-treated samples are used, but consistency is advised.)

References

1. Attwood PV (2013) Histidine kinases from bacteria to humans. Biochem Soc Trans 41:1023–1028
2. Fuhs SR, Hunter T (2017) pHisphorylation: the emergence of histidine phosphorylation as a reversible regulatory modification. Curr Opin Cell Biol 45:8–16
3. Attwood PV, Wieland T (2015) Nucleoside diphosphate kinase as protein histidine kinase. Naunyn Schmiedeberg's Arch Pharmacol 388:153–160
4. Fuhs SR, Meisenhelder J, Aslanian A, Ma L, Zagorska A, Stankova A et al (2015) Monoclonal 1- and 3-phosphohistidine antibodies: new tools to study histidine phosphorylation. Cell 162:198–210
5. Kee JM, Oslund RC, Perlman DH, Muir TW (2013) A pan-specific antibody for direct detection of protein histidine phosphorylation. Nat Chem Biol 9:416–421
6. Hohenester UM, Ludwig K, Krieglstein J, Konig S (2010) Stepchild phosphohistidine: acid-labile phosphorylation becomes accessible by functional proteomics. Anal Bioanal Chem 397:3209–3212
7. Oslund RC, Kee JM, Couvillon AD, Bhatia VN, Perlman DH, Muir TW (2014) A phosphohistidine proteomics strategy based on elucidation of a unique gas-phase phosphopeptide fragmentation mechanism. J Am Chem Soc 136:12899–12911
8. Potel CM, Lin MH, Heck AJR, Lemeer S (2018) Widespread bacterial protein histidine phosphorylation revealed by mass spectrometry-based proteomics. Nat Methods 15:187–190
9. Wei YF, Matthews HR (1990) A filter-based protein kinase assay selective for alkali-stable protein phosphorylation and suitable for acid-labile protein phosphorylation. Anal Biochem 190:188–192
10. Attwood PV, Piggott MJ, Zu XL, Besant PG (2007) Focus on phosphohistidine. Amino Acids 32:145–156
11. Tan E, Lin Zu X, Yeoh GC, Besant PG, Attwood PV (2003) Detection of histidine kinases via a filter-based assay and reverse-phase thin-layer chromatographic phosphoamino acid analysis. Anal Biochem 323:122–126
12. Huang JM, Wei YF, Kim YH, Osterberg L, Matthews HR (1991) Purification of a protein histidine kinase from the yeast Saccharomyces cerevisiae. The first member of this class of protein kinases. J Biol Chem 266:9023–9031
13. Besant PG, Attwood PV, Piggott MJ (2009) Focus on phosphoarginine and phospholysine. Curr Protein Pept Sci 10:536–550
14. Besant PG, Lasker MV, Bui CD, Turck CW (2000) Phosphohistidine analysis using reversed-phase thin-layer chromatography. Anal Biochem 282:149–153
15. Wei YF, Matthews HR (1991) Identification of phosphohistidine in proteins and purification of protein-histidine kinases. Methods Enzymol 200:388–414

Chapter 5

Analysis of 1- and 3-Phosphohistidine (pHis) Protein Modification Using Model Enzymes Expressed in Bacteria

Alice K. M. Clubbs Coldron, Dominic P. Byrne, and Patrick A. Eyers

Abstract

Despite the discovery of protein histidine (His) phosphorylation nearly six decades ago, difficulties in measuring and quantifying this unstable post-translational modification (PTM) have limited its mechanistic analysis in prokaryotic and eukaryotic signaling. Here, we describe reliable procedures for affinity purification, cofactor-binding analysis and antibody-based detection of phosphohistidine (pHis), on the putative human His kinases NME1 (NDPK-A) and NME2 (NDPK-B) and the glycolytic phosphoglycerate mutase PGAM1. By exploiting isomer-specific monoclonal N1-pHis and N3-pHis antibodies, we describe robust protocols for immunological detection and isomer discrimination of site-specific pHis, including N3-pHis on His 11 of PGAM1.

Key words NME1, NME2, NDPK-A, NDPK-B, PGAM1, Histidine phosphorylation, Western blotting, Immunoblotting, Differential scanning fluorimetry, N1-phosphohistidine, N3-phosphohistidine

1 Introduction

Protein phosphorylation is a common post-translational modification (PTM) that plays a vital role in all strands of life. The addition of a phosphate group to a protein can change its fundamental properties, including the regulation of catalysis and modification of the protein interactome [1]. Such regulation is the basis of a large number of essential cellular processes such as cell division, cell growth and programmed cell death [2]. Canonical phosphorylation occurs through phosphoester bond formation with free hydroxyl groups on Ser/Thr and Tyr amino acids in proteins, and can readily be analyzed using chemical, immunological, and/or mass spectrometry (MS)-based approaches [3, 4]. Recent findings have revealed even more complex phosphorylation patterns in human cells, including "non-canonical" phosphorylation of amino acids such as histidine [5–7]. Historically, His phosphorylation (pHis) was originally identified in a bovine preparation by

Claire E. Eyers (ed.), *Histidine Phosphorylation: Methods and Protocols*, Methods in Molecular Biology, vol. 2077,
https://doi.org/10.1007/978-1-4939-9884-5_5,

Boyer and colleagues in the early 1960s [8]. Subsequent work has confirmed fundamental roles for pHis as an enzyme intermediate, and as part of signal relay systems in prokaryotic and plant-based signal transduction [9, 10].

However, pHis remains largely unexplored in models of mammalian cell signaling or as a biomarker, due to the relative instability of the phosphoramidate (N-P) bond, which is labile at high temperature and acidic pH [5, 10]. Rapid hydrolysis of this bond during sample preparation has made it very difficult to determine the prevalence and importance of pHis in eukaryotic signaling, in marked difference to the identification of many thousands of sites of Ser/Thr/Tyr phosphorylation in proteomes [11]. Phosphohistidine is unique in that two biologically relevant isomers can occur. Both imidazole nitrogens can be phosphorylated [1], forming 1-phosphohistidine (1-pHis) or 3-phosphohistidine (3-pHis). NDPK-A (Nucleoside diphosphate kinase A, termed here NME1) and NDPK-B (nucleoside diphosphate kinase B, termed here NME2) are best known as nucleotide kinases [12]. NME protein family members catalyze the transfer of a γ-phosphate from nucleoside triphosphates to nucleoside diphosphates through a 1-pHis118 enzyme intermediate, and have also been reported to function as protein His kinases [13–15]. Phosphohistidine also functions as an enzyme intermediate in the glycolytic pathway, including a 3-pHis11 intermediate on the metabolic enzyme phosphoglycerate mutase 1 (PGAM1), which serves as the phosphate donor during the interconversion of 2-phosphoglycerate for 3-phosphoglycerate [16]. A related PGAM (PGAM5) has recently been proposed to function as a tumor-suppressive His phosphatase in mammals [17].

The recent development, exploitation and commercial release of individual 1-pHis and 3-pHis specific monoclonal antibodies [5] can now help in the identification of pHis-containing protein substrates, and cellular analysis of pHis in immunological, proteomic and biological assays. In particular, this allows the simple analysis of pHis in proteins through the use of modified immunoblotting procedures. This chapter focuses on methods developed in our laboratory to purify and exploit recombinant versions of human NME1/2 and PGAM1, which are pHis-containing proteins under certain experimental conditions. They are suitable for use in enzyme assays (to measure both *in cis* and *in trans* His phosphorylation of substrates), thermal stability assays (TSAs) for ligand-binding analysis [18], and as reliable experimental controls for SDS-PAGE and immunodetection-based analysis of either 1-pHis on NME1 and NME2 or 3-pHis on PGAM1.

2 Materials

Prepare all solutions with highest-quality deionized water at room temperature. Store all reagents at room temperature unless otherwise indicated. Nucleotides (−20 °C) and purified enzymatically cleaved enzymes (−80 °C) are stored frozen prior to use.

2.1 PCR and Agarose Gel Electrophoresis

1. CloneAmp HiFi PCR Premix (Takara).
2. 5× Infusion HD Enzyme Premix (Takara).
3. Midori Green marker dye.
4. TAE buffer (10×): 400 mM Tris–HCl, pH 8.0, 200 mM acetate, 10 mM EDTA. Weigh 48.4 g of Tris and 3.7 g of EDTA and transfer to a 1 L graduated cylinder. Add ~800 mL of water and 11.4 mL of acetic acid. Adjust to pH 8.0 if necessary and make up to a final volume of 1 L with water.
5. Low-melting temperature agarose: 1% (w/v) agarose. Weigh 1 g of Agarose and add 100 mL of 1× TAE buffer. Heat to dissolve and once cool, add 6 μL of Midori Green.
6. Nuclease-free water.
7. Primers suitable for His to Ala conversion in the protein of interest at a 100 μM stock made up in nuclease-free water (NFW), following manufacturer's instructions (*see* Table 1).
8. Template plasmid DNA in NFW.
9. Restriction enzymes for plasmid linearization (we use KpnI/ HindIII for pOPINJ).
10. QIAprep Spin MiniPrep Kit (QIAGEN).
11. QIAquick Gel Extraction Kit (QIAGEN).
12. Benchtop centrifuge.
13. PCR machine.
14. Bio-Rad Gel tanks and PAGE apparatus, including PowerPac™ HC high-current power supply suitable for mini-gel electrophoresis.

2.2 Protein Expression and Purification

The following pOPINJ and pET-based expression plasmids are employed to encode the following human proteins, which are expressed and purified to homogeneity (*see* Fig. 1): pOPINJ human NME1 (N-terminal 6His, GST, 3C-cleavable full-length amino acids 1–152), pOPINJ human NME2 (N-terminal 6His, GST, 3C-cleavable full-length amino acids 1–152), pOPINJ human PGAM1 (N-terminal 6His, GST, 3C-cleavable full-length amino acids 1–254), pOPINJ human PGAM1 His11Ala (N-terminal 6His, GST, 3C-cleavable full-length amino acids 1–254), pET28a human ABL (N-terminal 6His tag, amino acids

Table 1
Oligonucleotide primer sequences for site-directed mutagenesis

	PGAM1 primer
Forward External Primer (5′–3′)	AAGTTCTGTTTCAGGGCCCGATGGCCGCCTACAAACTGG
Reverse External Primer (5′–3′)	ATGGTCTAGAAAGCTTTATCACTTCTTGGCCTTGCCC
H11A PGAM1 Forward Mutagenic Primer (5′–3′)	CAAACTGGTGCTGATCCGG**GCC**GGCGAGAGCGCA TGGAAC
H11A PGAM1 Reverse Mutagenic Primer (5′–3′)	GTTCCATGCGCTCTCGCC**GGC**CCGGATCAGCACCAG TTTG

The mutated codon in PGAM1 is highlighted in bold in the primer pair

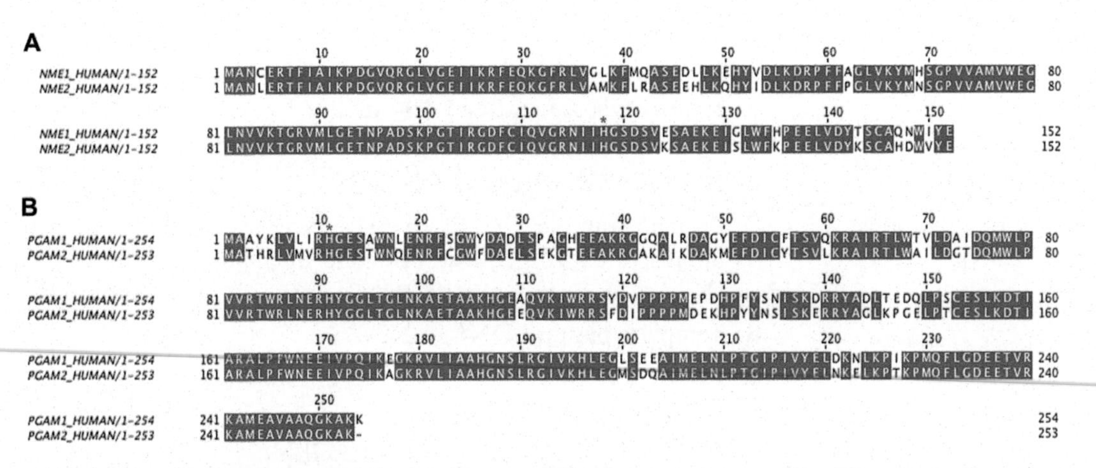

Fig. 1 Amino acid conservation in human NME1/2 and PGAM1/2 proteins used for pHis analysis. Sequence alignment of human NME1 (NDPK-A) and NME2 (NDPK-B) and human PGAM1 and PGAM2. (**a**) Amino acid (aa) comparison of NME1 (full-length, 152 aa) and NME2 (full-length, 152 aa) and (**b**) PGAM1 (full-length, 254 aa) and PGAM2 (full-length, 253 aa). Human NME1 and NME2 exhibit 88% amino acid identity, with 94% similarity. PGAM1 and PGAM2 exhibit 81% amino acid identity, with 90.5% similarity. Residues highlighted in blue exhibit 100% identity. The phosphorylated His residue is highlighted with a red asterisk for each enzyme. This is amino acid His118 on NME1 and NME2 and His11 on PGAM1 and PGAM2. NME1 UniProt ID: P15531, NME2 UniProt ID: P22392, PGAM1 UniProt ID: P18669, and PGAM2 UniProt ID: P15259

46–515), pET30 human Aurora A (N-terminal 6His-tag, amino acids 1–403).

1. Chemically competent *E. coli* cells: Top10, BL21 (DE3) pLys cells. Cells can be made chemically competent for plasmid transformation in-house using standard RbCl procedures [19], or purchased from an appropriate vendor.
2. Lysogeny broth (LB) agar plates: Weigh 3.7 g of LB agar and add 100 mL water to make ~5 agar plates. Autoclave to sterilize

and add the appropriate antibiotics (once cool) and pour liquid agar into a cylinder plate and leave to set.

3. LB (broth): Weigh 2.5 g of LB (broth) and add 100 mL water to a sterile conical flask. Autoclave to sterilize.
4. Antibiotics selected according to the requirements of the plasmid being used to express the protein of interest: chloramphenicol—final concentration 35 μg/mL; ampicillin—final concentration 50 μg/mL. These should be made up as 1000× stocks in ethanol and water respectively.
5. Super Optimal broth with Catabolite repression (SOC).
6. Isopropyl β-D-1-thiogalactopyranoside (IPTG): 100 mM. Weigh 1.19 g of IPTG and add 50 mL water. Mix to dissolve and pass through a 0.45 μM syringe. Store at −20 °C.
7. cOmplete™ EDTA-free Protease Inhibitor tablets.
8. Lysis buffer: 50 mM Tris–HCl, pH 7.4, 300 mM NaCl, 1 mM DTT, 0.1 mM EDTA, 0.1 mM EDTA, 1% (v/v) Triton X-100, 10% (v/v) glycerol, 10 mM Imidazole (*see* **Note 1**). Add 10 mL of glycerol to a 100 mL cylinder and add 5 mL of 1 M Tris–HCl, pH 7.4, 10 mL of 3 M NaCl, 100 μL of 1 M DTT, 100 μL of 100 mM EDTA, 100 μL of 100 mM EGTA, 1 mL of 1 M Imidazole, and 1 mL of Triton X-100 (*see* **Note 2**). Add water to a final volume of 100 mL and dissolve cOmplete™ EDTA-free Protease Inhibitor tablets (1 per 100 mL lysis buffer).
9. Gel filtration buffer: 50 mM Tris–HCl, pH 7.4, 100 mM NaCl, 1 mM DTT, 10% (v/v) glycerol. Add 100 mL of glycerol to a 1 L cylinder and add 50 mL of 1 M Tris–HCl, pH 7.4, 33 mL 3 M NaCl, and 1 mL DTT and mix. Transfer to a 1 L duran (*see* **Note 3**).
10. High-salt wash buffer: 50 mM Tris–HCl, pH 7.4, 500 mM NaCl, 20 mM Imidazole. Add 5 mL of 1 M Tris–HCl, pH 7.4, 16.67 mL of 3 M NaCl, 2 mL of 1 M imidazole to a 100 mL graduated cylinder. Add water to a final volume of 100 mL and mix.
11. Low-salt wash buffer: 50 mM Tris–HCl, pH 7.4, 100 mM NaCl, 20 mM imidazole. Add 5 mL of 1 M Tris–HCl, pH 7.4, 3.33 mL of 3 M NaCl, and 2 mL of 1 M imidazole to a 100 mL graduated cylinder. Add water to a final volume of 100 mL and mix.
12. Elution buffer: 50 mM Tris–HCl, pH 7.4, 100 mM NaCl, 1 mM DTT, 10% (v/v) glycerol, 500 mM imidazole. Add 10 mL of glycerol to a 100 mL graduated cylinder and add 5 mL of 1 M Tris–HCl, pH 7.4, 3.33 mL of 3 M NaCl, and 100 μL of 1 M DTT and add water to a final volume of 80 mL. Weigh out 3.4 g of imidazole and transfer to the cylinder. Adjust the pH with HCl to pH 7.4 and make up to 100 mL with water.

13. Shaking incubator with temperature control suitable for holding 8 × 2 L conical flasks.
14. Reusable plastic chromatography column (20 mL).
15. Nickel-NTA resin for IMAC.
16. Superdex 200 (16/60) column (GE Healthcare Life Sciences).
17. Automatic liquid chromatography system (AKTA or similar, although disposable gravity columns can suffice for IMAC).

2.3 SDS-PAGE

1. Resolving gel buffer: 1.5 mM Tris–HCl, pH 8.8. Add 100 mL water to a 1 L graduated cylinder. Weigh 181.7 g Tris base and transfer to the cylinder. Add water to a volume of 900 mL. Mix and adjust to pH 8.8 with HCl; make up to 1 L with water.
2. Stacking gel buffer: 0.5 mM Tris–HCl, pH 6.8. Weigh 60.6 g Tris base and prepare a 1 L solution as in previous step.
3. Bis–acrylamide solution: 29.2:0.8 acrylamide–bis ratio.
4. Ammonium persulfate: 10% (w/v) in water.
5. *N,N,N,N′*-tetramethyl-ethylenediamine (TEMED).
6. SDS-PAGE running buffer (10×): 25 mM Tris, 192 mM glycine, 3.5 mM SDS. Weigh 60.57 g Tris, 288.3 g glycine and 20 g SDS in 2 L of water and mix. Dilute 100 mL of 10× SDS-PAGE running buffer in 900 mL water to make a 1× solution for use.
7. SDS loading sample buffer (LSB) (5×): 250 mM Tris–HCl, pH 6.8, 10% (w/v) SDS, 50% (v/v) glycerol, 500 mM DTT, 0.5% (v/v) bromophenol blue (*see* **Note 4**). Add 2.5 mL 1 M Tris–HCl, pH 6.8, 1 g SDS, 5 mL glycerol, 0.7 g DTT, 0.5 g bromophenol blue to a 15 mL falcon tube and make up to 10 mL with water. Vortex to mix. Light heating may be required to dissolve the SDS.
8. Prestained molecular mass standards.
9. Gel plates, combs, and electrophoresis system for SDS-PAGE.
10. Coomassie blue: Add 2250 mL methanol, 50 g of Brilliant Blue, 500 mL acetic acid to a 5 L beaker and make up to 5 L with water. Mix well.
11. Destaining solution: Add 1 L methanol, 500 mL acetic acid to a 5 L beaker and make up to 5 L with water. Mix well.
12. Plastic container for gel staining.

2.4 Enzyme Assays

1. 6His-tagged 3C protease: 30 μM stock for removal of tandem 6His-GST tags on NME1/2 and PGAM1.
2. Purified (uncleaved) NME1, NME2, PGAM1, or PGAM1 H11A for phosphotransferase assays.
3. Adenosine triphosphate (ATP), adenosine diphosphate (ADP), guanosine triphosphate (GTP), guanosine diphosphate

(GDP), 2, 3-diphosphoglycerate (DPG). Make up 10 mM stock solutions in water and adjust to pH 8.0 with NaOH.

4. $MgCl_2$: 100 mM. Dissolve 9.5 mg $MgCl_2$ in 100 mL of water.
5. Assay buffer: 50 mM Tris–HCl, 100 mM NaCl, pH 8.0.
6. Microcentrifuge tubes.

2.5 Thermal Stability Assay

1. StepOnePlus Real-Time PCR System (Thermo Fisher).
2. SYPRO Orange: 1:1000 final dilution (Invitrogen), as described in [20].
3. Purified (3C-cleaved) NME1, NME2, PGAM1, or PGAM1 H11A; ATP, ADP, AMP, GTP, GDP, GMP, cyclic AMP (adenosine 3′-5′ cyclic monophosphate) and cyclic GMP (guanosine 3′-5′ cyclic monophosphate), 2, 3-diphosphoglycerate (DPG). Stock solutions (10 mM) of all nucleotides should be made in water and adjusted to pH 8.0 with NaOH.
4. 96-well clear microplates and adhesive cover.

2.6 Antibodies

1. Prestained molecular mass markers.
2. Rabbit anti-N1-pHis monoclonal primary antibody, clone SC50-3; N3-pHis monoclonal primary antibody, clone SC39-6 (Sigma-Aldrich) [5].
3. Mouse anti-6His HRP-linked primary antibody (Sigma-Aldrich).
4. Goat anti-rabbit HRP secondary antibody (Sigma-Aldrich).
5. Goat anti-mouse HRP secondary antibody (Cell Signaling Technology).
6. Mouse monoclonal pTyr antibody (P-Tyr-100) (Cell Signaling Technology).
7. Rabbit pT288 Aurora A antibody [21].

2.7 Immunoblotting and ECL Reagents

1. SDS Loading Sample Buffer (LSB) (5×) pH 8.8, as subheading 2.3, **step 7** but adjust pH to 8.8 (*see* **Note 5**).
2. Nitrocellulose membranes: 0.45 μm, cut to the size of the SDS-polyacrylamide gel.
3. Blotting paper: eight sheets per gel, cut to the size of the gel.
4. Transfer buffer: 25 mM Tris, 192 mM glycine, 20% (v/v) methanol. Weigh out 4.54 g of Tris base, 21.6 g of glycine and transfer to a 2 L beaker. Add 1.2 L of water and 300 mL methanol. Mix to dissolve at 4 °C for 1 h.
5. Tris-buffered saline with tween (TBST): 20 mM Tris–HCl, pH 7.6, 137 mM NaCl, 0.1% (w/v) Tween 20. Weigh out 4.84 g of Tris base, 16.0 g of NaCl and transfer to a 2 L beaker.

Add 1.8 L of water and adjust the pH to 7.6 with HCl. Add 2 mL of Tween 20 and water to make a final volume of 2 L.

6. Blocking solution: 5% (w/v) nonfat milk powder in TBST. Weigh 5 g powdered milk and transfer to a conical flask. Add 100 mL TBST buffer and mix to dissolve. Store at 4 °C.
7. ECL reagents (Amersham).
8. Cold room (4 °C) and ice bath to maintain low temperature and minimize pHis loss during electrophoresis.

3 Methods

3.1 Mega Primer Site-Directed Mutagenesis

This protocol is compatible for the pOPIN infusion vector suite. The example given here is for the creation of a His11 to Ala substitution in PGAM1. For the analysis of other pHis-containing proteins, specific primers can be designed to mutate His to Ala (or another nonmodifiable amino acid) using these principles.

3.1.1 First Round PCR

1. Assemble a 12.5 μL PCR reaction as follows: 25 ng template plasmid DNA, 2.5% DMSO, 0.3 μM forward external primer, 0.3 μM reverse mutagenic primer, and 6.25 μL Clone Amp reaction mix. Adjust the volume to 12.5 μL with nuclease-free water (NFW).
2. Perform PCR amplification using the following parameters: 98 °C for 10 s, 55 °C for 10 s, 72 °C for 1–2 min, followed by a final extension at 72 °C for 10 min (*see* **Note 6**).
3. Analyze the PCR product by agarose gel electrophoresis (*see* **Note 7**).
4. Excise the PCR product (forward megaprimer) from the agarose gel using a scalpel blade and purify the DNA fragment using the QIAquick gel extraction kit according to the manufacturer's instructions.
5. Collect in nuclease-free water (NFW).

3.1.2 Second Round PCR

1. Perform the second round of PCR amplification as described in subheading 3.1.1, substituting the forward external primer for the reverse external primer. Use the forward megaprimer extracted from the first round of PCR as the mutagenic primer.
2. Excise the resultant PCR product from the agarose gel using a scalpel blade and purify the DNA fragment from the gel using the QIAquick gel extraction kit according to the manufacturer's instructions.
3. Collect in NFW.

3.1.3 pOPINJ Infusion Reaction

1. Assemble a 2.5 μL infusion reaction as follows: 12.5 ng linearized vector (KpnI/HindIII restricted pOPINJ), 0.5 μL 5× In-fusion HD enzyme premix, and 12.5 ng excised PCR product (from Subheading 3.1.2, **step 3**). Adjust the volume to 2.5 μL with NFW.
2. Incubate at 50 °C for 15 min (*see* **Note 8**).
3. Place the infusion reaction on ice before bacterial transformation.

3.1.4 Transform DNA Plasmid into Competent Top 10 E. coli Cells

1. Add 1 μL of plasmid DNA (~50 ng) to 50 μL of competent *E. coli* cells and mix gently by rotating the pipette tip (do not mix by pipetting up and down). Incubate on ice for 30 min (*see* **Notes 9** and **10**).
2. Heat-shock the cells for 30 s at 42 °C. Place back on ice for 2 min.
3. Add 250 μL of SOC or LB medium and incubate at 37 °C in a shaker at 240 rpm for 30 min to 1 h.
4. Spread the transformation reaction evenly onto 50 μg/mL antibiotic-containing LB agar plates and incubate overnight at 37 °C.

3.1.5 Make a Starter Culture for Plasmid Extraction

1. Add 5 mL of the LB broth containing the appropriate antibiotic to a 15 mL falcon tube.
2. Pick a colony from the agar plate using a sterile inoculation loop and mix in the falcon tube. Incubate at 37 °C overnight at 240 rpm. Ensure the falcon tube is at an angle to maximise aeration.

3.1.6 MiniPrep: Plasmid Extraction

1. Pellet the cells at 5000 × *g* for 10 min and decant the supernatant.
2. Isolate the plasmid DNA using a QIAprep Spin MiniPrep kit (QIAGEN) according to the manufacturer's instructions.

3.2 Protein Expression and Purification

All buffers for protein purification should be prechilled to 4 °C. All protein samples should be kept on ice.

3.2.1 Transformation of Bacteria for Protein Expression

1. Add 1 μL of the appropriate plasmid (~50 ng) to 50 μL competent pLys BL21 (DE3) *E. coli* cells.
2. Carry out a transformation following standard protocol (as described in Subheading 3.1.4) and prepare an overnight cell culture of 100 mL (as described in Subheading 3.1.5) with a final concentration of 50 μg/mL Ampicillin and 35 μg/mL Chloramphenicol.

3.2.2 Bacterial Protein Expression of His-GST Tagged Proteins from pOPINJ

1. Prepare 8 × 2 L conical flasks with sterile 700 mL LB broth.
2. Add 5 mL of the overnight culture of plasmid-containing *E. coli* (from Subheading 3.2.1) to each flask along with the appropriate antibiotics (50 μg/mL of ampicillin and/or 35 μg/mL of chloramphenicol).
3. Incubate the flasks at 37 °C (240 rpm) for ~2 h until the OD_{600} reaches 0.6–0.8. Reduce the temperature of the incubator to 18 °C and induce protein expression by addition of 0.4 mM IPTG and leave overnight (~18 h).
4. Collect the cells by centrifugation at 5000 × *g* for 10 min, 4 °C, decant the supernatant and transfer the cell pellet into a 100 mL glass beaker (*see* **Note 11**).

3.2.3 Lysis and Sonication of the Bacteria

1. Resuspend the *E. coli* pellet in ice-cold lysis buffer by pipetting up and down. The sample can be vortexed if necessary to aid resuspension.
2. Sonicate on ice at an amplitude of 16 for 30 s. Leave on ice for 1 min. Repeat this 6–10 times as required to lyse the cells (*see* **Note 12**).
3. Centrifuge the lysate at 43,000 × *g* for 60 min at 4 °C (*see* **Note 13**).
4. Collect the supernatant and pass through a 0.45 μM syringe, retaining the clarified supernatant on ice for immediate purification.

3.2.4 Protein Purification

1. Equilibrate 1 ml of Ni-NTA agarose beads in lysis buffer by resuspending and washing the Ni-NTA agarose beads with 2 mL lysis buffer. Repeat this three times.
2. Add the Ni-NTA agarose beads to the cell lysate and incubate at 4 °C with gentle agitation using a magnetic stirrer for 1 h.
3. Load the lysate and the Ni-NTA agarose beads onto a reusable plastic chromatography column, collecting the protein-bound beads as the flow-through lysate passes through. Keep the lysate on ice during this step.
4. Wash the Ni-NTA agarose beads in the column with high-salt wash buffer in 5 mL increments (~50 mL in total).
5. Add 5 × 1 mL of elution buffer to the beads in the column, collecting 500 μL samples of eluent in microcentrifuge tubes.
6. Analyze the eluent fractions by SDS-PAGE and Coomassie staining to confirm the presence of protein (*see* **Note 14**). It is recommended that samples of total cell pellet, clarified lysate and the flow-through (non-binding fraction) from the Ni-NTA agarose column be analyzed alongside for comparison.

7. Centrifuge the protein-containing eluent fractions for 20 min at 4 °C, at 16,000 × *g* using a benchtop centrifuge to remove any aggregated protein.
8. Equilibrate a Superdex 200 (16/60) column with de-gassed gel filtration buffer at a flow rate of 1 mL/min for 200 mL (0.5 mPa). Purify the protein by size exclusion chromatography, collecting 1.5 mL fractions on ice.
9. Analyze the fractions by SDS-PAGE and stain the gel with Coomassie to determine which fractions contain the purified protein.

3.2.5 His-GST 3C Cleavage

1. Repeat Subheading 3.2.4, **steps 1** and **2**.
2. Resuspend the beads in the column with 10 mL low-salt wash buffer and transfer 5 mL to a 5 mL microcentrifuge tube.
3. Centrifuge at 1000 × *g* for 5 min at 4 °C and repeat for remaining 5 mL until all beads have been collected. Wash the beads 3–4 times with 4 mL low-salt wash buffer, centrifuging as above, removing the supernatant each time.
4. After all washes are complete, collect a small proportion of the wash supernatant as a sample for SDS-PAGE analysis.
5. Add an appropriate amount of 3C protease in a total volume of 500 μL low-salt wash buffer to facilitate on-bead cleavage, and incubate at 4 °C on a roller for ~3 h (*see* **Note 15**).
6. Centrifuge at 1000 × *g* for 5 min, 4 °C. Collect the supernatant containing the His-GST-cleaved protein.
7. Wash the beads with high-salt wash buffer and centrifuge again at 1000 × *g* for 5 min.
8. Remove the supernatant and resuspend the beads in 1 mL elution buffer. Leave for 1 min.
9. Centrifuge the beads at 1000 × *g* for 5 min. Collect the supernatant (eluent) and analyze a sample by SDS-PAGE (*see* **Note 16**).

3.2.6 SDS-PAGE

1. Add 2.5 μL of 5× gel loading sample buffer, pH 6.8 to 10 μL of each sample.
2. Heat samples at 95 °C for 5 min and separate by SDS-PAGE at 200 V for ~30 min or until the dye front reaches the bottom of the gel.
3. To visualize NME1/2 and PGAM1 (Fig. 2), stain the gel with Coomassie blue and use several changes of destaining solution to destain gel background.

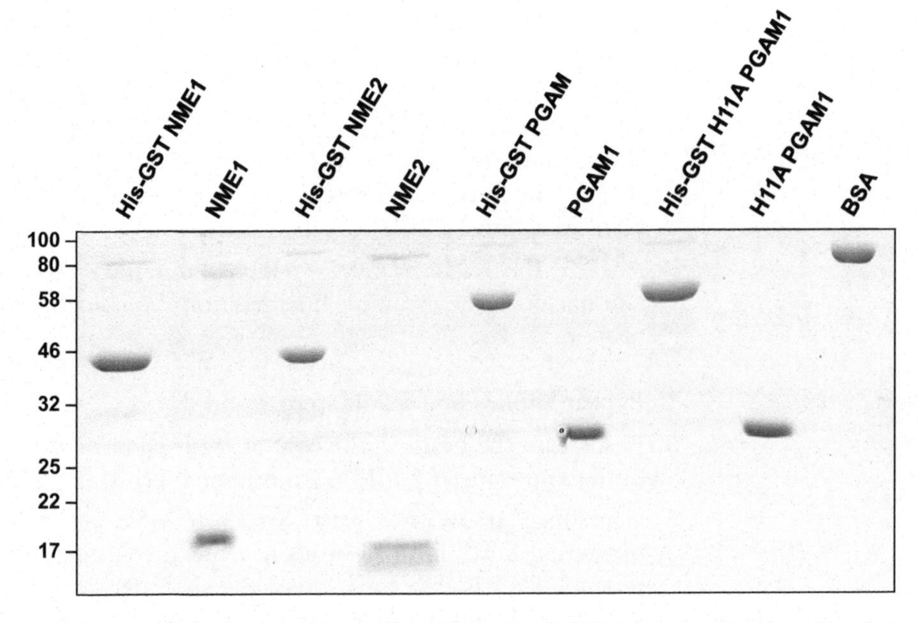

Fig. 2 SDS-PAGE analysis of cleaved and uncleaved NME1/2 and PGAM1 proteins. Analysis of purified proteins before and after 3C cleavage, which removes the N-terminal His-GST affinity tag from the protein of interest. The 3C protease was removed by IMAC. Two microgram of purified cleaved and uncleaved NME1, NME2, PGAM1, and H11A PGAM1 were processed for SDS-PAGE and visualized using Coomassie blue staining and destaining. A known amount of BSA 2 μg was used as a standard

3.3 Differential Scanning Fluorimetry

A more detailed explanation of differential scanning fluorimetry (DSF) can be found in [22]; specific procedures for the analysis of NME1/2 and PGAM1 are described in Fig. 3.

1. DSF experiments are performed using an Applied Biosystems StepOnePlus Real-Time PCR instrument, using a thermal ramping procedure developed and validated for canonical protein kinases and pseudokinases [22–25].
2. For each reaction, the final volume is 25 μL, which is dispensed in duplicate (ideally triplicate) into individual wells of a MicroAmp Fast Optical 96-well reaction plate (Applied Biosystems).
3. For each reaction, the appropriate amount of NME1/2 or PGAM1 in gel filtration buffer (final concentration 5 μM protein) is assembled alongside a mixture of nucleotide (typically 1 mM final) or divalent cation (typically 10 mM final) or both additives and SYPRO Orange dye (final dilution 1:1000 of supplied stock).
4. Set up a series of control reactions lacking nucleotide or divalent cations. In these tubes add a suitable volume of buffer to compensate to ensure a final volume of 25 μL.

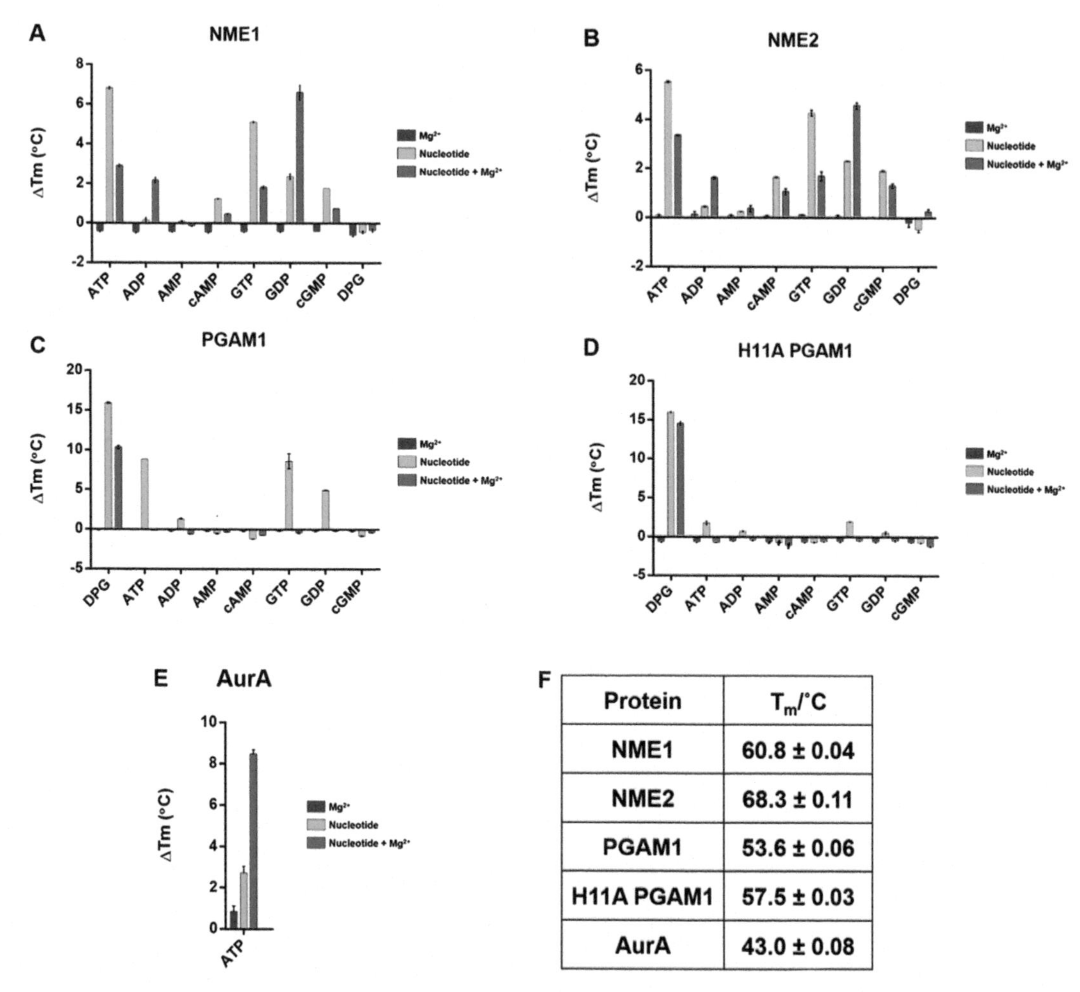

Protein	T_m/°C
NME1	60.8 ± 0.04
NME2	68.3 ± 0.11
PGAM1	53.6 ± 0.06
H11A PGAM1	57.5 ± 0.03
AurA	43.0 ± 0.08

Fig. 3 Differential scanning fluorimetry (DSF) to analyze thermal effects of ligand-binding on putative Histidine kinases. Thermal stability assay (TSA) using purified 3C-cleaved (**a**) NME1 (**b**) NME2 (**c**) PGAM1 or (**d**) H11A PGAM1 and the well-characterized Ser/Thr protein kinase Aurora A (**e**) in the presence or absence of 10 mM $MgCl_2$ ions (Mg^{2+}) (blue) 1 mM nucleotide alone (green) or both Mg^{2+} and nucleotide (red). Thermal-shift unfolding assays were performed using SYPRO Orange dye and thermal ramping. All proteins were diluted to a final concentration of 2–5 μM (optimally 5 μM) in the presence or absence of the indicated concentration of nucleotides [18] as previously described [22]. Both NME1 and NME2 bind to ATP and GTP in the absence of Mg^{2+} ions. Binding is inhibited by Mg ions, although for ADP and GDP, an enhancement was seen. Very low levels of thermal shift were seen with cAMP and cGMP or with AMP or DPG. No detectable thermal effects were seen in the presence of Mg ions alone. For PGAM1, DPG binding was independent of Mg ions, and although ATP, GTP, and GDP binding could be detected, this was completely abolished in H11A PGAM1, in contrast to DPG binding. The canonical protein kinase Aurora A bound to ATP in a Mg-dependent manner, as expected. (**f**) T_m values for each protein are reported from duplicate measurements ($n = 2$) and represent the mean ± SD. Note the increased thermal stability of H11A PGAM1 compared to PGAM1

5. Seal plates with optical adhesive cover and centrifuge at 500 × *g* for 10 s.
6. Set RT-PCR machine to measure fluorescence at 530 nm.

7. Measure fluorescence emission at a set rate of 0.3 °C/min, beginning at 25 °C and ending at 95 °C.
8. Normalize raw data exported from PCR machine using GraphPad Prism software.
9. In order to generate sigmoidal denaturation curves, fit normalized data to the Boltzmann Equation, and calculate average T_m (the temperature at which 50% fluorescence is measured, equivalent to 50% protein unfolded) and ΔT_m (the change in T_m values between experimental conditions) values (*see* **Note 17**).

3.4 Detection of Histidine Phosphorylation by Western Blotting

3.4.1 Autophosphorylation Assay and SDS-PAGE

1. Incubate 400 ng recombinant histidine kinase protein with 10 μM of the appropriate phosphate donor: ATP, DPG, or nucleotide of interest, buffered with 50 mM Tris–HCl (pH 8.0) and 100 mM NaCl in a total volume of 40 μL. Leave for 5 min at room temperature. Perform this reaction in the presence and absence of 10 mM $MgCl_2$ in a microcentrifuge tube (*see* **Note 18**).
2. Terminate the in vitro kinase reaction by addition of 10 μL 5× LSB, pH 8.8.
3. Divide the assay in two: heat one half of the reaction to 95 °C for 5 min to promote pHis hydrolysis (*see* **Note 19**), retain the other half on ice.
4. Separate 100 ng each of the heated and nonheated reaction mixtures on each of two separate SDS polyacrylamide gels as previously described (*see* **Notes 20** and **21**) to independently probe for N1 and N3 histidine phosphorylation by immunoblotting (Subheading 3.4.2). To maintain histidine phosphorylation, use running buffer prechilled to 4 °C, applying a voltage of 100 V. Maintain at 4 °C throughout by performing the electrophoresis in a cold room.

3.4.2 Immunoblotting

1. Transfer the proteins from the polyacrylamide gel to a nitrocellulose membrane for ~540 Vh (at 30 V). Precool the transfer buffer to 4 °C prior to use and maintain at 4 °C throughout by running on an ice tray in the cold room.
2. Remove the nitrocellulose membrane from the blotting apparatus and block with TBST containing nonfat milk powder for 1 h at room temperature, or overnight at 4 °C, to prevent nonspecific antibody binding.
3. Place the membrane in a 50 mL falcon tube (face-up) containing 5 mL of 5% milk/TBST with a 1 in 5000 dilution (1 μL) of the desired primary N1-pHis or N3-pHis antibody.
4. Incubate at room temperature for 1 h on a roller.
5. Wash with 10 mL TBST for 3 × 10 min.
6. Repeat **steps 3–5** using the secondary antibody.

3.4.3 Immunodetection

1. Treat the membrane with an immunodetection reagent such as the ECL reagent (Amersham) system as per manufacturer's instruction.
2. Place membrane in a plastic envelope.
3. Expose to X-ray film for between 5 and 20 min (*see* **Note 22**).
4. Develop and fix the film (*see* **Note 23**).

4 Notes

1. Make up the lysis buffer using stock buffers as follows: 1 M Tris–HCl, pH 7.4—dissolve 121.14 g of Tris base in 1 L water, adjust to pH 7.4; 3 M NaCl—dissolve 175.32 g of NaCl in 1 L water; 1 M Imidazole—dissolve 68.07 g imidazole in 1 L water; 1 M DTT—dissolve 15.43 g DTT in 10 mL water and make 1 mL aliquots to be stored at −20 °C; 100 mM EGTA—dissolve 19.0 g in 500 mL water; 100 mM EDTA—dissolve 14.6 g in 500 mL water.
2. As Triton X-100 is viscous, cut off the point of the pipette tip to aid accurate pipetting of the required volume.
3. Gel filtration buffer must be degassed for at least 1 h prior to use.
4. "Regular" (pH 6.8) SDS sample buffer can only be used if pHis is not being directly monitored in the procedure (e.g., Coomassie staining in Fig. 2).
5. When pHis is being assessed by immunoblotting, all reactions are placed on ice during assay preparation and after completion, and a pH 8.8 SDS sample buffer is used, without boiling, unless indicated. This helps preserve the P-N bond in pHis-containing proteins.
6. Annealing temperature and extension time can be adjusted for specific primers or for PCR optimization.
7. Prepare 100 mL of agarose gel solution with 1% agarose in 1× TAE buffer. Dissolve the agarose by heating and after cooling to room temperature add 6 μL of Midori Green. Pour into an appropriate gel cast and insert a gel comb.
8. Incubate the pOPINJ infusion reaction in a water bath.
9. *E. coli* cells should be kept on ice at all times to maintain cell competency.
10. Transform the entire 2.5 μL infusion reaction into competent *E. coli*. This should be done under sterile conditions, either in a laminar flow hood or using a Bunsen burner.
11. Cell pellet can be stored at −20 °C until use for up to 1 week prior to cell lysis.

12. Sonicate cells until a homogeneous lysate solution is obtained.
13. The centrifuge should be precooled to 4 °C prior to use.
14. We typically use 10% acrylamide gels of 0.75 mm thickness for SDS-PAGE, loading approximately 5 μL of sample per well.
15. 3C cleavage is relatively inefficient, since it takes place "on-bead." We typically use 100 μg of purified 6His-3C to prepare a batch of up to 10 mg of cleaved NME1/2 or PGAM1. The advantage of this procedure is that the 6His-3C is retained on the beads, and does not contaminate purified cleaved proteins.
16. For SDS-PAGE analysis of His-GST proteins pre- and post-3C cleavage, it is recommended that you also analyze a sample of 3C protease to ensure there is no 3C remaining in the protein sample (*see* **Note 15**, above). 3C can be detected by protein staining and comparison with protein markers, or by immunoblotting with a commercial 6His antibody.
17. The mean change in thermal stability (ΔT_m) is calculated by subtracting the control T_m value (protein + buffer) from the measured T_m value in the presence of additive (e.g., protein + nucleotide, metal ion, or small molecule).
18. Also include a sample of protein processed in the absence of phosphate donor as a negative control. This will confirm that NME1/2 and PGAM1 are not phosphorylated at N1 or N3 positions during expression or after purification.
19. Do not heat the other half of the reaction, as this will remove any pHis present.
20. In order to load the maximum possible amount of sample, SDS-PAGE can be performed using a 10% (v/v) acrylamide gels with a thickness of 1.0 or 1.5 mm.
21. Stripping and reprobing nitrocellulose membranes is not recommended when analyzing pHis, since boiling (or extended experimental handling) leads to loss of pHis from proteins. Instead, run duplicate gels and samples and ensure that all samples that might contain pHis are analyzed as quickly as possible.
22. These timeframes are suitable for the immunodetection of heat-sensitive N1-pHis and N3-pHis on NME1/2 and PGAM1 respectively (Fig. 4), and also to evaluate the comparative effects of different nucleotides on N1-pHis and N3-pHis incorporation (Fig. 5).
23. If the signal is faint, reapply ECL reagents and repeat exposure for a longer period of time.

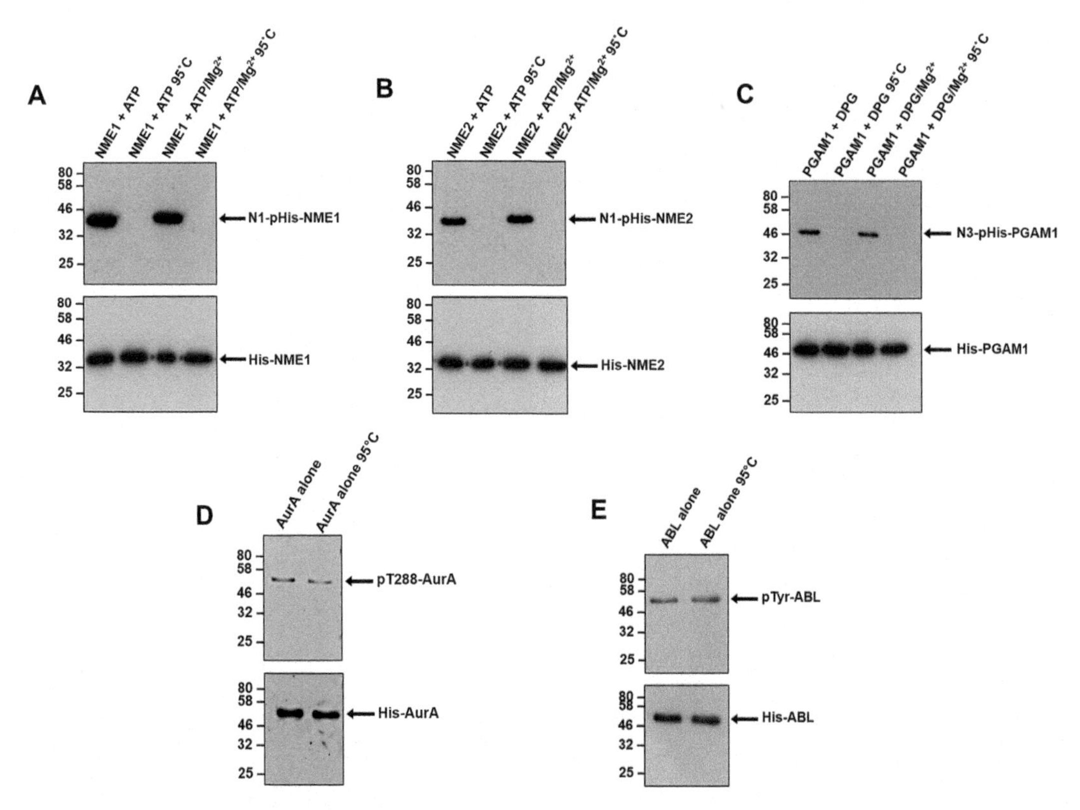

Fig. 4 Exploitation of NME1/2 and PGAM1 for detection of heat-labile N1 and N3 phosphorylation in situ. Two hundred nanogram of uncleaved NME1 (**a**), NME2 (**b**) or PGAM1 (**c**) were preincubated with the NME1/2 phosphate-donor, ATP or ATP and Mg^{2+} ions, or the PGAM1 phosphate-donor DPG for 5 min at room temperature. After this time, samples were either heated to 95 °C or left on ice, prior to SDS-PAGE and immunochemical analysis. After transfer to nitrocellulose (in duplicate) NME1 and NME2 were incubated with either anti-N1pHis antibody (pHis) or 6His antibody (total protein), *see* **Note 14**. After processing, PGAM1 immunoblots were incubated with anti-N3pHis antibody or 6His antibody, *see* **Note 14**. To confirm the thermostability of pThr and pTyr, heated or unheated recombinant Aurora A (200 ng) was immunoblotted with anti-pT288 Aurora A or 6His antibody (**d**) or heated or unheated ABL kinase (SH3-SH2 and catalytic domain) was immunoblotted with anti-pTyr antibody (**e**). For NME1, NME2, and PGAM1, 1 or 3-pHis is not detected after heat treatment, confirming the heat lability of the posttranslational modification, in contrast to pThr and pTyr

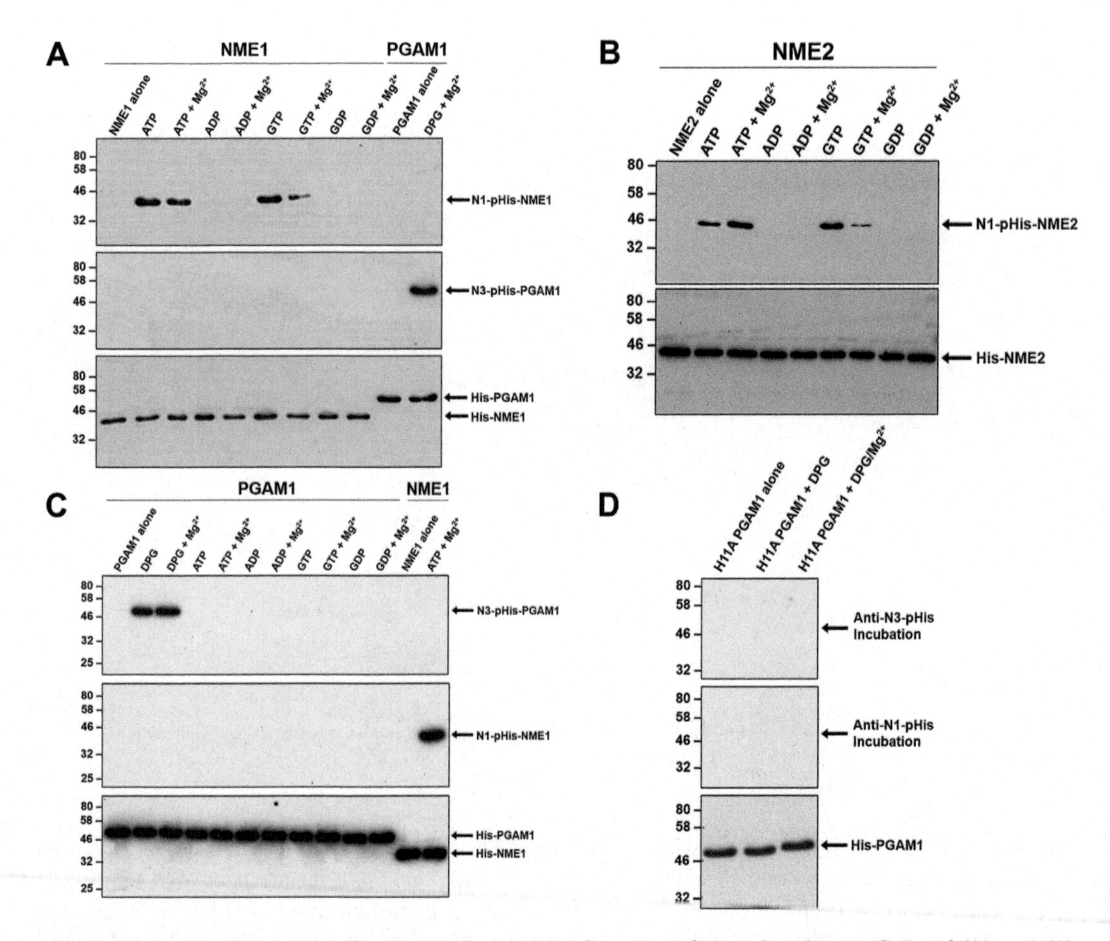

Fig. 5 Enzymatic His protein phosphorylation and confirmation of the phosphospecificity of N1 and N3 antibodies. (**a**) Uncleaved NME1 was incubated with the nucleotides ATP, ADP, GTP, and GDP +/− $MgCl_2$. Uncleaved PGAM1 was incubated in the presence and absence of DPG and $MgCl_2$. Unheated samples were processed for immunoblotting with either N1-pHis (top panel), N3-pHis (middle panel), or boiled, then immunoblotted for total protein (bottom panel). (**b**) Uncleaved NME2 was incubated with ATP, ADP, GTP, and GDP +/− $MgCl_2$ and processed for immunoblotting without heating with N1-pHis antibody (top panel) or for total protein after boiling (bottom panel). These data confirm that NME1 and NME2 autophosphorylate at the N1-His, but not the N3-His, position, and that this only occurs in the presence of the phosphate donors ATP or GTP. Moreover, this does not require the presence of Mg ions, consistent with DSF (Fig. 3). The nucleotides ADP and GDP do not serve as a phosphate donor for histidine phosphorylation on NME1 and NME2, despite exhibiting marked binding (Fig. 3). (**c**, **d**) PGAM1 was preincubated with DPG, ATP, ADP, GTP, and GDP +/− $MgCl_2$ and analyzed for the presence of N1 and N3-pHis. NME1 served as a positive control for N1-pHis. PGAM1 autophosphorylated on H11 at the N3 position, but only in the presence of the phosphate donor DPG, despite binding to ATP and GTP when analyzed using TSA (Fig. 3). Critically, the mutation of H11 to Ala (H11A PGAM1) prevented detection by the N3-pHis antibody, confirming antibody phosphospecificity

References

1. Hunter T (2012) Why nature chose phosphate to modify proteins. Philos Trans R Soc Lond Ser B Biol Sci 367:2513–2516
2. Cohen P (2002) The origins of protein phosphorylation. Nat Cell Biol 4:E127–E130
3. Haydon CE, Eyers PA, Aveline-Wolf LD et al (2003) Identification of novel phosphorylation sites on Xenopus laevis Aurora A and analysis of phosphopeptide enrichment by immobilized metal-affinity chromatography. Mol Cell Proteomics 2:1055–1067
4. Schweppe RE, Haydon CE, Lewis TS et al (2003) The characterization of protein post-translational modifications by mass spectrometry. Acc Chem Res 36:453–461
5. Fuhs SR, Meisenhelder J, Aslanian A et al (2015) Monoclonal 1- and 3-phosphohistidine antibodies: new tools to study histidine phosphorylation. Cell 162:198–210
6. Kee JM, Oslund RC, Perlman DH et al (2013) A pan-specific antibody for direct detection of protein histidine phosphorylation. Nat Chem Biol 9:416–421
7. Hardman G et al. (2017) Extensive non-canonical phosphorylation in human cells revealed using strong-anion exchange-mediated phosphoproteomics. bioRxiv. doi: 10.1101/202820
8. Boyer PD, Deluca M, Ebner KE et al (1962) Identification of phosphohistidine in digests from a probable intermediate of oxidative phosphorylation. J Biol Chem 237: PC3306–PC3308
9. Kee JM, Muir TW (2012) Chasing phosphohistidine, an elusive sibling in the phosphoamino acid family. ACS Chem Biol 7:44–51
10. Puttick J, Baker EN, Delbaere LT (2008) Histidine phosphorylation in biological systems. Biochim Biophys Acta 1784:100–105
11. Olsen JV, Blagoev B, Gnad F et al (2006) Global, in vivo, and site-specific phosphorylation dynamics in signaling networks. Cell 127:635–648
12. Chen Y, Gallois-Montbrun S, Schneider B et al (2003) Nucleotide binding to nucleoside diphosphate kinases: X-ray structure of human NDPK-A in complex with ADP and comparison to protein kinases. J Mol Biol 332:915–926
13. Srivastava S, Panda S, Li Z et al (2016) Histidine phosphorylation relieves copper inhibition in the mammalian potassium channel KCa3.1. elife 5:pii: e16093
14. Attwood PV, Muimo R (2018) The actions of NME1/NDPK-A and NME2/NDPK-B as protein kinases. Lab Investig 98:283–290
15. Attwood PV, Wieland T (2015) Nucleoside diphosphate kinase as protein histidine kinase. Naunyn Schmiedeberg's Arch Pharmacol 388:153–160
16. Vander Heiden MG, Locasale JW, Swanson KD et al (2010) Evidence for an alternative glycolytic pathway in rapidly proliferating cells. Science 329:1492–1499
17. Panda S, Srivastava S, Li Z et al (2016) Identification of PGAM5 as a mammalian protein histidine phosphatase that plays a central role to negatively regulate CD4(+) T cells. Mol Cell 63:457–469
18. Byrne DP, Li Y, Ngamlert P et al (2018) New tools for evaluating protein tyrosine sulfation: tyrosylprotein sulfotransferases (TPSTs) are novel targets for RAF protein kinase inhibitors. Biochem J 475:2435–2455
19. Green R, Rogers EJ (2013) Transformation of chemically competent *E. coli*. Methods Enzymol 529:329–336
20. Byrne DP, Vonderach M, Ferries S et al (2016) cAMP-dependent protein kinase (PKA) complexes probed by complementary differential scanning fluorimetry and ion mobility-mass spectrometry. Biochem J 473:3159–3175
21. Scutt PJ, Chu ML, Sloane DA et al (2009) Discovery and exploitation of inhibitor-resistant aurora and polo kinase mutants for the analysis of mitotic networks. J Biol Chem 284:15880–15893
22. Foulkes DM, Byrne DP, Yeung W et al (2018) Covalent inhibitors of EGFR family protein kinases induce degradation of human Tribbles 2 (TRIB2) pseudokinase in cancer cells. Sci Signal 11:pii: eaat7951
23. Murphy JM, Zhang Q, Young SN et al (2014) A robust methodology to subclassify pseudokinases based on their nucleotide-binding properties. Biochem J 457:323–334
24. Mohanty S, Oruganty K, Kwon A et al (2016) Hydrophobic core variations provide a structural framework for tyrosine kinase evolution and functional specialization. PLoS Genet 12: e1005885
25. Rudolf AF, Skovgaard T, Knapp S et al (2014) A comparison of protein kinases inhibitor screening methods using both enzymatic activity and binding affinity determination. PLoS One 9:e98800

Chapter 6

Determination of Phosphohistidine Stoichiometry in Histidine Kinases by Intact Mass Spectrometry

Lauren J. Tomlinson, Alice K. M. Clubbs Coldron, Patrick A. Eyers, and Claire E. Eyers

Abstract

Protein histidine phosphorylation has largely remained unexplored due to the challenges of analyzing relatively unstable phosphohistidine-containing proteins. We describe a procedure for determining the stoichiometry of histidine phosphorylation on the human histidine kinases NME1 and NME2 by intact mass spectrometry under conditions that retain this acid-labile protein modification. By characterizing these two model histidine protein kinases in the absence and presence of a suitable phosphate donor, the stoichiometry of histidine phosphorylation can be determined. The described method can be readily adapted for the analysis of other proteins containing phosphohistidine.

Key words NME1, NME2, Histidine phosphorylation, Intact mass spectrometry

1 Introduction

Protein phosphorylation is an important post-translational modification (PTM) found in eubacteria, archaebacteria, prokaryotes, and eukaryotes, typically mediated by target-specific protein kinases [1]. Phosphorylation plays vital roles in signal transduction, mediating the cellular response to external stimuli, and in the regulation of diverse cellular processes including transcription, metabolism, cell cycle progression, apoptosis and differentiation [2, 3].

Histidine (His) phosphorylation has been well-studied in prokaryotes, in particular in bacteria due to their crucial involvement in two-component signaling systems (TCS), which comprise a receptor His kinase and an effector protein. Two-component systems were first reported in bacteria in 1980 and their widespread relevance in bacterial signaling systems is now accepted. They are thought to be critically important in the response of cells to environmental factors (sensory responses). However, TCS are not limited to prokaryotes, and can be found in major eukaryotic

Claire E. Eyers (ed.), *Histidine Phosphorylation: Methods and Protocols*, Methods in Molecular Biology, vol. 2077, https://doi.org/10.1007/978-1-4939-9884-5_6,

kingdoms, including plants and fungi [4, 5]. More recently, phosphohistidine (pHis) has been recognized to have roles in vertebrate signaling [6–8], and there is growing evidence that this acid-labile PTM may function in a similar cell signaling capacity as described for the much-better studied phosphorylation of serine, threonine and tyrosine. However, pHis analysis is challenging, due to the high free energy of hydrolysis (~12 kcal/mol) of the phosphoramidate (N-P) bond [9], meaning that loss of the phosphate group occurs rapidly at high temperatures and at low pH [10], conditions often employed for biochemical analysis. The two isomers of phosphohistidine (1-pHis, or 3-pHis) arising due to phosphate group addition on the nitrogen at either the one or three positions of the histidine imidazole ring, also have a different ΔG° of hydrolysis, with 1-pHis believed to be more labile [11].

NME1 and NME2 are known mammalian histidine kinases and belong to the family of nucleoside diphosphate kinases (NDPKs), a family of proteins encoded by the *nme* (nonmetastatic cells) genes. NDPKs are responsible for catalyzing the transfer of γ-phosphate from nucleoside triphosphate to nucleoside diphosphate. This mechanism of catalytic transfer is associated with the generation of a "high-energy" phosphohistidine intermediate [12, 13].

Mass spectrometry (MS)-based proteomic methods represent a leading technique for protein identification and characterization of both the sites and stoichiometry of phosphorylation [14–18]. Analysis of intact proteins by MS permits the mass of the phosphorylated protein to be defined, and consequently the average number of phosphorylation sites per molecule to be identified, enhancing peptide-driven analyses (which are discussed in Chapter 15). In general, MS-based analysis is a label-free analytical strategy that overcomes any requirement for radioactive ($^{32/33}$P) labelling (as presented in Chapters 3 and 4), or for PTM-specific antibodies (as in Chapter 12), and is thus a simple method for evaluating His protein kinase activity.

We present a straightforward MS method to identify phosphorylation-induced mass shifts in two model histidine phosphorylated proteins that can be used to determine relative phosphorylation stoichiometry. To do this, we employ His autophosphorylation of the putative protein histidine kinases NME1 and NME2 as exemplars. This method can be readily adapted to evaluate autophosphorylation of other histidine kinases, or enzyme-mediated histidine phosphorylation of substrate proteins, and for the analysis of proteins containing other unstable phosphoamino acids [19–22].

2 Materials

Prepare all solutions and buffers using HPLC grade water and acetonitrile where applicable. Store all reagents at room temperature. Nucleotides can be stored frozen (−20 °C) prior to use. Protein solutions should be maintained on ice, unless otherwise stated. For longer-term storage, proteins should be kept at −20 °C in a glycerol-containing buffer of suitable pH.

2.1 Phosphorylation Assay

1. Histidine kinases: NME1 and NME2. Enzymes should be ~0.1–0.3 mg/mL in 50 mM Tris–HCl (pH 7.4), 100 mM NaCl, 1 mM DTT, 10% (v/v) glycerol (*see* **Notes 1** and **2**).
2. Low-bind microcentrifuge tubes.
3. HPLC water.
4. 1 M Tris–HCl (pH 8.0): Weigh out 121 g Tris base. Add to a graduated cylinder with ~800 mL of water. Mix and adjust to a pH of 8.0 with HCl (*see* **Note 3**). Make up to 1 L with water.
5. 1 M NaCl: Weigh out 58 g of NaCl. Add to a graduated cylinder and make up to 1 L with water.
6. Nucleotides (10 mM): Make up a stock of adenosine triphosphate (ATP) in water and adjust to pH 8.0 with NaOH (*see* **Note 4**).
7. Assay buffer: 50 mM Tris–HCl (pH 8.0), 100 mM NaCl. Add 500 μL of 1 M NaCl and 1 mL of 1 M Tris–HCl (pH 8.0) to 8.5 mL of water. Check that the pH is still at 8.0 (*see* **Note 5**).
8. Assay buffer with nucleotides: 50 mM Tris–HCl (pH 8.0), 100 mM NaCl, 10 μM nucleotide. Add 500 μL of 1 M NaCl, 1 mL of 1 M Tris–HCl (pH 8.0), and 10 μL of the appropriate phosphate donor (*see* **Note 4**) to a low-bind microcentrifuge tube. Make up to 10 mL with water. Check that the pH is still at 8.0 (*see* **Note 5**).

2.2 Liquid Chromatography–Mass Spectrometry (LC-MS)

1. HPLC Water.
2. Buffer A: 0.1% (v/v) formic acid in HPLC grade water. Add 1 mL of formic acid to 1 L HPLC-grade water in a clean glass bottle.
3. Buffer B: 0.1% (v/v) formic acid in HPLC grade acetonitrile. Add 1 mL of formic acid to 1 L acetonitrile in a clean glass bottle.
4. TFA (0.1% v/v): Add 0.5 mL of TFA to 500 mL of HPLC grade water in a clean glass bottle.
5. (Glu1)-Fibrinopeptide B calibrant (GluFib): 100 fmol/μL in 0.1% formic acid.
6. Myoglobin: 500 fmol/μL in buffer A.

7. MS sample vials (*see* **Note 6**).
8. LC-MS instrument: We use a Waters Synapt G2-S*i* in-line with a Nano Acquity UPLC system (*see* **Note 7**).
9. MassPREP Micro Desalting Column (Waters).
10. MassLynx data analysis software (*see* **Note 8**).

3 Methods

3.1 MS Calibration

1. To calibrate the time-of-flight (ToF) mass analyzer, infuse the Lockspray Flow Control with GluFib. Start at 20 μL/min and decrease to 10 μL/min once a strong signal is observed at m/z 785.8.
2. Calibrate the ToF using the Intellistart program, acquiring data for GluFib over a m/z range of 200–4000, in resolution mode.
3. Repeat if necessary to ensure a mass accuracy of 1 ppm or below (*see* **Note 9**).

3.2 LC-MS Setup

1. Wash the UPLC column with 50% buffer A/50% buffer B for 1 min at a flow rate of 8 mL/min.
2. Program the LC-MS gradient as shown in Table 1.
3. Define the MS settings as listed in Table 2.
4. To evaluate instrument configuration, inject 0.5 μL of 500 fmol/μL of myoglobin into the sample loop and acquire mass spectra over the period of reverse-phase chromatographic separation (*see* **Note 10**).

3.3 Auto-phosphorylation Assay

1. Label two low-bind tubes for each enzyme being analyzed, one for the assay, and the other for the negative control.
2. To each tube add 200 ng of recombinant enzyme (*see* **Note 11**). To the assay tube, make up to 20 μL with assay buffer containing the appropriate nucleotide/phosphate donor. To the negative control tube, make up to 20 μL with assay buffer (no phosphate donor) (*see* **Notes 12** and **13**).
3. Incubate all vials at room temperature for 5 min (or longer as required) and place back on ice (*see* **Note 14**).

3.4 Intact Mass Spectrometry Analysis

1. Immediately after the in vitro protein kinase assay, transfer each reaction mixture to a clean MS sample vial.
2. Place the vials in the auto sampler (*see* **Note 15**) and analyze the samples using the LC-MS parameters as defined in Subheading 3.2.

Table 1
LC/MS gradient

Time (min)	Flow (μL/min)	Buffer A (%)	Buffer B (%)	Curve
Initial	25	95	5	6
0.10	40	95	5	6
5.10	40	95	5	6
5.20	25	95	5	6
6.00	25	95	5	6
6.10	25	95	5	6
7.60	25	10	90	6
7.90	25	95	5	6
8.60	25	10	90	6
8.90	25	95	5	6
9.60	25	10	90	6
11.60	25	10	90	6
11.70	25	95	5	6

Details of the LC-MS gradient used for intact protein separation by C4 reverse-phase chromatography. Flow rate and buffer composition (% A and % B) are listed

Table 2
MS instrument settings

Source	Capillary (kV) Sampling cone Source offset	3.0 40 80
Temperature (°C)	Source Desolvation	100 250
Gas flow settings	Cone gas (L/h) Desolvation gas (L/h) Nebulizer (bar)	50 300 6.0

Acquisition parameters for intact protein analysis using the Synapt G2-*Si* (Waters)

3.5 Data Analysis

Data analysis should be performed using instrument-specific software. In our case, using the Synapt G2-*Si* (Waters), all data analysis was done using MassLynx software.

1. Extract the chromatographic peak of interest from the Total Ion Chromatogram (TIC).
2. Combine the mass spectra for the protein over the chromatographic elution window and view the averaged mass spectrum.

3. Process the MS data using the MaxEnt 1 function to generate a deconvoluted (zero charge state) mass spectrum. Adjust the mass range according to the size of the protein being analyzed.
4. Amend the MaxEnt 1 deconvolution settings according to the experimental parameters. We typically use: 0.5 Da/channel resolution; Uniform Gaussian distribution with a width at half height of 0.500 Da; Minimum intensity ratios for left and right at 33% (*see* **Note 16**).
5. The deconvoluted spectrum will open in a new window. Compare the generated MaxEnt 1 mock data with the originator mass spectrum to evaluate how well the deconvolution has worked (*see* **Note 17**).
6. Compare the MS data from the control reaction (with no phosphate donor) with the phosphate donor-containing kinase assay reaction mixture. Examples of the MS data generated for NME1 and NME2 following in vitro autophosphorylation in the absence and presence of ATP are presented in Figs. 1 and 2.
7. Calculate the stoichiometry of histidine phosphorylation by comparing the relative heights of the phosphorylated and

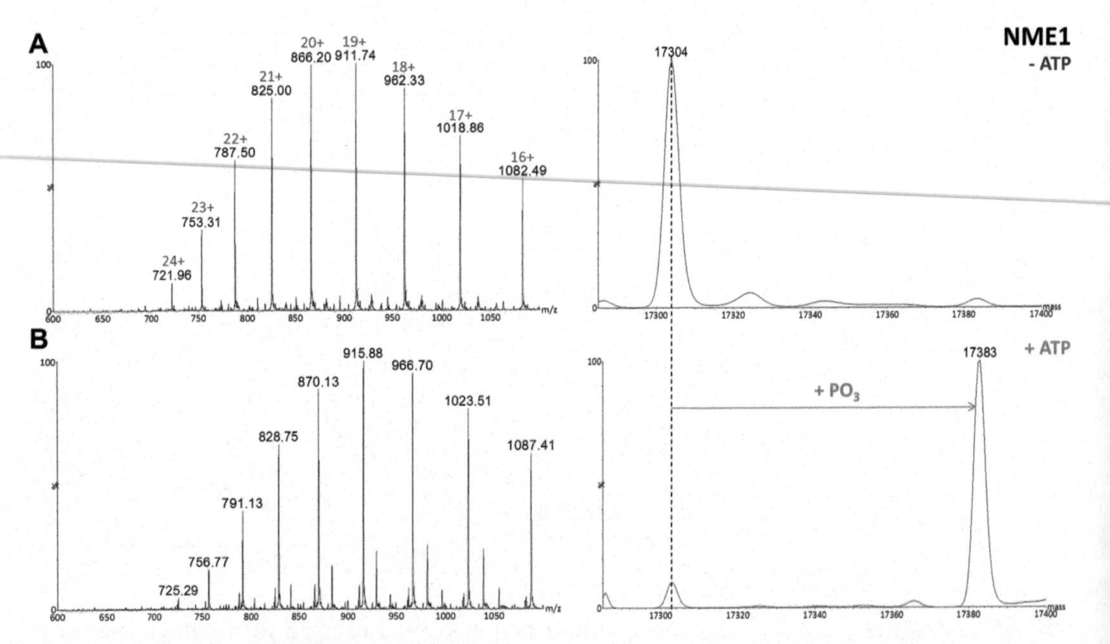

Fig. 1 Intact mass analysis of NME1 in the absence and presence of ATP. In vitro kinase assays was carried out for 5 min at room temperature with 200 ng recombinant protein and 10 μM ATP. The reaction mixtures were then analyzed by LC-MS using a Waters G2-S*i* Synapt. MS data were deconvoluted using MaxEnt 1 within MassLynx. (**a**) *Left* Combined mass spectra for NME1 (no ATP) over the period of LC elution. Numbers in red indicate the observed charge states; *Right* zero charge state mass spectrum for NME1. (**b**) *Left* Combined mass spectra for NME1 following incubation with ATP; *Right* zero charge state mass spectrum for phosphorylated NME1. A difference in mass due to phosphorylation (compare panels on the right) can be clearly observed and a stoichiometry of 90% phosphorylation can be observed

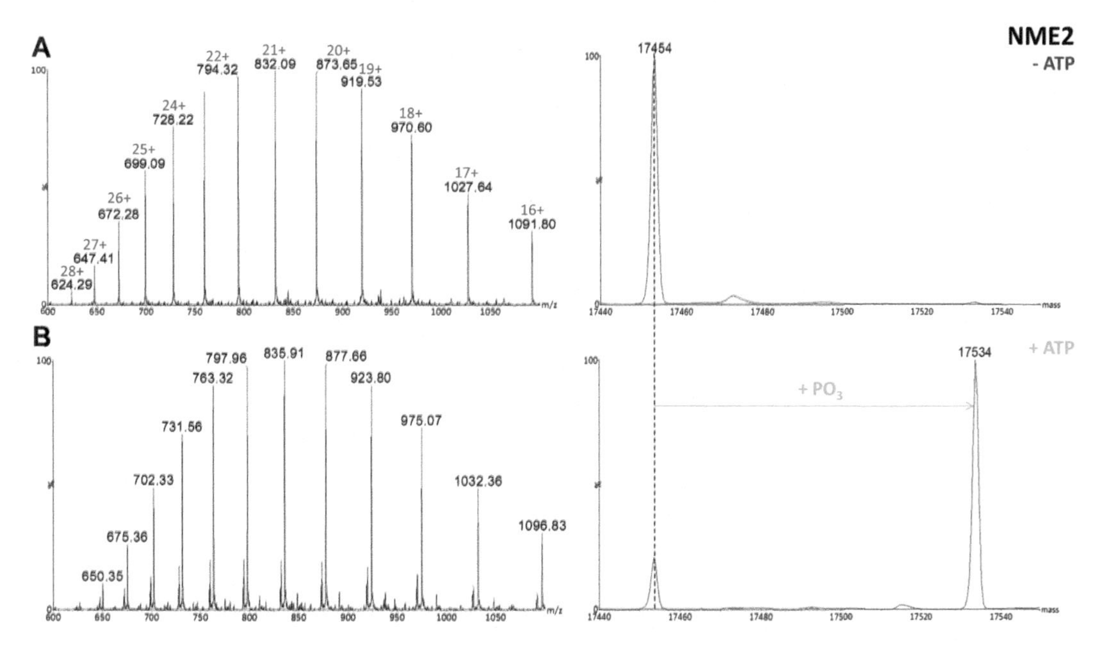

Fig. 2 Intact mass analysis of NME2 in the absence and presence of ATP. In vitro kinase assays was carried out for 5 min at room temperature with 200 ng recombinant protein and 10 μM ATP. The reaction mixtures were then analyzed by LC-MS using a Waters G2-S*i* Synapt. MS data were deconvoluted using MaxEnt 1 within MassLynx. (**a**) *Left* Combined mass spectra for NME2 (no ATP) over the period of LC elution. Numbers in red indicate the observed charge states; *Right* zero charge state mass spectrum for nonphosphorylated NME2. (**b**) *Left* Combined mass spectra for NME2 following incubation with ATP; *Right* zero charge state mass spectrum for phosphorylated NME2. A difference in mass due to phosphorylation (compare panels on the right) can be clearly observed and a stoichiometry of 80% phosphorylation can be observed

nonphosphorylated protein in the reaction mixture containing the phosphate donor.

8. To validate that the change in mass is due to heat-labile histidine phosphorylation, heat the samples at 80 °C for 10 min and repeat the LC-MS analysis as outlined in Subheading 3.4.

4 Notes

1. Histidine kinases can be made by recombinant protein expression and purification (*see* Chapters 2 and 5), or purchased.
2. A buffer suitable to maintain the activity and stability of the protein should be used. We find that these proteins are generally stable in the buffer used for the final purification step (*see* Chapter 5): 50 mM Tris–HCl (pH 7.4), 100 mM NaCl, 1 mM DTT, 10% (v/v) glycerol.
3. Make sure that the Tris base is dissolved completely in the water prior to adjusting the pH. 1 M HCl can be used initially,

although a lower concentration of HCl should be used for final pH adjustment.

4. Phosphate donor type and concentration may need to be adapted according to the specific requirements of the recombinant enzyme. For example, PGAM uses DPG as a phosphate donor instead of ATP.
5. The pH can be checked using indicator paper. If the pH requires minor adjustment at this point, use dilute concentrations of HCl or NaOH as appropriate.
6. We use TruView™ LCMS sample vials which minimize sample loss.
7. Any LC-MS instrumentation can be used that is capable of mass resolution of 20,000 or greater.
8. Data analysis software used will be dependent on the make of LC-MS instrumentation.
9. If the mass accuracy is greater than 1 ppm during ToF calibration, calibration should be repeated.
10. To minimize the time the sample is kept on ice after in vitro phosphorylation, the mass spectrometer should be calibrated and the LC-MS system set up and evaluated prior to the start of the enzyme assay.
11. Ensure the protein is kept on ice prior to starting the assay.
12. Some proteins may have residual phosphorylation prior to the start of the assay. It is essential to evaluate a control sample of the enzyme in the absence of a phosphate donor to determine the change in phosphorylation stoichiometry arising as a result of the assay.
13. This assay configuration assumes a single time point for analysis. If undertaking a time course is of interest, increase the relative volume of the assay and components, and remove 20 μL to a clean Eppendorf tube kept on ice.
14. The intact protein MS analysis should be performed as soon as possible after the end of the in vitro assay.
15. Set the autosampler to remain at 4 °C throughout the experiment.
16. It is often useful to generate a coarse-grained survey spectrum using a resolution of 10–25 Da/channel to define the major protein components in the mixture and allow the mass range to be suitably adjusted.
17. The channel resolution settings can be further adjusted if required to between ~0.25 and 1 Da to gain a zero charge mass spectrum of high signal to noise.

References

1. Loomis WF, Shaulsky G, Wang N (1997) Histidine kinases in signal transduction pathways of eukaryotes. J Cell Sci 110:1141–1145
2. Manning G, Whyte DB, Martinez R, Hunter T, Sudarsanam S (2002) The protein kinase complement of the human genome. Science 298(5600):1912
3. Cohen P (2002) The origins of protein phosphorylation. Nat Cell Biol 4(5):127–130
4. Adam K, Hunter T (2018) Histidine kinases and the missing phosphoproteome from prokaryotes to eukaryotes. Pathol Focus 98:233–247
5. Kee JM, Muir TW (2012) Chasing phosphohistidine, an elusive sibling in the phosphoamino acid family. ACS Chem Biol 7(1):44–51
6. Klumpp S, Krieglstein J (2009) Reversible phosphorylation of histidine residues in proteins from vertebrates. Sci Signal 2(61):1–4
7. Fuhs SR et al (2015) Monoclonal 1- and 3-phosphohistidine antibodies: new tools to study histidine phosphorylation. Cell 162 (1):198–210
8. Srivastava S et al (2016) Histidine phosphorylation relieves copper inhibition in the mammalian potassium channel KCa3.1. eLife 5:pii: e16093
9. Puttick J, Baker E, Delbaere L (2008) Histidine phosphorylation in biological systems. Biochim Biophys Acta 1784(1):100–105
10. Attwood PV (2013) P/N bond protein phosphatases. Biochim Biophys Acta 1834:470–478
11. Attwood PV, Muimo R (2018) The actions of NME1/NDPK-A and NME2/NDPK-B as protein kinases. Lab Investig 98:283–290
12. Attwood PV, Wieland T (2015) Nucleoside diphosphate kinase as protein histidine kinase. Naunyn Schmiedeberg's Arch Pharmacol 388 (2):153–160
13. Wu Z, Tiambeng TN, CaI W, Chen B, Lin Z, Zachery R, Gregorich ZR, Ge Y (2018) Impact of phosphorylation on the mass spectrometry quantification of intact phosphoproteins. Anal Chem 90(8):4935–4939
14. Haydon CE, Eyers PA, Aveline-Wolf LD, Resing KA, Maller JL (2003) Identification of novel phosphorylation sites on Xenopus laevis Aurora A and analysis of phosphopeptide enrichment by immobilized metal-affinity chromatography. Mol Cell Proteomics 2 (10):1055–1067
15. Schweppe RE, Haydon CE, Lewis TS, Resing KA, Ahn NG (2003) The characterization of protein post-translational modifications by mass spectrometry. Acc Chem Res 36 (6):453–456
16. Johnson H, Eyers CE, Eyers PA, Beynon RJ, Gaskell SJ (2009) Rigorous determination of the stoichiometry of protein phosphorylation using mass spectrometry. J Am Soc Mass Spectrom 20(12):2211–2220
17. Byrne DP, Vonderach M, Ferries S, Brownridge PJ, Eyers CE, Eyers PA (2016) cAMP-dependent protein kinase (PKA) complexes probed by complementary differential scanning fluorimetry and ion mobility-mass spectrometry. Biochem J 473(19):3159–3175
18. Ferries S, Perkins S, Brownridge PJ, Campbell A, Eyers PA, Jones AR, Eyers CE (2017) Evaluation of parameters for confident phosphorylation site localization using an orbitrap fusion tribrid mass spectrometer. J Proteome Res 16(9):3448–3459
19. Attwood PV, Besant PG, Piggott MJ (2011) Focus on phosphoaspartate and phosphoglutamate. Amino Acids 40(4):1035–1051
20. Besant PG, Attwood PV, Piggott MJ (2009) Focus on phosphoarginine and phospholysine. Curr Protein Pept Sci 10(6):536–550
21. Hardman G, Perkins S, Ruan Z, Kannan N, Brownridge P, Byrne DP, Eyers PA, Jones AR, Eyers CE (2017) Extensive non-canonical phosphorylation in human cells revealed using strong-anion exchange-mediated phosphoproteomics. bioRxiv 202820. https://doi.org/10.1101/202820
22. Hardman G, Perkins S, Brownridge PJ, Clarke CJ, Byrne DP, Campbell AE, Anton Kalyuzhnyy A, Myall A, Eyers PA, Jones AR, Eyers CE (2019) Strong anion exchange-mediated phosphoproteomics reveals extensive human non-canonical phosphorylation. The EMBO Journal. https://doi.org/10.15252/embj.2018100847

Chapter 7

Protein Phosphohistidine Phosphatases of the HP Superfamily

Daniel J. Rigden

Abstract

Histidine phosphorylation of proteins is increasingly recognised as an important regulatory posttranslational modification in eukaryotes as well as prokaryotes. The HP (Histidine Phosphatase) superfamily, named for a key catalytic His residue, harbors two known groups of protein phosphohistidine phosphatases (PPHPs). The bacterial SixA protein acts as a regulator of His-Asp phosphorelays with two substrates characterized in vitro and/or in vivo. The recently characterized eukaryotic PHPP PGAM5 only has one currently known substrate, NDPK-B, through which it helps regulate T-cell signaling. SixA and PGAM5 appear to share no particular sequence or structural features relating to their PPHP activity suggesting that PHPP activity has arisen independently in different lineages of the HP superfamily. Further members of the HP superfamily may thus harbor (additional) unsuspected PHPP activity.

Key words Protein phosphohistidine phosphatase, Histidine phosphatase superfamily, Posttranslational modification, SixA, PGAM5

1 Introduction

The histidine phosphatase (HP) superfamily [1] is a very large and functionally diverse group of enzymes. They are named for a key His residue which becomes transiently phosphorylated during the catalytic cycle. As the name suggests, a large majority of the characterized enzymes in the HP superfamily are phosphatases. Ironically and unfortunately, however, the first characterized enzyme was cofactor-dependent phosphoglycerate mutase (dPGM), a glycolytic enzyme that interconverts 2- and 3-phosphoglycerate [2] and earlier nomenclatures for the group often made use of this activity. As a result, both computational and even human annotations have regularly been biased toward mutase rather than phosphatase activities. The human phosphatase discussed extensively later is a particularly egregious example, bearing the gene locus name PGAM5, despite PGAM being a common abbreviation for phosphoglycerate mutase [3, 4]. A related difficulty in the

Claire E. Eyers (ed.), *Histidine Phosphorylation: Methods and Protocols*, Methods in Molecular Biology, vol. 2077, https://doi.org/10.1007/978-1-4939-9884-5_7,

superfamily is its functional diversity: this has led to the production and propagation of much overannotation and misannotation, especially in the early days of automated function prediction based on BLAST [1].

Catalysis in the HP superfamily centres on a conserved "phosphate pocket" [1] containing two His residues and two Arg residues. One of each, along with a conserved but not invariant Gly residue, forms an easily recognised RHG motif that lies at the end of the first β-strand of the HP domain. With the frequent addition of other residues, in an enzyme-specific way, the phosphate pocket is a strongly positively charged cavity that recognizes the phospho group to be hydrolyzed (in the case of phosphatases) or moved (in the case of mutases) before, during and after its covalent linkage to the central His residue. In crystal structures, the phosphate pocket is often occupied by tetrahedral ions present in the crystallization solution such as phosphate, sulphate or tungstate (e.g., [5–7]). Catalysis also requires the presence of a proton donor, an Asp or Glu residue that is located in at least three different positions in the HP domain in different enzymes [7–10].

The HP superfamily can be divided into two branches, the homology of which is only clear at a sequence level when sensitive, iterative database searches are done (Fig. 1). Structures of the two branches unambiguously reveal their relationship and show how a minimal fold, represented by the branch 1 protein SixA, can be extensively decorated by insertions—many long and rich in regular secondary structure—or extensions, particularly at the C-terminus [1]. These enzyme-specific accretions form the distinct substrate-binding sites for each activity, and are so substantial that the largest HP structures are almost three times the size of the smallest. Branches 1 and 2 are represented by Pfam [14] entries PF00300 and PF00328, respectively. Branch 1 is more common in prokaryotes than eukaryotes while this distribution is reversed in branch 2. Branch 1 is the larger by about fivefold.

Although the HP superfamily is most notable for harboring an impressive variety of activities relating to small organic molecules containing phosphate groups [1], a number of activities on phosphorylated protein substrates have been characterized. Protein phosphotyrosine phosphatases exist in both branches of the superfamily. For example, STS-1 (also known as TULA-2, and encoded by the Ubash3b gene) in Branch 1 targets tyrosine phosphorylated, ubiquitinated proteins within TCR signaling pathways [15–17]. In Branch 2, prostatic acid phosphatase can dephosphorylate the receptor tyrosine kinases ErbB2 [18] and EGFR [19]. Notably, phosphatase activities on small molecules and protein substrates are not mutually exclusive. For example, prostatic acid phosphatase has also been characterized as an ectonucleosidase [20] and a lysophospholipase [21], each activity having plausible physiological significance. Similarly, before the protein phosphatase role of STS-1

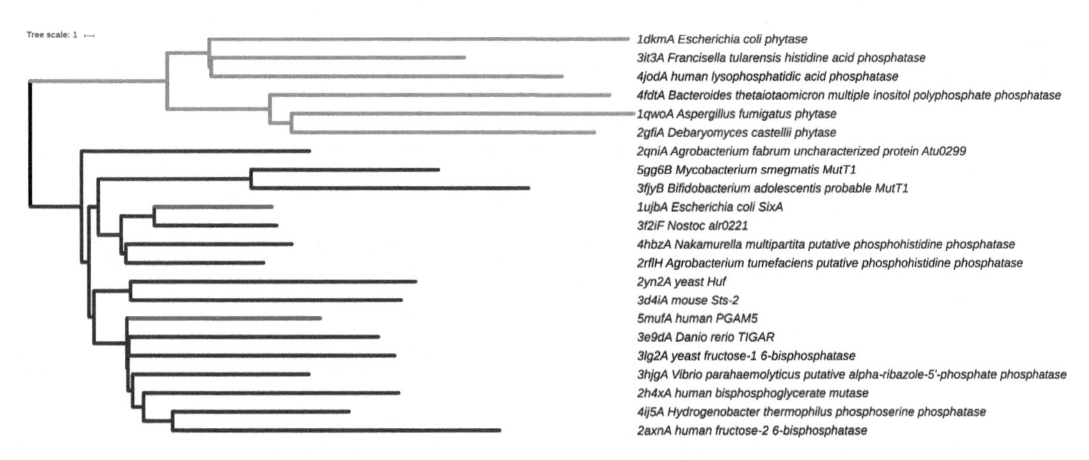

Fig. 1 A dendrogram representation of structural relationships between HP superfamily members. Protein structures are derived from a redundancy-reduced version of the PDB [11] in which no pair of structures share more than 25% sequence identity. The tree is calculated on the basis of all-against-all DALI scores [12]. Branch 1 is shown in purple, with Branch 2 in green. PHPPs discussed in detail in the text are shown in red. The tree was formatted at the iTOL server [13]

was discovered, it was shown to dephosphorylate ecdysteroid and steroid phosphates [22].

Protein phosphohistidine phosphatase (PPHP) activity [23] has been associated with a number of distinct families and more may yet be discovered. The HP superfamily contains two distinct lineages with this proven activity: SixA, characterized in *Escherichia coli* in 1998 [24], and eukaryotic PGAM5 proteins—previously considered as solely protein phosphoserine/phosphothreonine phosphatases—whose PHPP activity was reported as recently as 2016 [25]. This chapter summarizes what is known of the biological functions of these two groups, revealing some surprising limits of our knowledge, and asks whether protein phosphohistidine phosphatases of the HP superfamily share any characteristic features.

2 SixA

The *E. coli* SixA protein is a short, 161 residue member of the HP superfamily. It was characterized as a PPHP [24] by a screening approach looking for proteins affecting the His-Asp multistep phosphorelay downstream of the sensory kinase ArcB. The His-Asp phosphorelay is a common signaling mechanism in plants and microbes [26, 27], by which the phospho group, transferred to a protein by a sensory His kinase, is progressively transferred along a series of His or Asp residues to its ultimate target. Recognition of the characteristic HP superfamily RHG motif in the SixA sequence led to the hypothesis that this protein acts to dephosphorylate ArcB

[24]. This hypothesis was confirmed in vitro, with the target site for dephosphorylation determined to be the His residue of the HPt domain (rather than the His or Asp sites elsewhere in ArcB), and the SixA PPHP activity shown to be dependent on the core His residue of the RHG motif. A SixA gene deletion provided further evidence that SixA acts as a regulator of His-Asp phosphorelay [24]. Interestingly, a variety of protein phosphoaspartate phosphatases act on components of these phosphorelays [28, 29] but SixA seems to be the only PPHP yet characterized in this context. An *Arabidopsis thaliana* protein, ARR22, has been described as a PPHP [30, 31], yet it lacks a recognizable catalytic site, containing instead a Response Regulator domain. Its undoubted impact on signaling seems better described as resulting from competition for phosphotransfer from upstream His-Asp phosphotransfer proteins.

Very recent data [32] have shed some doubt on the in vivo significance of the ArcB dephosphorylation activity and have suggested an alternative substrate for SixA. The work of Schulte and Goulian demonstrated that *sixA* deletion, although producing a growth defect, did not affect ArcA/ArcB-dependent gene expression in vivo [32] arguing against the regulation of ArcB by SixA dephosphorylation. By screening for suppressors of the growth defect, a link between SixA and the nitrogen-related phosphotransferase system (PTS^{Ntr}) [33] was detected. This system contains three components which are sequentially phosphorylated on His residues and cross talk with the similar PTS^{sugar} system. SixA (but not an active site mutant) reduced phosphorylation levels of the final component, $EIIA^{Ntr}$, in a manner dependent on the presence of another component, NPr. Given this NPr-dependence of SixA function, and since SixA is a known PHPP, it was hypothesized that SixA directly dephosphorylates the phosphorylated NPr protein.

Crystal structures of *E. coli* SixA were published in 2005 [7] (Table 1) revealing a minimal version of the HP domain fold and confirming earlier predictions [9] that the likely catalytic proton donor in SixA and relatives was positioned differently in the catalytic site than in HP enzymes structurally characterized hitherto. The structure of the enzyme in complex with that of its substrate, the HPt domain of ArcB [35], was manually modeled to position the dephosphorylated substrate residue, His717, appropriately in the catalytic site [7] (approximately reconstructed in Fig. 2a since the coordinates are not available). As previously predicted [9], the ability of SixA to accommodate a large protein globular domain as substrate was related to the absence of significant elaborations of the core HP domain structure, allowing the catalytic site to be presented in the centre of a large flat platform [7].

A key open question is how many of the homologs of SixA, broadly distributed among proteobacteria, actinobacteria, and cyanobacteria [38], also function as PPHPs in their respective species. As mentioned, pure similarity-based function annotation in the HP

Table 1
Properties of HP superfamily enzymes known or proposed to dephosphorylate phospho-His residues

Protein group	Species	Construct	PDB code	Annotation at the PDB	Resolution (Å)	Physiologically relevant ligands or analogs	Publication of structure	UniProt entry name	Biological substrate (s) containing phosphorylated His residue
SixA	*Escherichia coli*	Full-length	1ujb	SixA	2.06	-	[7]	SIXA_ECOLI	ArcB, NPr
			1ujc		1.9	Tungstate, occupies phospho pocket at the catalytic site			
	Agrobacterium fabrum (A. tumefaciens)	Full-length	2rfl	Putative phosphohistidine phosphatase SixA	2.35	-	Unpublished	A9CJ67_AGRFC	Unknown
	Nakamurella multipartita	Full-length	4hbz	Putative phosphohistidine phosphatase SixA	1.55	-	Unpublished	C8XA43_NAKMY	Unknown
PGAM5	*Homo sapiens*	ΔN90 construct.	3mxo	Human phosphoglycerate mutase family member 5 (PGAM5)	1.7	Apo structure	[34]	PGAM5_HUMAN	NDPK-B
		ΔN90 construct.	3o0t		1.9	Phosphate occupies phospho pocket at the catalytic site. Catalytic residues in the “off” conformation			
		ΔN54 construct.	5muf		3.1	Phosphate, occupies phospho pocket at the catalytic site. Catalytic residues in the “on” conformation			

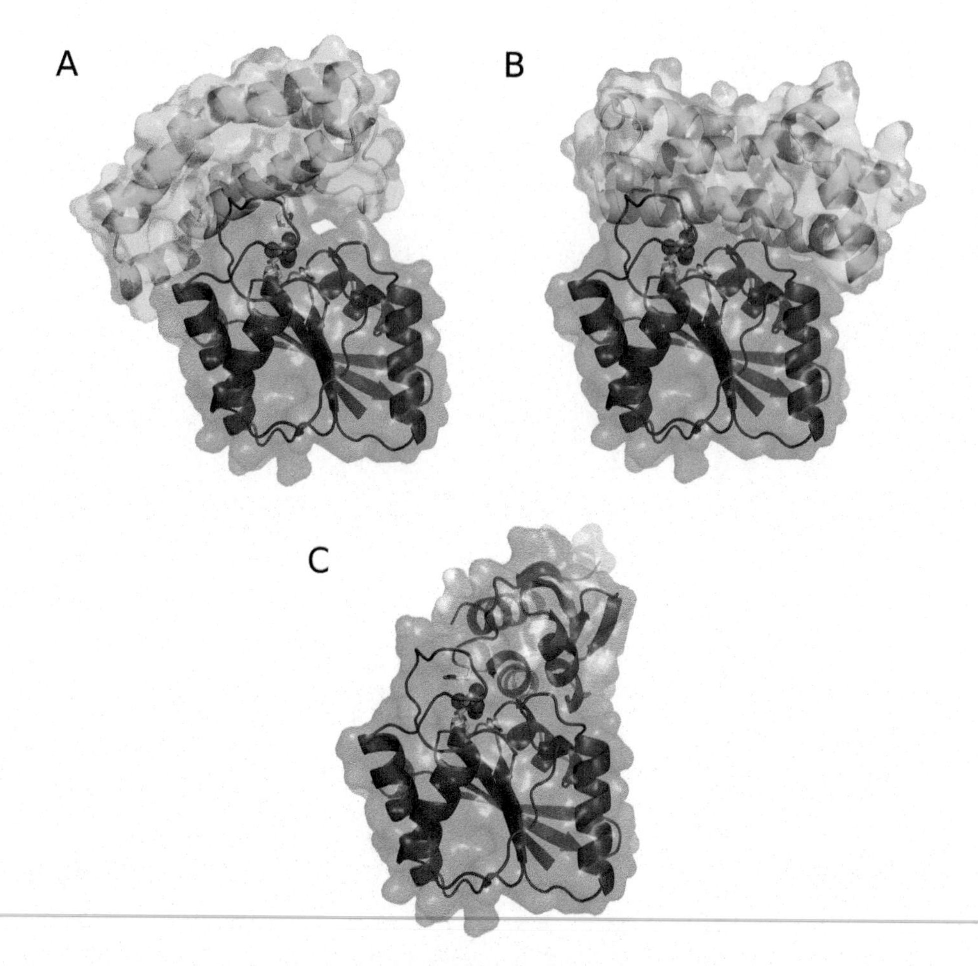

Fig. 2 Predicted docking models of SixA with substrate. The manually obtained docking mode of SixA (PDB code 1ujc; [7]) with ArcB HPt domain (PDB code 1A0B; [35]) is shown in (**a**) while the ClusPro [36] prediction for the same proteins is shown in (**b**). (**c**) Shows the ClusPro predicted interaction mode of SixA with NPr protein (PDB code 5t1n; [37]). In each case SixA is shown in grey in the same orientation with side chains shown for residues of its "phosphate pocket" and spheres shown for the bound tungstate ion. Sticks are shown for the His residues of substrates (number 717 in ArcB HPt domain; number 16 in NPr) whose phosphorylations are removed by SixA and which must therefore lie close to the SixA catalytic site

superfamily is complicated by its sheer functional diversity: without further support it is dangerous to assume that any except the closest homologs necessarily share the same activity. In principle, the link between the minimal SixA domain structure, with a flat binding site, and protein substrate should be very valuable. Indeed, this feature was used to suggest that a distinct group of HP enzymes, centered on the Ais protein, should have a large substrate, possibly—given the putative regulatory function of Ais—a phosphorylated protein [9]. However, one of the proteins in question, *Salmonella enterica* PmrG, was later shown to hydrolyze the Hep (II) phosphate of large lipopolysaccharide molecules [39] and thereby confer iron and aluminum resistance on the bacterium.

While the numerous precedents of HP enzymes with dual small molecule and protein specificities mean that an additional PPHP function cannot be ruled out, it is certainly now clear that not every small HP superfamily member is necessarily a PPHP.

A crystal structure of SixA in complex with a substrate would undoubtedly assist in functional annotation of homologs by revealing substrate-binding residues and potential specificity determinants. While the manual docking mentioned above [7] was used to predict that a number of SixA residues were involved in electrostatic and other hydrogen-bonding interactions with HPt domain substrate, that docking pose was not reproduced among the top results of the current best-performing [40] protein–protein docking server ClusPro [36]. Filtering results using the knowledge that His717 must lie suitably close to the tungstate bound in the phosphate pocket of the SixA crystal structure, the best scoring docking pose is shown in Figure 2b. Compared to the previous manual pose (Fig. 2a), the substrate domain is rotated by around 180°. Interestingly, the computationally predicted pose in Fig. 2b is only number 15 in the list of results (using the default "balanced" scoring); better scoring poses place His717 well away from the SixA catalytic site. This result is surprising since these computational methods generally perform well when the docked structures are experimentally determined, as here. One possibility is that either SixA or its substrate undergoes a significant conformational change upon binding to its partner, such that the protein interfaces of the isolated structures are less complementary than expected. Similar results were obtained when ClusPro was used to dock the crystal structure of *E. coli* NPr protein [37] to that of SixA. Again the knowledge that the phosphorylated residue His16 must lie within the SixA catalytic site was used to filter residues. Once more, however, the only remaining candidate pose (Fig. 2c) scored relatively poorly, being number 13 in the results scored by a hydrophobic-favoring scoring scheme. Taken together, therefore, it seems that protein structural data on the enzyme–substrate interaction are not of high enough quality to guide the prediction of interactions between homologs.

Other potential sources of guidance to predict SixA-like activities in other proteins and cognate substrates are also uninformative here. Subunits of complexes or components of pathways can often be linked, especially in microbes, by genome context: the tendency of functionally linked genes to be found in similar genomic neighborhoods across species [41]. Similarly, coordinated patterns of presence or absence of proteins across species can be used to infer connected functions [42]. However, the STRING database [43], one of the main resources for determining such functional linkages, does not link SixA with either of its characterized substrates, ArcB or NPr, suggesting that this information will be of limited utility to predict novel substrates.

Taken together, these considerations paint a surprisingly limited picture of PPHP activities in microbes. In vitro and/or in vivo evidence shows that *E. coli* SixA acts on ArcB Hpt domain and the NPr protein. However, it is not known if these are its only substrates. Are other *E. coli* proteins containing Hpt or PTS-HPr domains also regulated by SixA in this way? Can phosphorylated His residues borne by other domains be substrates? Is SixA even multifunctional, acting on nonproteic substrates as well? The picture is even less clear in other organisms: in the absence of information on substrate-binding residues, it remains unclear what proportion of SixA homologs catalyze PPHP activity. The two unpublished putative "SixA-like" protein crystal structures in the PDB (Table 1) are cases in point: they share less than 30% sequence identity with *E. coli* SixA and between themselves, sufficient only to infer that they share a broad functional similarity as phosphatases, but far from enough to assure that they catalyze the same PHPP reaction [44]. A full picture of the molecular roles of SixA and homologs evidently awaits further elucidation.

3 PGAM5

The official name of the protein encoded by the PGAM5 locus is "Human phosphoglycerate mutase family member 5." PGAM is a common abbreviation for phosphoglycerate mutase, the first activity characterized in the HP superfamily, but while PGAM1 is well-characterized [45, 46], and PGAM2 and PGAM4 are presumed to encode genuine phosphoglycerate mutases (PGAM3 was realized to be the same as PGAM4), PGAM5 bears only around 22% sequence similarity with them and so, with hindsight, its naming was speculative and misleading. Current evidence, discussed below, suggests that this protein functions solely to dephosphorylate pSer, pThr, and pHis residues.

Early published studies on PGAM5 reported its binding to Bcl-X_L, lending it the temporary name "Bcl-XL-binding protein v68" and tentatively linking it to the regulation of apoptosis effected by members of the Bcl-2 protein family [47]. Later work confirmed that interaction, and characterized PGAM5 as a substrate for Keap1-regulated ubiquitination and subsequent degradation [3]. The interaction with Keap1 was shown by site-directed mutagenesis to be mediated by a conserved motif in PGAM5 similar to that characterized in the transcription factor Nrf2. Interestingly, both transcriptional and posttranscriptional mechanisms were shown to mediate the common Keap1-mediated regulation of PGAM5 and Nrfl in the response to oxidative stress. This work also demonstrated for the first time the existence of two isoforms, PGAM5-L and -S (for long or short) which, though identical up to residue 239, differ thereafter as a result of alternative splicing.

At that time, PGAM5 was suggested to mediate protein–protein interactions only rather than functioning as an enzyme. In 2008, PGAM5 was found to exist in a ternary complex with Keap1 and Nrf2, localized to the outer mitochondrial membrane by an N-terminal targeting sequence [48]. Subsequent work has reported the presence of PGAM5 in the mitochondrial inner membrane (e.g., [49, 50]) and has suggested that PGAM5 is released into the cytosol as a result of loss of integrity of the outer mitochondrial membrane under stress conditions [50].

Phosphatase activity was first assigned to PGAM5 by Takeda et al. [4], who showed that although PGAM5 lacked phosphoglycerate mutase activity, it was capable of activating ASK1 kinase by dephosphorylating inhibitory pSer/pThr sites. They also showed that PGAM5 orthologs from *Drosophila melanogaster* and *Caenorhabditis elegans* similarly activated ASK1 counterparts in their respective species [4]. Subsequent work from multiple groups has collectively placed PGAM5 at the heart of the regulation of mitochondrial dynamics, including roles in necrosis [51], apoptosis [52] and most recently mitophagy, the process by which dysfunctional mitochondria are selectively degraded by a specialized form of autophagy [53, 54–56]. Interestingly, not all its functions appear to be dependent on its catalytic activity [53].

PGAM was first characterized as a PHPP by Panda et al. [25], who identified its natural substrate as NDPK-B—specifically its catalytic His118 residue—by coimmunoprecipitation and mutational analysis. NDPK-B is a member of the NME family of nucleoside diphosphate kinases [57] which transfer γ-phospho groups from nucleoside triphosphates to nucleoside diphosphates. In addition to this nucleotide homeostatic role, the NME proteins are implicated in a variety of other cellular processes and two of them, NDPK-A and NDPK-B, function as protein histidine kinases [58]. One substrate identified for NDPK-B is the calcium-activated potassium channel, KCa3.1, which regulates membrane potential and calcium signaling. Autophosphorylated NDPK-B transfers its phospho group to KCa3.1 channels thereby activating K^+ efflux [59]. Thus, PGAM5 has an overall inhibitory effect on T-cell receptor-stimulated Ca^{2+} influx and hence cytokine production in $CD4^+$ T-cells. (Coincidentally, PGAM5's distant relatives in the HP superfamily STS-1 and -2 are also implicated in regulation of T-cell signaling: indeed, their names derive from *S*uppressor of *T*-cell receptor *S*ignaling [15].) As mentioned, the best characterized functions of PGAM5 relate to mitochondria, but mitochondrial membrane-resident PGAM5 can undergo proteolytic cleavage by rhomboid protease PARL [60] to remove the N-terminal 24 residues and thereby liberate a soluble form into the cytosol. The cleaved version of PGAM5-L, but not of PGAM5-S, binds and dephosphorylates NDPK-B [25]. Contributing to this picture of highly specific interaction, neither NDPK-A or -C which share 88%

and 66% sequence identity respectively with NDPK-B, interacted with PGAM5 [25]. Knockdown or overexpression of PGAM5 (wild-type or H105A) in cells in culture confirmed the relevance of PGAM5 PHPP activity to measured KCa3.1 channel activity. Finally, T-cells from PGAM knockout mice showed the expected higher levels of phosphorylation of NDPK-B and KCa3.1, as well as elevated calcium influx and cytokine production [25]. Interestingly, the system is also finely regulated by the unrelated PHPP enzyme PHPT-1 which, although it dephosphorylates and inhibits KCa3.1, does not act to dephosphorylate NDPK-B [61].

Structural information on human PGAM5 comes from the work of Chaikuad et al. [34] who determined crystal structures of two different N-terminal deletion mutants, with or without bound phosphate (Table 1). Through a series of biophysical experiments they determined that the C-terminal region of PGAM5 is required for dimerization and addressed the roles of the conserved WDXNWD motif, previously noted to be required for further oligomerization and allosteric regulation [62]. They showed that the motif interacted with other residues within the dimer, but also packed between dimers enabling the assembly of an unusual dodecameric structure [34]. However, whether this or other large oligomeric states are physiologically relevant remains to be determined, since the experiments were performed using truncated recombinant proteins. Interestingly, the three crystal structures differed in the conformation of their catalytic sites, especially between the two phosphate-bound structures, labeled "on" and "off." The former shows the enzyme in its likely catalytically competent state, while in the latter the key phosphorylatable His230 and the likely proton donor Glu177 are both in "out" positions that are presumably, by comparison with the consensus picture across the superfamily [1], incompatible with catalysis.

As with the interaction between *E. coli* SixA and its substrate, experimental protein structures are available for PGAM5 [34] and its substrate NDPK-B [63]. Inspection of the structures (Fig. 3), however, immediately suggests that substantial conformational changes must occur in one or both for a catalytically competent complex to form, since both the PGAM5 catalytic site and the dephosphorylated His118 of NDPK-B lie at the bottom of cavities. In the case of PGAM5, the flanking loops around 105–115 and 180–195 would need to part, while the region around 43–70 of NDPK-B (Fig. 3b, right) would need to change conformation to expose His118 for dephosphorylation. Movement of these regions is plausible, however: residues 181–190 are disordered in the PGAM5 apo structure [34] suggesting intrinsic flexibility, while the subdomain corresponding to residues 43–70 of NDPK-B, containing two helices joined by a turn, has been observed to adopt different conformations in different crystal structures, again indicative of flexibility [63].

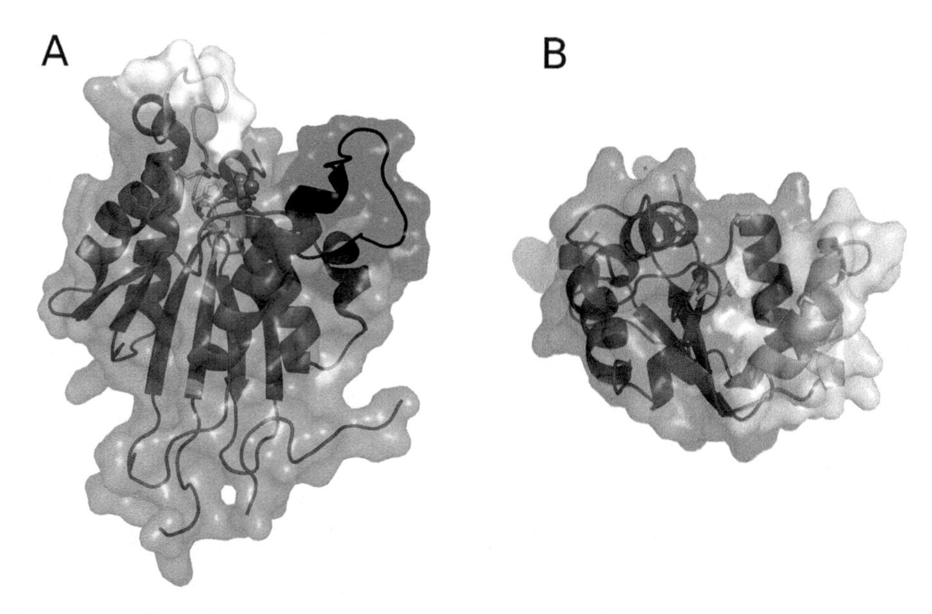

Fig. 3 Structures of (**a**) PGAM5 (PDB code 5muf; [34]) and (**b**) NDPK-B (PDB code 1nsk; [63]). The catalytic site of PGAM5 is marked by the four residues of the phospho pocket (sticks) and the bound phosphate group (spheres), and lies in a cavity. His118 of NDPK-B is shown in yellow and is similarly positioned at the bottom of a pocket. Specific regions mentioned in the text are colored as follows: PGAM5 105–115, white; PGAM5 180–195, black; NDPK-B 43–70, light pink

4 Conclusions

The history of PHPPs in the HP superfamily encompasses an interesting contrast between the bacterial SixA example, characterized as early as 1998 [24], to the very recent discovery that eukaryotic PGAM5 has PHPP activity [25]. Mirroring discoveries of protein Tyr phosphatase activities in the same superfamily, where proteins were previously characterized or predicted as having a distinct small molecule phosphatase activity, or to be catalytically inactive (e.g., [18, 22]), the discovery of PGAM5 PHPP activity followed an initial prediction of a noncatalytic function for this protein [3] followed by nine years of work characterizing its activity on substrates phosphorylated on Ser or Thr [4]. These precedents encourage speculation that further members of the HP superfamily may harbor PHPP activity. Prediction of these with bioinformatics methods would be possible if SixA and PGAM5 contained distinctive sequence or structural features that could be screened for across the superfamily. However, a structure-based alignment of a representative set of HP superfamily members is discouraging. Figure 1 shows a dendrogram resulting from an all-against-all comparison of a nonredundant (at a 25% sequence identity threshold) set of HPs generated with DALI [12] and displayed with the iTOL server

[13]. Clearly, the two known PHPP enzymes bear no strong overall structural similarity. Furthermore, their catalytic sites are significantly different: the deep active site cleft observed in the PGAM5 crystal structures [34] contrasts sharply with the open, flat active site of SixA [7]. Moreover, at the sequence level there do not appear to be any shared features that could be inferred as correlating with their common PHPP activity. In the DALI structure-based sequence alignment of the two, only 19 SixA residues are identical in PGAM5, a mere 12% shared sequence identity. These 19 residues lie very largely at the catalytic site and at neighboring buried positions: elsewhere there is no cluster of conserved residues that could represent a common substrate protein-binding interface. While available data are therefore little guide in predicting which other HP superfamily members may harbor unsuspected PHPP activity, the encoding of PHPP activity in two so diverse proteins should encourage further exploration of the HP superfamily for additional PHPP functionality.

References

1. Rigden DJ (2008) The histidine phosphatase superfamily: structure and function. Biochem J 409:333–348
2. Fothergill LA, Harkins RN (1982) The amino acid sequence of yeast phosphoglycerate mutase. Proc R Soc Lond B Biol Sci 215:19–44
3. Lo SC, Hannink M (2006) PGAM5, a Bcl-XL-interacting protein, is a novel substrate for the redox-regulated Keap1-dependent ubiquitin ligase complex. J Biol Chem 281:37893–37903
4. Takeda K, Komuro Y, Hayakawa T, Oguchi H, Ishida Y, Murakami S, Noguchi T, Kinoshita H, Sekine Y, Iemura S, Natsume T, Ichijo H (2009) Mitochondrial phosphoglycerate mutase 5 uses alternate catalytic activity as a protein serine/threonine phosphatase to activate ASK1. Proc Natl Acad Sci U S A 106:12301–12305
5. Rigden DJ, Littlejohn JE, Henderson K, Jedrzejas MJ (2003) Structures of phosphate and trivanadate complexes of Bacillus stearothermophilus phosphatase PhoE: structural and functional analysis in the cofactor-dependent phosphoglycerate mutase superfamily. J Mol Biol 325:411–420
6. Rigden DJ, Walter RA, Phillips SE, Fothergill-Gilmore LA (1999) Sulphate ions observed in the 2.12 A structure of a new crystal form of S. cerevisiae phosphoglycerate mutase provide insights into understanding the catalytic mechanism. J Mol Biol 286:1507–1517
7. Hamada K, Kato M, Shimizu T, Ihara K, Mizuno T, Hakoshima T (2005) Crystal structure of the protein histidine phosphatase SixA in the multistep His-Asp phosphorelay. Genes Cells 10:1–11
8. Lin K, Li L, Correia JJ, Pilkis SJ (1992) Glu327 is part of a catalytic triad in rat liver fructose-2,6-bisphosphatase. J Biol Chem 267:6556–6562
9. Rigden DJ (2003) Unexpected catalytic site variation in phosphoprotein phosphatase homologues of cofactor-dependent phosphoglycerate mutase. FEBS Lett 536:77–84
10. Ostanin K, Van Etten RL (1993) Asp304 of Escherichia coli acid phosphatase is involved in leaving group protonation. J Biol Chem 268:20778–20784
11. wwPDB consortium (2018) Protein Data Bank: the single global archive for 3D macromolecular structure data. Nucleic Acids Res 47 (D1): D520–D528
12. Holm L, Laakso LM (2016) Dali server update. Nucleic Acids Res 44:W351–W355
13. Letunic I, Bork P (2016) Interactive tree of life (iTOL) v3: an online tool for the display and annotation of phylogenetic and other trees. Nucleic Acids Res 44:W242–W245
14. El-Gebali S, Mistry J, Bateman A, Eddy SR, Luciani A, Potter SC, Qureshi M, Richardson LJ, Salazar GA, Smart A, Sonnhammer ELL, Hirsh L, Paladin L, Piovesan D, Tosatto SCE, Finn RD (2018) The Pfam protein families

database in 2019. Nucleic Acids Res 47(D1): D427–D432
15. Carpino N, Turner S, Mekala D, Takahashi Y, Zang H, Geiger TL, Doherty P, Ihle JN (2004) Regulation of ZAP-70 activation and TCR signaling by two related proteins, Sts-1 and Sts-2. Immunity 20:37–46
16. Carpino N, Chen Y, Nassar N, Oh HW (2009) The Sts proteins target tyrosine phosphorylated, ubiquitinated proteins within TCR signaling pathways. Mol Immunol 46:3224–3231
17. Mikhailik A, Ford B, Keller J, Chen Y, Nassar N, Carpino N (2007) A phosphatase activity of Sts-1 contributes to the suppression of TCR signaling. Mol Cell 27:486–497
18. Meng TC, Lin MF (1998) Tyrosine phosphorylation of c-ErbB-2 is regulated by the cellular form of prostatic acid phosphatase in human prostate cancer cells. J Biol Chem 273:22096–22104
19. Lin MF, Clinton GM (1988) The epidermal growth factor receptor from prostate cells is dephosphorylated by a prostate-specific phosphotyrosyl phosphatase. Mol Cell Biol 8:5477–5485
20. Zylka MJ, Sowa NA, Taylor-Blake B, Twomey MA, Herrala A, Voikar V, Vihko P (2008) Prostatic acid phosphatase is an ectonucleotidase and suppresses pain by generating adenosine. Neuron 60:111–122
21. Tanaka M, Kishi Y, Takanezawa Y, Kakehi Y, Aoki J, Arai H (2004) Prostatic acid phosphatase degrades lysophosphatidic acid in seminal plasma. FEBS Lett 571:197–204
22. Davies L, Anderson IP, Turner PC, Shirras AD, Rees HH, Rigden DJ (2007) An unsuspected ecdysteroid/steroid phosphatase activity in the key T-cell regulator, Sts-1: surprising relationship to insect ecdysteroid phosphate phosphatase. Proteins 67:720–731
23. Fuhs SR, Hunter T (2017) pHisphorylation: the emergence of histidine phosphorylation as a reversible regulatory modification. Curr Opin Cell Biol 45:8–16
24. Ogino T, Matsubara M, Kato N, Nakamura Y, Mizuno T (1998) An Escherichia coli protein that exhibits phosphohistidine phosphatase activity towards the HPt domain of the ArcB sensor involved in the multistep His-Asp phosphorelay. Mol Microbiol 27:573–585
25. Panda S, Srivastava S, Li Z, Vaeth M, Fuhs SR, Hunter T, Skolnik EY (2016) Identification of PGAM5 as a mammalian protein histidine phosphatase that plays a central role to negatively regulate CD4(+) T cells. Mol Cell 63:457–469
26. Mizuno T (1998) His-Asp phosphotransfer signal transduction. J Biochem 123:555–563
27. Sakakibara H, Taniguchi M, Sugiyama T (2000) His-Asp phosphorelay signaling: a communication avenue between plants and their environment. Plant Mol Biol 42:273–278
28. Mueller JP, Sonenshein AL (1992) Role of the Bacillus subtilis gsiA gene in regulation of early sporulation gene expression. J Bacteriol 174:4374–4383
29. Ohlsen KL, Grimsley JK, Hoch JA (1994) Deactivation of the sporulation transcription factor Spo0A by the Spo0E protein phosphatase. Proc Natl Acad Sci U S A 91:1756–1760
30. Kiba T, Aoki K, Sakakibara H, Mizuno T (2004) Arabidopsis response regulator, ARR22, ectopic expression of which results in phenotypes similar to the wol cytokinin-receptor mutant. Plant Cell Physiol 45:1063–1077
31. Horak J, Grefen C, Berendzen KW, Hahn A, Stierhof YD, Stadelhofer B, Stahl M, Koncz C, Harter K (2008) The Arabidopsis thaliana response regulator ARR22 is a putative AHP phospho-histidine phosphatase expressed in the chalaza of developing seeds. BMC Plant Biol 8:77–2229-8-77
32. Schulte JE, Goulian M (2018) The phosphohistidine phosphatase SixA targets a phosphotransferase system. MBio 9. https://doi.org/10.1128/mBio.01666-18
33. Pfluger-Grau K, Gorke B (2010) Regulatory roles of the bacterial nitrogen-related phosphotransferase system. Trends Microbiol 18:205–214
34. Chaikuad A, Filippakopoulos P, Marcsisin SR, Picaud S, Schroder M, Sekine S, Ichijo H, Engen JR, Takeda K, Knapp S (2017) Structures of PGAM5 provide insight into active site plasticity and multimeric assembly. Structure 25:1089–1099.e3
35. Kato M, Mizuno T, Shimizu T, Hakoshima T (1997) Insights into multistep phosphorelay from the crystal structure of the C-terminal HPt domain of ArcB. Cell 88:717–723
36. Kozakov D, Hall DR, Xia B, Porter KA, Padhorny D, Yueh C, Beglov D, Vajda S (2017) The ClusPro web server for protein-protein docking. Nat Protoc 12:255–278
37. Strickland M, Stanley AM, Wang G, Botos I, Schwieters CD, Buchanan SK, Peterkofsky A, Tjandra N (2016) Structure of the NPr:EIN (Ntr) complex: mechanism for specificity in paralogous phosphotransferase systems. Structure 24:2127–2137
38. Hakoshima T, Ichihara H (2007) Structure of SixA, a histidine protein phosphatase of the

ArcB histidine-containing phosphotransfer domain in Escherichia coli. Methods Enzymol 422:288–304

39. Nishino K, Hsu FF, Turk J, Cromie MJ, Wosten MM, Groisman EA (2006) Identification of the lipopolysaccharide modifications controlled by the Salmonella PmrA/PmrB system mediating resistance to Fe(III) and Al(III). Mol Microbiol 61:645–654
40. Vajda S, Yueh C, Beglov D, Bohnuud T, Mottarella SE, Xia B, Hall DR, Kozakov D (2017) New additions to the ClusPro server motivated by CAPRI. Proteins 85:435–444
41. Lathe WC,3rd, Snel B, Bork P (2000) Gene context conservation of a higher order than operons. Trends Biochem Sci 25:474–479
42. Skunca N, Dessimoz C (2015) Phylogenetic profiling: how much input data is enough? PLoS One 10:e0114701
43. Szklarczyk D, Gable AL, Lyon D, Junge A, Wyder S, Huerta-Cepas J, Simonovic M, Doncheva NT, Morris JH, Bork P, Jensen LJ, von Mering C (2018) STRING v11: protein-protein association networks with increased coverage, supporting functional discovery in genome-wide experimental datasets. Nucleic Acids Res 47(D1):D607–D613
44. Devos D, Valencia A (2000) Practical limits of function prediction. Proteins 41:98–107
45. Wang Y, Wei Z, Liu L, Cheng Z, Lin Y, Ji F, Gong W (2005) Crystal structure of human B-type phosphoglycerate mutase bound with citrate. Biochem Biophys Res Commun 331:1207–1215
46. Durany N, Joseph J, Cruz-Sanchez FF, Carreras J (1997) Phosphoglycerate mutase, 2,3-bisphosphoglycerate phosphatase and creatine kinase activity and isoenzymes in human brain tumours. Br J Cancer 76:1139–1149
47. Hammond PW, Alpin J, Rise CE, Wright M, Kreider BL (2001) In vitro selection and characterization of Bcl-X(L)-binding proteins from a mix of tissue-specific mRNA display libraries. J Biol Chem 276:20898–20906
48. Lo SC, Hannink M (2008) PGAM5 tethers a ternary complex containing Keap1 and Nrf2 to mitochondria. Exp Cell Res 314:1789–1803
49. Lu W, Karuppagounder SS, Springer DA, Allen MD, Zheng L, Chao B, Zhang Y, Dawson VL, Dawson TM, Lenardo M (2014) Genetic deficiency of the mitochondrial protein PGAM5 causes a Parkinson's-like movement disorder. Nat Commun 5:4930
50. Yamaguchi A, Ishikawa H, Furuoka M, Yokozeki M, Matsuda N, Tanimura S, Takeda K (2019) Cleaved PGAM5 is released from mitochondria depending on proteasome-mediated rupture of the outer mitochondrial membrane during mitophagy. J Biochem 165:19–25
51. Wang Z, Jiang H, Chen S, Du F, Wang X (2012) The mitochondrial phosphatase PGAM5 functions at the convergence point of multiple necrotic death pathways. Cell 148:228–243
52. Ishida Y, Sekine Y, Oguchi H, Chihara T, Miura M, Ichijo H, Takeda K (2012) Prevention of apoptosis by mitochondrial phosphatase PGAM5 in the mushroom body is crucial for heat shock resistance in Drosophila melanogaster. PLoS One 7:e30265
53. Lenhausen AM, Wilkinson AS, Lewis EM, Dailey KM, Scott AJ, Khan S, Wilkinson JC (2016) Apoptosis inducing factor binding protein PGAM5 triggers mitophagic cell death that is inhibited by the ubiquitin ligase activity of X-linked inhibitor of apoptosis. Biochemistry 55:3285–3302
54. Chen G, Han Z, Feng D, Chen Y, Chen L, Wu H, Huang L, Zhou C, Cai X, Fu C, Duan L, Wang X, Liu L, Liu X, Shen Y, Zhu Y, Chen Q (2014) A regulatory signaling loop comprising the PGAM5 phosphatase and CK2 controls receptor-mediated mitophagy. Mol Cell 54:362–377
55. Lu W, Sun J, Yoon JS, Zhang Y, Zheng L, Murphy E, Mattson MP, Lenardo MJ (2016) Mitochondrial protein PGAM5 regulates mitophagic protection against cell necroptosis. PLoS One 11:e0147792
56. Wu H, Xue D, Chen G, Han Z, Huang L, Zhu C, Wang X, Jin H, Wang J, Zhu Y, Liu L, Chen Q (2014) The BCL2L1 and PGAM5 axis defines hypoxia-induced receptor-mediated mitophagy. Autophagy 10:1712–1725
57. Boissan M, Dabernat S, Peuchant E, Schlattner U, Lascu I, Lacombe ML (2009) The mammalian Nm23/NDPK family: from metastasis control to cilia movement. Mol Cell Biochem 329:51–62
58. Attwood PV, Wieland T (2015) Nucleoside diphosphate kinase as protein histidine kinase. Naunyn Schmiedebergs Arch Pharmacol 388:153–160
59. Srivastava S, Li Z, Ko K, Choudhury P, Albaqumi M, Johnson AK, Yan Y, Backer JM, Unutmaz D, Coetzee WA, Skolnik EY (2006) Histidine phosphorylation of the potassium channel KCa3.1 by nucleoside diphosphate kinase B is required for activation of KCa3.1 and CD4 T cells. Mol Cell 24:665–675
60. Sekine S, Kanamaru Y, Koike M, Nishihara A, Okada M, Kinoshita H, Kamiyama M,

Maruyama J, Uchiyama Y, Ishihara N, Takeda K, Ichijo H (2012) Rhomboid protease PARL mediates the mitochondrial membrane potential loss-induced cleavage of PGAM5. J Biol Chem 287:34635–34645

61. Srivastava S, Zhdanova O, Di L, Li Z, Albaqumi M, Wulff H, Skolnik EY (2008) Protein histidine phosphatase 1 negatively regulates CD4 T cells by inhibiting the K+ channel KCa3.1. Proc Natl Acad Sci U S A 105:14442–14446

62. Wilkins JM, McConnell C, Tipton PA, Hannink M (2014) A conserved motif mediates both multimer formation and allosteric activation of phosphoglycerate mutase 5. J Biol Chem 289:25137–25148

63. Webb PA, Perisic O, Mendola CE, Backer JM, Williams RL (1995) The crystal structure of a human nucleoside diphosphate kinase, NM23-H2. J Mol Biol 251:574–587

Chapter 8

In Vitro Assays for Measuring Protein Histidine Phosphatase Activity

Brandon S. McCullough and Amy M. Barrios

Abstract

In order to obtain a detailed kinetic characterization, identify inhibitors, and elucidate the biological roles of an enzyme, it is advantageous to have a facile, sensitive enzyme assay protocol. Here we present a brief overview of the techniques available to monitor histidine phosphatase activity and provide protocols for measuring the activity and inhibition of PHPT1 in vitro using the fluorescent probe 6,8-difluoro-4-methylumbelliferyl phosphate (DiFMUP). This assay uses small quantities of commercially available materials, making its use feasible for most laboratories.

Key words In vitro enzyme assay, Enzyme kinetics, Dephosphorylation, Phosphohistidine, Phosphatase activity, Fluorogenic substrate

1 Introduction

Although protein histidine phosphorylation was discovered in 1962 [1], the enzymes responsible for dephosphorylation of phosphohistidine (pHis) in mammals were not discovered until much more recently [2–8]. One of the major challenges in the field is the lability of pHis under relatively mild conditions [6–8], making the adduct much more difficult to study than phosphorylated serine, threonine, and tyrosine residues. In order to understand the biological roles of pHis it is necessary to identify and characterize the kinases and phosphatases that coordinate the regulated addition and removal of phosphate from histidine residues. This chapter provides an overview of the methods that have been used to study histidine phosphatase activity in vitro (*see* Table 1 for summary) along with a detailed protocol for a facile, commercially available histidine phosphatase assay that should be readily applicable in most laboratories.

Proteins containing phosphorylated histidine (pHis) residues provide the most biologically relevant substrates for the histidine phosphatases. Consequently, histidine phosphatase activity can be

Claire E. Eyers (ed.), *Histidine Phosphorylation: Methods and Protocols*, Methods in Molecular Biology, vol. 2077, https://doi.org/10.1007/978-1-4939-9884-5_8, © Springer Science+Business Media, LLC, part of Springer Nature 2020

Table 1
Summary of methods available to assay histidine phosphatase activity in vitro

Technique	Substrate	Substrate concentration	Enzyme	Enzyme concentration	Detection method	Qualitative or quantitative	Special notes	Reference
Radiolabeling	Protein with ^{32}P-labeled pHis	NA	Cell extracts	NA	Radiography	Qualitative/ semiquantitative	Requires radioisotopes	[3, 13–15]
Immunoblotting	Cell extracts	NA	LHPP	NA	Western blot	Qualitative	Requires anti-pHis antibodies	[17]
HPLC	pHis peptide	7 mM	PHPT1	NA	Absorbance	Quantitative	Discontinuous assay	[2, 18]
MS	pHis peptide	2.4 mM	PHPT1	370 nM	ESI-MS	Semiquantitative	Discontinuous assay	[19]
NMR	pHis peptide	5 mM	PHPT1	14 μM	NMR	Quantitative	Requires deuterated buffer	[21]
Colorimetric assay	pHis peptide	10–80 mM	PHPT1	20–50 nM	Absorbance	Quantitative	pHis lability requires special protocol	[24, 25]
Paper electrophoresis	pHis	10 mM	LHPP	5.1 μM	Staining	Qualitative		[27]
Colorimetric assay	*p*-NPP	1–50 mM	PHPT1	2.4 μM	Absorbance	Quantitative	Can be run as a continuous or discontinuous assay	[28]
Fluorescence spectroscopy	DiFMUP	2–0.02 mM	PHPT1	110 nM	Fluorescence	Quantitative	Highly sensitive, continuous assay	[29]

monitored using pHis-containing proteins from cell lysates or with purified proteins. Use of a ^{32}P radiolabel has been a common method to detect pHis in proteins [1, 9–15], providing a highly sensitive means of detecting changes in protein phosphorylation. More recently, antibodies to pHis have been developed that can be used in immunoblotting assays [16].

The recent development of anti-pHis antibodies has paved the way for more precise characterization of histidine kinases and phosphatases [6, 16]. For example, the primary anti-pHis antibody SC44-1 was recently used to aid in the identification of the histidine phosphatase LHPP as a tumour suppressor [17]. The availability of anti-pHis antibodies is expected to facilitate significant advances in our understanding of the roles of histidine phosphorylation in health and disease.

Like pHis-containing proteins, histidine-phosphorylated peptides also provide convenient substrates for assaying histidine phosphatase activity; they are advantageous as they can be readily made synthetically and should have well characterized phosphorylation sites. High performance liquid chromatography (HPLC) and/or mass spectrometry (MS)-based methods can be used detect changes in histidine phosphorylation levels in peptides that do not contain ^{32}P. Typically, histidine kinase/phosphatase activity is evaluated by quantifying the relative amounts of the phosphorylated and nonphosphorylated peptide substrates, either using a UV detector (and an appropriate UV-absorbing substrate peptide) [2, 18], or by MS [19] which measures the mass change due to the presence or absence of the phosphate group.

Small changes in the levels of histidine phosphorylation on a peptide substrate can also be evaluated using either ^{31}P or ^{1}H nuclear magnetic resonance spectroscopy (NMR) as demonstrated by studies with histone H4 [20, 21].

Malachite green is a commonly used and well-validated reagent for measuring free phosphate concentration in phosphatase assays [22, 23]. However, in the context of studying histidine phosphatases, the use of malachite green is significantly more challenging due to the highly acidic conditions required for assay quenching, which results in hydrolysis of the P–N bond of pHis [6, 7], and thus erroneous evaluation of phosphatase activity. To overcome this problem, a protocol has been developed that compares the total amount of substrate used with the amount of phosphorylated substrate remaining after enzymatic hydrolysis of a peptide substrate [24, 25]. Following ion exchange chromatography to remove the free phosphate resulting from enzymatic activity, the total concentration of peptide was determined by measuring the absorbance at 320 nm. The eluted peak was then incubated with malachite green for 2 h to release and quantify the acid labile phosphate from the pHis. By subtracting the amount of phosphate released from the amount of total substrate, the enzymatic activity of PHPT1 could be determined. This approach has also been used to monitor dephosphorylation of the histone H4 protein, histone H1 protein, and a 30 kDa polylysine peptide [24, 26].

Small molecule substrates have often been used to monitor phosphatase activity in vitro. Although they are not as representative of the biological substrates of the histidine phosphatases as pHis-containing proteins and peptides, they often provide significantly more sensitive and facile assays that are convenient for many applications. Paper electrophoresis, along with absorbance spectroscopy and fluorescence spectroscopy using the substrates *para*-nitrophenyl phosphate (*p*-NPP), or 6,8-difluoro-4-methylumbelliferyl phosphate (DiFMUP), are all used to quantify levels of pHis following incubation with the phosphatase of interest [19, 27–29].

Assays that take advantage of an increase in fluorescence upon dephosphorylation of the substrate provide an excellent signal/noise ratio and improved sensitivity over colorimetric substrates. DiFMUP is a phosphotyrosine mimetic moiety whose fluorescence is quenched by phosphorylation of a coumarin core. Upon dephosphorylation, DiFMUP becomes strongly fluorescent with an excitation maximum at 350 nm and an emission band centred at 455 nm.

Encouraged by a preliminary report indicating that DiFMUP could serve as a substrate for monitoring histidine phosphatase activity [30], we developed an optimized protocol [29] and validated the use of DiFMUP as a substrate for PHPT1. Compared to *p*-NPP, using DiFMUP as a substrate to assay PHPT1 activity significantly improved the kinetic parameters (K_m of 220 mM and k_{cat}/K_m of 1900 M^{-1} s^{-1}). The assay is also much more sensitive, requiring only 20 μM DiFMUP and 110 nM PHPT1 to obtain satisfactory kinetic data. Additionally, this assay is useful in identifying enzyme inhibitors, as demonstrated by our discovery of the inhibitory nature of Cu^{2+} and Zn^{2+} on PHPT1 activity. The optimized protocol for measuring purified PHPT1 activity in vitro using DiFMUP as a substrate is described below.

2 Materials

Unless otherwise stated, use high purity Milli-Q water and HPLC grade DMSO. Perform all assays with 10% (v/v) DMSO.

2.1 PHPT1 Assay

1. Reaction buffer: 50 mM HEPES, 10 mM NaCl, 0.01% (w/v) Brij 35, pH 8.0. Add approximately 150 mL of water and a single stir bar to a 250 mL graduated cylinder (*see* **Note 1**). Weigh out 2.98 g HEPES free acid, 146 mg NaCl, and 25 mg Brij 35 and transfer to the graduated cylinder (*see* **Note 2**). After all components are fully dissolved, fill to 225 mL with water and adjust the pH to 8.0 as needed by careful addition of

NaOH (*see* **Note 3**). Finally, fill to 250 mL with water and filter the solution through a 0.45 μM PES membrane. Store at room temperature.

2. PHPT1 enzyme: Dissolve 100 μg of lyophilized PHPT1 in 1.316 mL water to make a 5 μM stock in buffer, as directed by the manufacturer. Divide into 22 μL aliquots. Store at −80 °C and minimize freeze–thaw cycles (*see* **Note 4**).
3. DiFMU stock solution and dilutions for standard curve: Dissolve 2.15 mg of DiFMU in 1 mL DMSO to make a 10 mM stock. Perform serial dilutions of this stock using DMSO to prepare 100 μM and 10 μM solutions. Store these at −20 °C.
4. DiFMUP stock solution and dilutions for Michaelis–Menten kinetics and inhibitor assays: Dissolve 3.51 mg of DiFMUP in 1 mL DMSO to make a 12 mM stock solution. Dilutions of the stock in DMSO are prepared as needed (*see* Table 3 for examples). Store at −20 °C.
5. DTT stock solution: Dissolve 7.7 mg of DTT in 500 μL of water to make a 100 mM stock solution (*see* **Note 5**). This solution should be prepared fresh daily and stored at 4 °C.
6. Inhibitor stock solution: Inhibitor should be dissolved in water or DMSO to create ~10 mM stock solution (*see* **Note 6**). Subsequent dilutions should be made with reaction buffer when possible (*see* **Note 7**). Store the stocks at a temperature appropriate for the inhibitor molecule (likely −20 °C).
7. Reaction vessel: 96-well half-area microplate (*see* **Note 8**).
8. Fluorescent plate reader: Temperature controllable 96-well fluorescent plate (*see* **Note 9**).
9. Plastic film.

3 Methods

Unless otherwise specified, carry out all steps at room temperature, with the exception of the enzyme assays that are performed at 37 °C (in a total volume of 50 μL).

3.1 Standard Curve

1. Set the excitation and emission wavelengths of the plate reader to 350 and 455 nm, respectively. If available, set the emission filter at a wavelength between 355 and 450 nm (*see* **Note 10**) and program the instrument to shake the plate for 5 s prior to the first measurement (*see* **Note 11**).
2. To a black 96-well half-area microplate, add 45 μL of reaction buffer, followed by DMSO and DiFMU according to Table 2. The maximum volume of DMSO + DiFMU should be 5 μL to ensure that the total concentration of DMSO does not exceed

Table 2
Summary of required reagents necessary to produce a DiFMU standard curve

DiFMU concentration in well, μM	Volume of DiFMU	Volume of DMSO, μL
10	5 μL of 100 μM stock	0
8	4 μL of 100 μM stock	1
6	3 μL of 100 μM stock	2
4	2 μL of 100 μM stock	3
2	1 μL of 100 μM stock	4
1	5 μL of 10 μM stock	0
0.8	4 μL of 10 μM stock	1
0.6	3 μL of 10 μM stock	2
0.4	2 μL of 10 μM stock	3
0.2	1 μL of 10 μM stock	4

10% (v/v). Each DiFMU concentration should be prepared in triplicate.

3. Insert the microplate into the plate reader and measure the relative fluorescence units (RFUs) of each well.
4. Average the three measured RFUs and determine the standard deviation and the coefficient of variation (CV) for each DiFMU concentration measurement. If the CV is not less than 0.1, repeat **steps 2** and **3** until this is achieved.
5. Obtain a standard curve: Perform a least squares linear regression analysis on the data, plotting the DiFMU concentration on the *x*-axis and the mean fluorescence of each triplicate on the *y*-axis. The R^2 value should be greater than 0.98. The linear equation generated from this method will be used to convert fluorescence intensity into product concentration (*see* **Note 12**).

3.2 Michaelis–Menten Kinetics Assay

1. Preheat the plate reader to 37 °C and adjust instrument settings as described in Subheading 3.1, **step 1**.
2. Remove one aliquot of enzyme from the −80 °C storage and place on ice to thaw (*see* **Note 13**). Thaw the required DiFMUP dilutions (*see* Table 3 for suggested dilutions) and keep at room temperature.
3. Add 73 μL of reaction buffer and 5 μL of the DTT stock solution to the enzyme (*see* **Note 14**). Mix thoroughly by pipetting up and down, or by vortexing (*see* **Note 15**). Store the enzyme stock solution on ice.

Table 3
Example of amounts of DiFMUP and DMSO to use when making a Michaelis–Menten curve

DiFMUP concentration in well, μM	Volume of DiFMU	Volume of DMSO, μL
1200	5 μL of 12 mM stock	0
1000	5 μL of 10 mM stock	0
500	2.5 μL of 10 mM stock	2.5
250	1.25 μL of 10 mM stock	3.75
100	5 μL of 1 mM solution	0
50	2.5 μL of 1 mM stock	2.5
30	1.25 μL of 1 mM stock	3.75
20	1 μL of 1 mM stock	4

4. Add 40 μL of reaction buffer, followed by DMSO and DiFMUP (according to Table 3) to the microplate as required according to the DiFMUP concentrations being tested (*see* **Note 16**). Each DiFMUP concentration should be assayed in triplicate.
5. Quickly transfer 5 μL of PHPT1 enzyme solution from **step 3** to each well to initiate the reaction (*see* **Note 17**).
6. Insert the microplate into the plate reader and measure the fluorescence value of each well every 60 s for 30 min. *See* Fig. 1 for a visualization of the data processing steps.
7. Using only the linear portion of the RFU vs. time curve generated in **step 6**, convert the increase in fluorescence into μM DiFMU for all data points according to the standard curve generated in Subheading 3.1.
8. Perform a least squares linear regression for the data in each well using time as the x-variable and μM DiFMU as the y-variable. The slope of each line will be the rate of the reaction at each concentration of DiFMUP.
9. Average the rates (μM DiFMU formed per second) obtained in **step 8** for each DiFMUP concentration triplicate to obtain the final rate of reaction at each concentration of DiFMUP.
10. To obtain the Michaelis–Menten parameters for the enzyme, plot the concentration of DiFMUP in μM on the x-axis and the rate obtained from DiFMUP concentration triplicate average (μM/s) on the y-axis. Perform a nonlinear curve fit using the equation:

$$y = \frac{ax}{b + x}$$

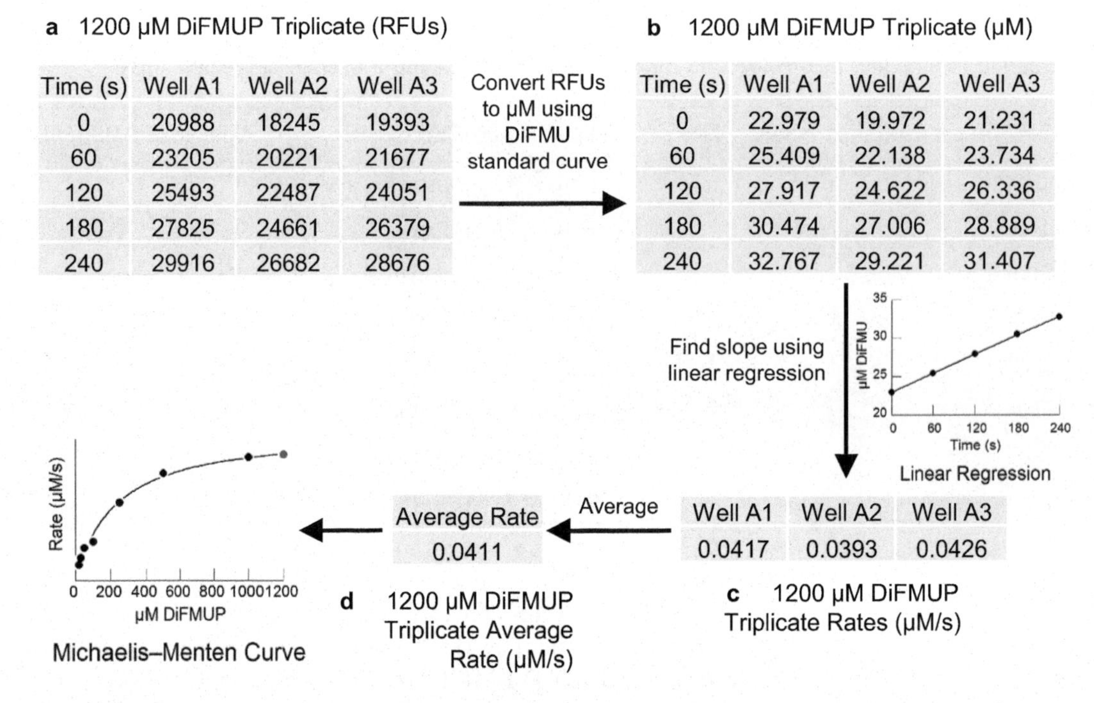

Fig. 1 Workflow of data processing to produce a Michaelis–Menten Curve. After collecting the data from the linear portion of the RFU *vs* time curve for a single DIFMUP concentration triplicate at the time points within the linear region (**a**), each fluorescence value should be converted into μM DiFMU using a prepared standard curve (**b**). For each well in the triplicate, perform a linear regression with time on the *x*-axis and μM DiFMU on the *y*-axis and determine the slope of the line (**c**). Average these slopes together to find the average enzyme activity rate for that DiFMUP concentration (**d**). Repeat this process for all DiFMUP concentrations to generate the data necessary for plotting a Michaelis–Menten curve

where a is V_{max} (μM/s), and b is K_m (μM) (*see* **Note 18**). The k_{cat} value can be calculated by dividing V_{max} by the enzyme concentration in the assay (110 nM in this protocol).

3.3 Inhibition Assay

1. Set-up the assay according to Subheadings 3.2, **steps 1** through **4**.
2. Add reaction buffer and inhibitor to each well such that their total volume is 40 μL (*see* **Note 19**). If the inhibitor is prepared in DMSO, be sure to add an appropriate volume of DMSO to each well such that the final DMSO concentration is 10% (v/v). Each inhibitor concentration should be tested in triplicate. Be sure to include a control with no inhibitor (10% (v/v) DMSO only) in triplicate as well.
3. Add 5 μL of PHPT1 enzyme solution to each well. Cover the microplate with a layer of plastic film and incubate at room temperature for 30 min (*see* **Note 20**).

4. Carefully remove the plastic film and quickly transfer 5 μL of 2 mM DiFMUP solution to each well to initiate the reaction (*see* **Note 17**).
5. Insert the microplate into the plate reader and measure the fluorescence of each well every 60 s for 30 min.
6. Using only data that falls within the linear portion of the curve produced in **step 5**, calculate the slope of this line for each well. Average the slopes for the triplicate measurements for each inhibitor concentration.
7. Divide the mean slope of each inhibitor triplicate by the mean slope of the control and multiply by 100. The resulting values are the percent inhibition at that inhibitor concentration.
8. Plot inhibitor concentrations (x-axis) against the percent inhibition (y-axis). Perform a nonlinear curve fit using the equation:

$$y = a + \frac{b - a}{1 + \left(\frac{x}{c}\right)^d}$$

where the a-value is the expected maximum inhibition value, the b-value is the expected minimum inhibition value, the c-value is the IC_{50}, the d-value is the Hill coefficient (*see* **Note 21**).

4 Notes

1. Addition of the water at this point ensures that the stir bar does not become trapped by HEPES as it is added to the cylinder.
2. While Brij 35 can be added directly, some may find it easier to prepare a concentrated solution separately, then add an appropriate volume to the graduated cylinder to achieve a final concentration of 0.01% (w/v).
3. When adjusting pH, it may be easier to start with a relatively high concentration of NaOH, then use lower concentrations to hone in on pH 8.0.
4. Enzyme activity decreases when subjected to freeze–thaw cycles. Therefore, to ensure maximum enzyme activity, is it critical to avoid excessive and/or unnecessary freeze–thaw cycles.
5. The values are provided as an example. Many will probably find it easier to weigh out some amount of material and then add an appropriate amount of solvent to obtain a final solution with the desired concentration.
6. If the inhibitor of interest is insoluble in water, it will be necessary to alter the preparation of DiFMUP stocks to ensure

the DMSO in each well remains at 10% (v/v). For example, a minimal amount of DMSO could be used to dissolve the DiFMUP, and water added to obtain stock solution with the desired DiFMUP concentration. Further dilutions could be performed using buffer. DiFMUP stock solutions in buffer will be less stable during storage than those made in DMSO alone.

7. As every molecule has different affinity for PHPT1, it is impossible to give the concentrations of inhibitor needed—this will have to be determined experimentally.
8. Black, opaque microplates are strongly recommended, as clear microplates may lead to fluorescence cross-contamination between wells. If half-area plates are not used, the amount of each reagent should be adjusted as appropriate.
9. The temperature of the plate reader should be controllable as the assays are carried out at 37 °C. The plate reader should have top read capabilities if using microplates without clear bottoms.
10. In our experience, it is best to use a filter as close as possible to the emission wavelength, without going over (e.g., a 450 nm cutoff filter for an emission of 450 nm).
11. If your instrument is not capable of shaking the plate, it will be necessary to pipette each well up and down manually prior to reading to ensure proper mixing of the components.
12. Although in a perfect system this linear equation should have no *b*-value (indicating that the fluorescence is zero when there is no fluorophore present), errors during preparation of the plate and measuring fluorescence intensity will typically result in a nonzero *b*-value.
13. Each 22 μL aliquot of enzyme will be sufficient for approximately 19–20 wells. If more wells are needed, multiple aliquots should be used.
14. To minimize potential error when using multiple aliquots, dilute each aliquot independently, mix thoroughly, combine all aliquots into a single container, and mix again.
15. In our experience, pipetting up and down results in slightly higher activity than vortexing.
16. The values listed in Table 3 are those which we used to produce a Michaelis–Menten curve for PHPT1 and required two aliquots of enzyme to test. The maximum volume of DMSO + DiFMUP should be 5 μL to keep the total DMSO concentration at 10% (v/v).
17. Addition of enzyme is best performed using a multichannel pipette to reduce the amount of time needed to transfer the enzyme into the wells.

18. Some programs require estimated *a*- and *b*-values. When possible, use the data to estimate these values. For V_{max}, this would be the highest *y*-value in the data set. For K_m, this would be the *x*-value at which the *y*-value is at half maximum. The data points should be sufficiently distributed throughout the modelled curve. In addition, the curve should appear to be reaching a horizontal asymptote at high DiFMUP concentrations. If both conditions are not met, the assay should be repeated with the number of DiFMUP concentrations increased or the existing concentrations adjusted accordingly.
19. We recommend at most 5 μL of inhibitor be used. If more inhibitor is required, prepare a higher concentration inhibitor solution.
20. We recommend 30 min as a starting point for inhibitor incubations. The exact duration may vary depending on the inhibitor, and time-dependent inhibition should be tested for.
21. Some programs require estimated *a*, *b*, *c*, and *d* values. When possible, use the data to estimate these values. For the relative IC_{50}, the *c* value should be the approximate x value halfway between the *a* and *b* values, NOT the x value at 50% inhibition. The Hill coefficient (*d*) is usually set to 1. The equation listed is a general method of curve fitting to inhibition data—more complex or unusual systems may require modified or different curve fit equations [31, 32].

Acknowledgments

This work was supported by a Teva Pharmaceuticals Mark A. Goshko Memorial Grant award (56426-TEV) and an NSF award (CHE 1308766) to A.M.B.

References

1. Boyer PD, Deluca M, Ebner KE, Hultquist DE, Peter JB (1962) Identification of phosphohistidine in digests from a probable intermediate of oxidative phosphorylation. J Biol Chem 237:PC3306–PC3308
2. Ek P, Pettersson G, Ek B, Gong F, Li J-P, Zetterqvist Ö (2002) Identification and characterization of a mammalian 14-kDa phosphohistidine phosphatase. Eur J Biochem 269:5016–5023
3. Klumpp S, Hermesmeier J, Selke D, Baumeister R, Kellner R, Krieglstein J (2002) Protein histidine phosphatase : a novel enzyme with potency for neuronal signaling. J Cereb Blood Flow Metab 22:1420–1424
4. Kim Y, Huang J, Cohen P, Matthews HR (1993) Protein phosphatases 1, 2A, and 2C are protein histidine phosphatases. J Biol Chem 268:18513–18518
5. Matthews HR, MacKintosh C (1995) Protein histidine phosphatase activity in rat liver and spinach leaves. FEBS Lett 364:51–54
6. Fuhs SR, Hunter T (2017) phisphorylation: the emergence of histidine phosphorylation as a reversible regulatory modification. Curr Opin Cell Biol 45:8–16
7. Kee JM, Muir TW (2012) Chasing phosphohistidine, an elusive sibling in the phosphoamino acid family. ACS Chem Biol 7:44–51

8. Gonzalez-Sanchez MB, Lanucara F, Helm M, Eyers CE (2013) Attempting to rewrite history: challenges with the analysis of histidine-phosphorylated peptides. Biochem Soc Trans 41:1089–1095
9. Chen CC, Bruegger BB, Kern CW, Lin YC, Halpern RM, Smith RA (1977) Phosphorylation of nuclear proteins in rat regenerating liver. Biochemistry 16:4852–4855
10. Motojima K, Goto S (1993) A protein histidine kinase induced in rat liver by peroxisome proliferators. In vitro activation by Ras protein and guanine nucleotides. FEBS Lett 319:75–79
11. Hegde AN, Das MR (1987) Ras proteins enhance the phosphorylation of a 38 kDa protein (p38) in rat liver plasma membrane. FEBS Lett 217:74–80
12. Noiman S, Shaul Y (1995) Detection of histidine phospho-proteins in animal tissues. FEBS Lett 364:63–66
13. Srivastava S, Zhdanova O, Di L, Li Z, Albaqumi M, Wulff H et al (2008) Protein histidine phosphatase 1 negatively regulates CD4 T cells by inhibiting the K+ channel KCa3.1. Proc Natl Acad Sci 105:14442–14446
14. Mäurer A, Wieland T, Meissl F, Niroomand F, Mehringer R, Krieglstein J et al (2005) The β-subunit of G proteins is a substrate of protein histidine phosphatase. Biochem Biophys Res Commun 334:1115–1120
15. Klumpp S, Bechmann G, Mäurer A, Selke D, Krieglstein J (2003) ATP-citrate lyase as a substrate of protein histidine phosphatase in vertebrates. Biochem Biophys Res Commun 306:110–115
16. Fuhs SR, Meisenhelder J, Aslanian A, Ma L, Zagorska A, Stankova M et al (2015) Monoclonal 1- and 3-phosphohistidine antibodies: new tools to study histidine phosphorylation. Cell 162:198–210
17. Hindupur SK, Colombi M, Fuhs SR, Matter MS, Guri Y, Adam K et al (2018) The protein histidine phosphatase LHPP is a tumour suppressor. Nature 555:678–682
18. Ma R, Kanders E, Beckman-Sundh U, Geng M, Ek P, Zetterqvist Ö et al (2005) Mutational study of human phosphohistidine phosphatase: effect on enzymatic activity. Biochem Biophys Res Commun 337:887–891
19. Martin DR, Dutta P, Mahajan S, Varma S, Stevens SM Jr (2016) Structural and activity characterization of human PHPT1 after oxidative modification. Sci Rep 6:1–12
20. Fujitaki JM, Fung G, Oh EY, Smith RA (1981) Characterization of chemical and enzymatic acid-labile phosphorylation of histone H4 using phosphorus-31 nuclear magnetic resonance. Biochemistry 20:3658–3664
21. Attwood PV, Ludwig K, Bergander K, Besant PG, Adina-Zada A, Krieglstein J et al (2010) Chemical phosphorylation of histidine-containing peptides based on the sequence of histone H4 and their dephosphorylation by protein histidine phosphatase. Biochim Biophys Acta 1804:199–205
22. Geladopoulos TP, Sotiroudis TG, Evangelopoulos AE (1991) A malachite green colorimetric assay for protein phosphatase activity. Anal Biochem 192:112–116
23. Feng J, Chen Y, Pu J, Yang X, Zhang C, Zhu S et al (2011) An improved malachite green assay of phosphate: mechanism and application. Anal Biochem 409:144–149
24. Beckman-Sundh U, Ek B, Zetterqvist Ö, Ek P (2011) A screening method for phosphohistidine phosphatase 1 activity. Ups J Med Sci 116:161–168
25. Inturi R, Wäneskog M, Vlachakis D, Ali Y, Ek P, Punga T et al (2014) A splice variant of the human phosphohistidine phosphatase 1 (PHPT1) is degraded by the proteasome. Int J Biochem Cell Biol 57:69–75
26. Ek P, Ek B, Zetterqvist Ö (2015) Phosphohistidine phosphatase 1 (PHPT1) also dephosphorylates phospholysine of chemically phosphorylated histone H1 and polylysine. Ups J Med Sci 120:20–27
27. Hiraishi H, Yokoi F, Kumon A (1998) 3-phosphohistidine and 6-phospholysine are substrates of a 56-kDa inorganic pyrophosphatase from bovine liver. Arch Biochem Biophys 349:381–387
28. Gong W, Li Y, Cui G, Hu J, Fang H, Jin C et al (2009) Solution structure and catalytic mechanism of human protein histidine phosphatase 1. Biochem J 418:337–344
29. McCullough BS, Barrios AM (2018) Facile, fluorogenic assay for protein histidine phosphatase activity. Biochemistry 57:2584–2589
30. Eerland M (2015) Design, synthesis and evaluation of PHP inhibitors. Dissertation, Technischen Universität Dortmund
31. Di Veroli GY, Fornari C, Goldlust I, Mills G, Koh SB, Bramhall JL et al (2015) An automated fitting procedure and software for dose-response curves with multiphasic features. Sci Rep 5:1–11
32. Beck B, Chen Y, Dere W, Devanarayan V, Eastwood BJ, Farmen MW et al (2004) Assay operations for SAR support. In: Assay guidance manual. Eli Lilly & Company and the National Center for Advancing Translational Sciences, Bethesda, MD. https://www.ncbi.nlm.nih.gov/books/NBK91994/. Accessed 19 Nov 2018

Chapter 9

Structural and Functional Characterization of Autophosphorylation in Bacterial Histidine Kinases

Laura Miguel-Romero, Cristina Mideros-Mora, Alberto Marina, and Patricia Casino

Abstract

Autophosphorylation of histidine kinases (HK) is the first step for signal transduction in bacterial two-component signalling systems. As HKs dimerize, the His residue is phosphorylated in *cis* or *trans* depending on whether the ATP molecule used in the reaction is bound to the same or the neighboring subunit, respectively. The *cis* or *trans* autophosphorylation results from an alternative directionality in the connection between helices α1 and α2 in the HK DHp domain, in such a way that α2 could be oriented almost 90° counterclockwise or clockwise with respect to α1. Sequence and length variability of this connection appears to lie behind the different directionality and is implicated in partner recognition with the response regulator (RR), highlighting its importance in signal transduction. Despite this mechanistic difference, HK autophosphorylation appears to be universal, involving conserved residues neighboring the phosphoacceptor His residue. Herein, we describe a simple protocol to determine both autophosphorylation directionality of HKs and the roles of the catalytic residues in these protein kinases.

Key words Histidine kinases, Two-component systems, *Cis–trans* autophosphorylation, Signal transduction, Heterodimer production and purification, X-ray crystallography

1 Introduction

Bacterial histidine kinases (HKs) are signal transduction enzymes involved in the sensing of different stimuli and transmitting changes in those stimuli to the inner cell [1]. Sensing and transmitting signals by HKs imply ATP uptake, autophosphorylation and phosphoryl transfer. In recent years, autophosphorylation of HKs has drawn attention as the structural and functional features governing this enzymatic activity have been better characterized. HKs are dimeric proteins and comprise a catalytic core formed by two domains joined by a flexible linker. One domain, called DHp (*D*imerization and *H*istidine *p*hosphotransfer), is formed by two

Laura Miguel-Romero and Cristina Mideros-Mora contributed equally to this work.

Claire E. Eyers (ed.), *Histidine Phosphorylation: Methods and Protocols*, Methods in Molecular Biology, vol. 2077, https://doi.org/10.1007/978-1-4939-9884-5_9, © Springer Science+Business Media, LLC, part of Springer Nature 2020

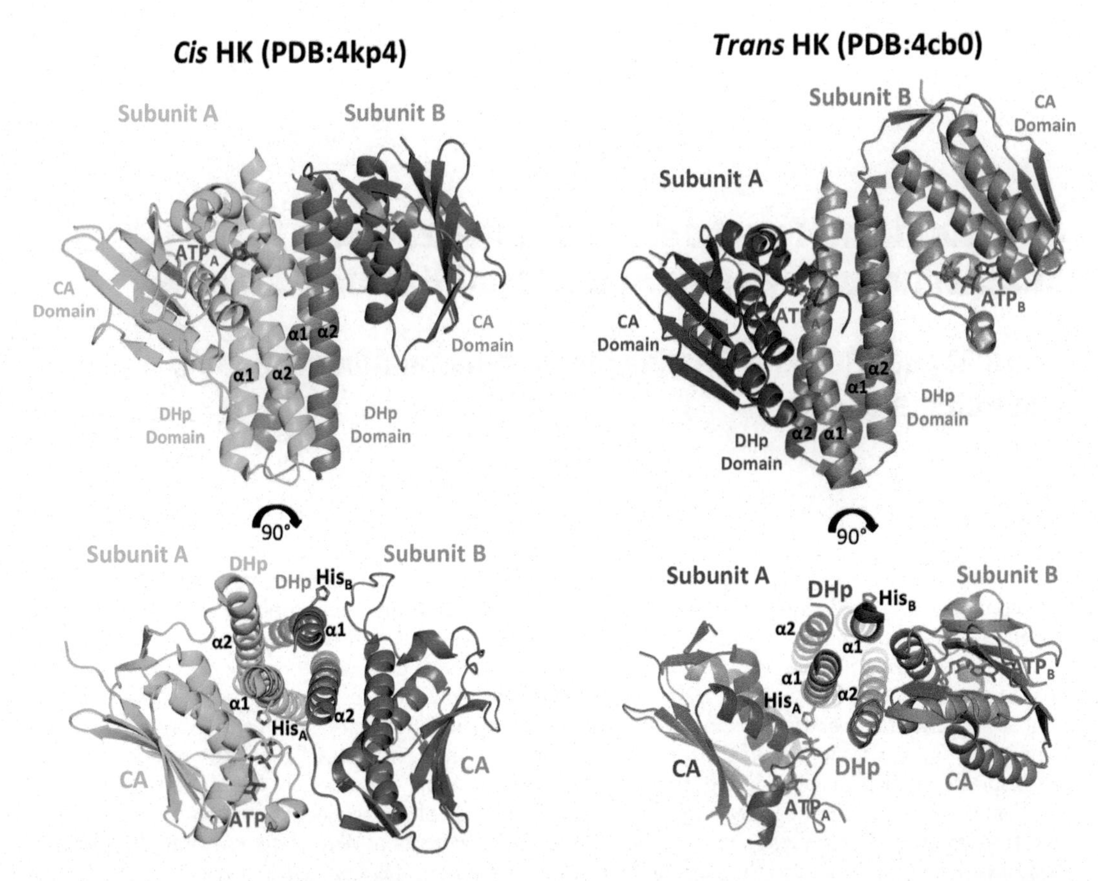

Fig. 1 *Cis* and *trans* autophosphorylation. 3D structures of the catalytic domains of HKs showing the conformations for autophosphorylation. Two orthogonal views of $EnvZ^{chim}$ (PDB:4kp4) showing *cis* autophosphorylation, and CpxA (PDB:4cb0) showing *trans* autophosphorylation are represented in cartoon. In the dimers, each subunit is represented in a different color: green and blue for the *cis* HK subunits, and orange and purple for the *trans* HK subunits. The ATP ligands are represented in stick with carbons in red. The DHp and CA domains conforming the HK catalytic domains are labelled, as well as the α1-α2 helices and the phosphoacceptor His in the DHp domains

antiparallel α-helices (α1 and α2) connected by a loop, while the other domain, CA (*C*atalytic and *A*TP-binding), is globular with an α/β sandwich fold that binds ATP [1–4] (Fig. 1). The DHp domain holds the phosphoacceptor His at helix α1 and mediates dimerization into a four helix bundle merging two subunits into a dimer (Fig. 1). In that way, the CA domain that binds ATP can approach the phosphoacceptor His in either subunit of the dimer, enabling autophosphorylation [1]. If the CA domain bound to ATP approaches the phosphoacceptor His within the same subunit, the HK adopts the so called "*cis*" autophosphorylation conformer, while if it approaches the His of the other subunit the HK adopts the "*trans*" autophosphorylation conformer [5–7] (Fig. 1). Structural and functional evidence suggests that the connection between

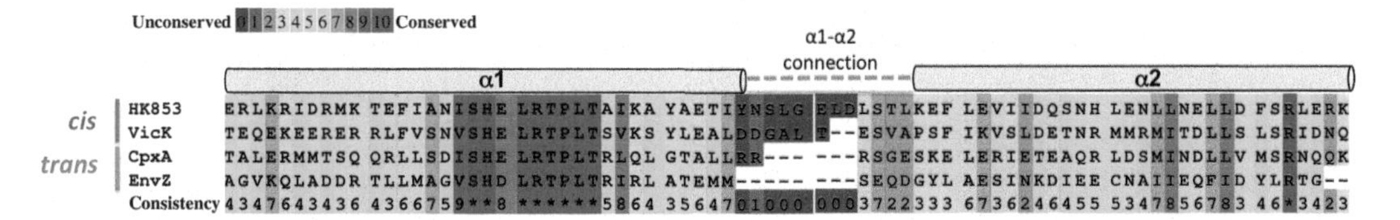

Fig. 2 Sequence alignment of the DHp. The sequences of DHp domains from HKs with *cis* (HK853 and VicK) and *trans* (EnvZ and CpxA) autophosphorylation mechanisms are aligned. The alignment was done with PRALINE webserver; the color code representing residue conservation is in the top of the figure. The secondary structure elements of the DHp are shown at the top of the alignment. The sequence and size variable connector between α1 and α2 helices is indicated

helices α1 and α2 in the DHp domain could play a key role regulating the directionality of autophosphorylation [7–9]. Therefore, it has been hypothesized that the length and residue composition of this link (Fig. 2) would select between *cis* or *trans* autophosphorylation [10]. Crystal structures of DHp domains from HKs whose autophosphorylation has been confirmed biochemically to occur in *cis* (i.e., HK853 from *Thermotoga maritima* [7]) or in *trans* (i.e., EnvZ and CpxA from *E. coli* [11, 12]) show that the helix α2 in the DHp domain is connected almost 90° counterclockwise or clockwise with respect to α1, respectively (Fig. 3). Interestingly, the alternative directionality adopted by HKs is independent of the catalytic mechanism for the autophosphorylation reaction, which has been demonstrated to be conserved [10]. Thus, the change in directionality at the connection between helices α1 and α2 could play a different role in two-component systems (TCS) signalling. Since the DHp α1–α2 connection is one of the regions recognized by the RR in the interaction with the HK, alternative directionalities could be critical in preserving the exquisite fidelity showed by the HK-RR pairs in TCS, avoiding cross-talk between phosphate donors and acceptors, which could compromise bacterial survival. Here, we will outline methodology that can be used to unveil if a HK works by a *cis* or *trans* autophosphorylation reaction, which should be directly related with the directionality of the connection between helices α1 and α2 at the DHp domain. Moreover, we will dissect the catalytic mechanism for the autophosphorylation reaction through the analysis of the most relevant catalytic residues. These methodologies will be broadly applicable for the characterization of structural and functional features of other HKs that govern their autophosphorylation.

2 Materials

2.1 Cloning

The proteins used for all the experiments should correspond to the catalytic domains of HKs, comprising only the DHp and CA domains. To locate the catalytic domains of the HK under study,

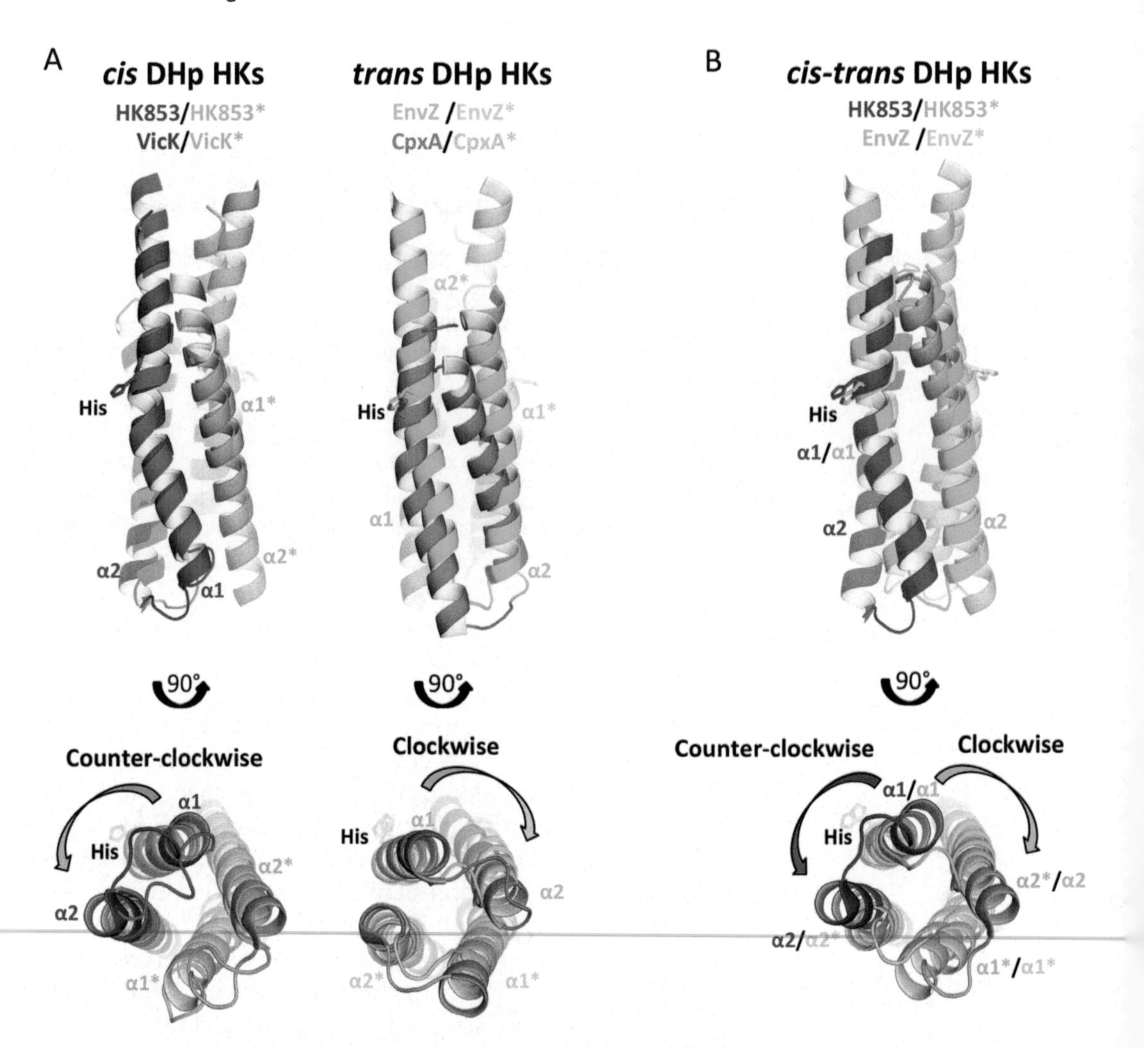

Fig. 3 Directionality of the $\alpha1-\alpha2$ DHp connection in HKs. (**a**) Superimposition of DHp domains from HK853 (red tones; PDB: 3DGE) and VicK (blue tones; PDB: 4I5S) HKs representative of *cis* autophosphorylation, and from EnvZ (green tones; PDB: 4CTI) and CpxA (orange tones; PDB: 4BIU) that show *trans* autophosphorylation. (**b**) Superimposition of DHp domains of *cis* and *trans* autophosphorylation HKs, HK853 and EnvZ respectively. An apical view of the superimposed structures showing counterclockwise or clockwise direction of the $\alpha1-\alpha2$ DHp connection is shown at the bottom of the panel

computational tools such as SMART [13] and Pfam [14] can be used. For example, the catalytic domain of HK853 of *Thermotoga maritima* comprises residues 232–489, while residues 224–450 constitute this domain in EnvZ from *E. coli*.

1. DNA primers for amplification of the coding sequence for the HK catalytic domain of interest: 10 μM final concentration in water or in Tris–EDTA (TE) buffer (TE buffer: 10 mM Tris–HCl pH 8.0 and 1 mM EDTA) (*see* **Notes 1** and **2**).

2. Bacterial expression plasmids: 5–50 ng dissolved in water or in Tris–EDTA buffer (*see* **Note 3**).
3. Agarose gel solution: 1% (w/v) agarose dissolved in 1× TAE buffer (40 mM Tris, 20 mM acetate, and 1 mM EDTA). Heat in the microwave until completely dissolved taking care not to overheat. For a 100 mL agarose gel add 2 μL of GreenSafe Premium (NZYtech, Portugal) before loading the agarose solution in the casting tray with a comb.
4. PCR or gel cleaning kit (e.g., NucleoSpin Gel and PCR Clean-up, Macherey-Nagel).
5. DpnI restriction enzyme (Clontech).
6. T4-DNA polymerase (*see* **Note 4**).
7. Competent cells of *E. coli* strain DH5α: use 50 μL of electro-competent cells prepared in 10% glycerol (*see* **Note 5**).
8. Luria-Bertani (LB) agar plates: Weigh 10 g of tryptone, 5 g of yeast extract, 10 g of NaCl and 15 g bacteriological agar and make up to 1 L in water. Autoclave and, when the medium is under 50 °C, supplement with the necessary antibiotics, typically ampicillin at 100 μg/mL, chloramphenicol 33 μg/mL and/or kanamycin 33 μg/mL, depending on the cell strain and the plasmid used in each case. Dispense in agar plates and leave to set.
9. Primers flanking the plasmid multicloning site to sequencing the resulting plasmid.

2.2 Site-Directed Mutagenesis

Site-directed mutagenesis can be performed using, for example, the QuikChange method developed by Stratagene (La Jolla, CA).

1. High-fidelity DNA polymerase (e.g., KOD Hot Start DNA polymerase (Merck Millipore)).
2. PCR machine.
3. DNase-free water.

2.3 Protein Expression

1. Luria-Bertani (LB) broth: weigh 10 g of tryptone, 5 g of yeast extract, 10 g of NaCl and make up to 1 L in water. Autoclave and, when the medium is under 50 °C, supplement with the necessary antibiotics, typically ampicillin at 100 μg/mL, chloramphenicol 33 μg/mL and/or kanamycin 33 μg/mL, depending on the cell strain and the plasmid used in each case.
2. Spectrophotometer to measure OD at 600 nm.
3. Isopropyl β-D-1-thiogalactopyranoside (IPTG): 1 M stock solution in water.

2.4 Protein Purification

1. Ultrasonic processor or emulsifier.

2. Lysis buffer (buffer A): 50 mM Tris–HCl pH 8.0 and 500 mM NaCl, containing 1 mM phenylmethylsulfonyl fluoride (PMSF) and 1 mM β-mercaptoethanol (β-ME) (*see* **Note 6**). pH of the lysis buffer should be adjusted prior to addition of PMSF and β-ME, according to the pI of the protein.
3. Affinity chromatography columns: The type of chromatography column will be dependent on the tag fused to the protein. Typically 1 mL or 5 mL His-Trap columns (GE Healthcare) are used for proteins tagged with His_6.
4. Elution buffer (buffer B): buffer A supplemented with a specific competitor ligand to dissociate protein bound to the affinity column (e.g., imidazole at 500 mM for His-Trap columns).
5. Gel filtration buffer: 50 mM Tris–HCl, pH 8.0 and 150 mM NaCl.
6. Gel filtration column: HiLoad 16/60 Superdex 200 (GE Healthcare).
7. Equipment and reagents for SDS-PAGE: it is recommended that a 15% SDS-polyacrylamide gel is used for a HK with a molecular mass of ~30 kDa.
8. SDS Loading buffer: mix 3.55 mL of deionized water, 1.25 mL of 0.5 M Tris pH 6.8, 2.5 mL of glycerol, 2 mL of 10% (w/v) SDS and 0.2 mL of 0.5% (v/v) bromophenol blue. Vortex to mix. Prior to use, transfer 950 μL loading buffer to a clean Eppendorf tube and add 50 μL of β-ME. The ratio of sample: SDS loading buffer is 2:1.
9. Centrifugal filters such as Amicon Ultra (Millipore).
10. ÄKTA Purification System (GE Healthcare).
11. Dialysis Tubing, 3.5 K MWCO.

2.5 HK Phosphorylation

Radioactive samples and waste should be handled behind Perspex screens. When handling radioactivity, it is imperative that you wear suitable protective clothing and use a personal radiation monitor (as mandated by local institutional regulations). A suitable portable radiation detector should be on hand to monitor contamination. All radioactive waste should be disposed of as mandated by local regulations.

1. Buffer A supplemented with 100 mM KCl_2 and 10 mM $MgCl_2$.
2. [γ^{32}]ATP: Mixture of cold (0.1–0.3 mM) ATP and hot (0.1 mCi/mL) ATP-[γ^{32}] ATP (3000 μCi/mmol) diluted in deionized water. It is recommended to prepare a 10× stock of ATP such that the same amount can be added to all samples.
3. Stop buffer: SDS loading buffer containing 50 mM EDTA and 4 % SDS.

4. Apparatus for gel blotting and drying: use 3MM blotting paper to transfer the gel followed by incubation into a gel drying system under vacuum (Bio-Rad Model 583 Gel Dryers and Hydrotech pump).
5. Phospho-imager: phosphorylated proteins are visualized by autoradiography using specific equipment, for example Fluoro Image Analyzer FLA-5000 (Fuji) and analyzed with the associated software (MultiGauge software, Fuji).

2.6 Crystallization and Crystal Freezing

1. Commercial crystallization screens (at least three). We recommend starting with JBScreen Classic HTS I + II (from Jena Bioscience) or MIDAS (from Molecular Dimensions) [15].
2. MRC 2- or 3-well crystallization plates (Molecular Dimensions) that can be used for crystallization of the protein alone and in the presence of ligands such as ATP, ADP, or the non-hydrolyzable ATP analogs AMP-PNP or AMP-PCP.
3. Liquid nitrogen.
4. Cryo loops for crystal mounting (MiTeGen or Hampton Research)).
5. X-ray data processing programs (e.g., IMosflm from CCP4 [17]).
6. Software suites for automated macromolecular structure determination by X-ray crystallography (e.g., CCP4 [17]or PHENIX [18]).
7. Macromolecular visualization software such as Coot [17] or PYMOL [21].

3 Methods

3.1 Determination of the Autophosphorylation Directionality in HKs

Due to the dimeric nature of HKs, the autophosphorylation reaction could take place within the same (intra-)subunit or between (inter-)subunits driving a *cis* or *trans* phosphorylation in the homodimer, respectively (Fig. 1). To find out which directionality is driving a specific HK, it is necessary to discern which subunit is being phosphorylated within the homodimer and which undertakes the phosphorylation. For this propose, heterodimers containing two subunits with different lengths, a Short (S) and a Long (L) version of the catalytic domains, should be produced. Furthermore, to unambiguously distinguish the phosphorylated subunit in the dimer, two additional types of mutant heterodimers are required: one type contains a mutation in the CA domain just in the L subunit, impairing ATP-binding ($L^{\emptyset}$; *kinase mutant*), and producing the $L^{\emptyset}S$ heterodimer; the other type contains subunit $L^{\emptyset}$ and a mutation at the phosphoacceptor His in the S subunit ($S^{\emptyset}$; *phosphoacceptor mutant*), impairing its capacity to be phosphorylated, and producing $L^{\emptyset}S^{\emptyset}$ heterodimer (Fig. 4a).

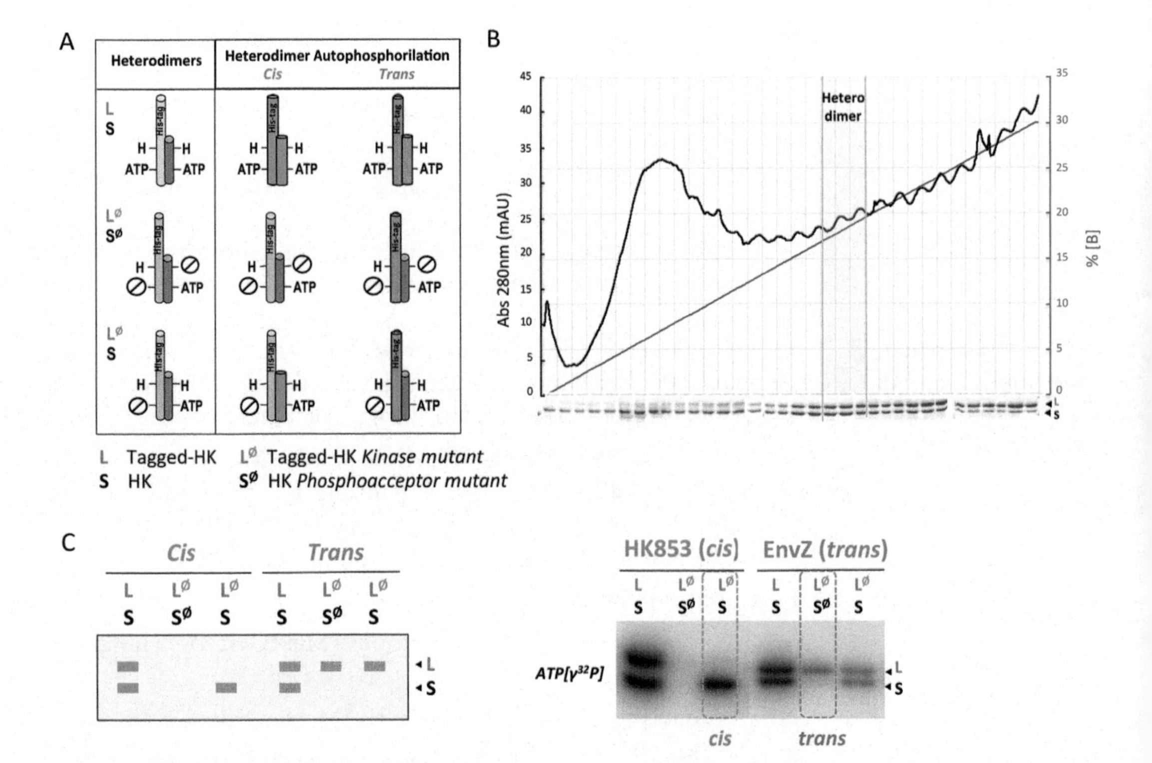

Fig. 4 Heterodimer formation. (**a**) *Left,* Schematic representation of heterodimer species. Subunits are labelled according to the absence (L and S) or presence (L^{Φ} and S^{Φ}) of mutation. *Right,* The phosphorylated subunits are colored in each heterodimer depending on whether the autophosphorylation reaction mechanism is *cis* (blue color) or in *trans* (orange color). (**b**) Example UV trace following heterodimer purification by affinity chromatography. The chromatogram corresponds to a 5 mL HisTrap FF column eluted with a gradient of buffer B (red line) in 20 column volumes. Samples from each 3 mL fraction were separated by SDS-PAGE. Fractions containing an equal ratio of L and S subunits as heterodimer (region in purple shadow) were selected for further analysis. (**c**) Visualization of heterodimer autophosphorylation. *Left,* Schematic representation of autophosphorylation patters with samples obtained in (**a**) in a hypothetical SDS-PAGE gel. *Right,* autoradiography of an experiment carried out with the heterodimers for HK853 (*cis*) and EnvZ (*trans*). Phosphorylation of L^{Φ} subunit in $L^{\Phi}S^{\Phi}$ heterodimer reveals *trans* autophosphorylation (orange) while phosphorylation of the S subunit in the $L^{\Phi}S$ heterodimer is characteristic of *cis* autophosphorylation (blue)

3.1.1 Cloning of HKs Constructs to Generate L and S Subunits

To generate subunits L and S with different length, genes for the specific HKs must be cloned into two plasmids, one containing a tag (i.e., His-tag) and the other without a tag (or with a tag of different size). Plasmids containing a tag, or the longest tag will produce a longer HK protein, defined by L, while the plasmids devoid of tag or with the shortest tag will produce shorter protein defined by S.

1. Amplify the genetic region of a specific HK construct by performing a Polymerase Chain Reaction (PCR) according to the protocol recommended for the High fidelity DNA polymerase used (e.g., for 25 μL, mix 5–50 ng of DNA template

with the polymerase Buffer 1×, specifically designed primers and dNTPs at 0.2 μM final concentration, 2 mM $MgSO_4$, and 1 U of High fidelity DNA polymerase. Incubate for 10 min at 95 °C for denaturation, followed by 35 cycles of: (a) 30 s at 95 °C, (b) 30 s at primer annealing temperature, (c) 30 s/kb extension at 72 °C. Leave for an additional 5–10 min at 72 °C (*see* **Note 1**).

2. Linearize the plasmid. The exact procedure for this step will vary depending on the cloning techniques used (e.g., LIC cloning, In-fusion, or enzyme restriction).
3. Load the PCR products and linearized plasmids onto a 1–2% agarose gel and visualize the size of the PCR amplified regions and linearized plasmids.
4. Cut the corresponding gel bands with a scalpel and purify the DNA with a gel cleaning kit following the manufacturer's instructions. Linearized plasmids could also be directly cleaned with a PCR clean-up kit without the electrophoretic step.
5. For LIC cloning, treat the amplified genes and linearized plasmid with T4 polymerase. For a 10 μL reaction use 5 μL of sample, 2.5 mM of corresponding dNTP, 0.1 mg/mL BSA, 1× T4 polymerase buffer, 5 mM DTT (1,4-dithiothreitol), and 0.5 μL of T4-DNA polymerase. Incubate the reaction in a PCR machine or thermoblock for 25 min at 22 °C, followed by 20 min 75 °C for enzyme inactivation. Store at 4 °C.
6. Mix the treated amplified gene and linearized plasmid in a 2:1 ratio (gene:plasmid). Incubate at room temperature (RT) for 5 min and place on ice. For In-fusion cloning, mix the amplified gene and plasmid in a 2:1 ratio with 5× In-Fusion HD Enzyme Premix, incubate for 50 °C for 15 min and then place on ice.
7. Immediately after **step 6**, transform 50 μL of DH5α cells with 3 μl of cloning mixture by electroporation. Add 1 mL of cold LB to the cells followed by incubation at 37 °C with gentle shaking (200 rpm) for 1 h.
8. Centrifuge the tube at 6000 × *g* for 2 min, discard the media without touching the cells. Resuspend the cells slowly in the same media; spread onto an agar plate with the appropriate antibiotics and grow overnight (o/n) at 37 °C.
9. Check if the cloning was successful by picking the colonies grown in the plate and performing a PCR assay (DNA template comes from the colony) using primers flaking the plasmid multicloning site. Confirm the correct cloning by sequencing the PCR products coming from those colonies which show bands of the correct molecular size.

10. For further mutagenesis or protein expression analysis, purify plasmid DNA of the positive colonies.

3.1.2 Mutant Design

The HK point mutations are a key factor in this methodology to generate subunits that cannot be phosphorylated and subunits that lack enzyme activity. To obtain a subunit that cannot be phosphorylated, point mutations at the phosphoacceptor His should be introduced in the plasmids devoid of tag or with the shortest tag (*phosphoacceptor mutant*, defined as construct $S^{Ø}$). To produce a construct that lacks enzymatic activity, mutations that impair ATP binding at the CA domain (*see* **Note 7**) could be introduced in the plasmids containing the tag or the longest tag (*kinase mutant*, defined as construct $L^{Ø}$) (*see* **Note 8**).

1. Perform the site directed mutagenesis using the QuikChange® method developed by Stratagene (La Jolla, CA). Carry out PCR with the wild type HK plasmids and the appropriate primers as described in the Subheading 3.1.1, **step 1**, but using longer extension times in order to introduce the mutation while amplifying the whole plasmid.
2. Digest the PCR product with 1 μL of DnpI at 37 °C for 2 hrs or the time indicated by supplier to remove the methylated parental plasmid.
3. Prepare the DNA and plasmid, and transform the *E. coli* as described in sections Continue with transformation as described in Subheadings 3.1.1, **steps 7–8**.
4. Check if the colonies contain the plasmid with the specific mutation by extracting the DNA and sequencing.

3.1.3 Homo and Heterodimer Production

To produce homodimer and heterodimer species of a HK, similar steps related with expression and purification are followed. However, to generate the heterodimer other methodologies, besides the coexpression method, can be used which take advantage of the spontaneous subunit exchange peculiarity of some HKs (*see* **Note 9**) [11].

1. For homodimer production, transform by electroporation 20 ng of the corresponding plasmid in 50 μL of expression cells. For heterodimer production, cotransform 20 ng of each plasmid in 50 μL of expression cells. To determine the autophosphorylation directionality using heterodimers, three different species of heterodimers should be generated: LS, $L^{Ø}S$, and $L^{Ø}S^{Ø}$. For that purpose, three different cotransformations should be performed for each HK using two different plasmids containing the appropriate constructs; i) (L) and (S) containing wild type constructs, ii) ($L^{Ø}$) and (S) containing mutation at the long construct, iii) ($L^{Ø}$) and ($S^{Ø}$) containing mutations at both long and short constructs (Fig. 4a). Notice that plasmids

for the L and S versions of the protein must confer resistance to different antibiotics (*see* **Note 10**).

2. Add 1 mL of cold LB to the cells followed by incubation at 37 °C with shaking for 1 h.
3. Spread 100–200 μL on a plate with the corresponding antibiotics (plasmid and cells) and grow o/n.
4. To generate a bacterial glycerol stock for future protein expressions, inoculate one colony in 5 mL of LB with the corresponding antibiotics and grow o/n. Next day, mix 800 μL of saturated culture with 400 μL of sterile 50% (v/v) glycerol and store at −80 °C.

3.1.4 Protein Expression

1. To recover the bacteria from the glycerol stock, scrape some of the frozen bacteria and streak it onto a LB plate with the corresponding antibiotics and incubate at 37 °C o/n.
2. Inoculate one colony from the LB-agar plate in 20 mL LB containing appropriate antibiotics and grow o/n in a shaker at 37 °C to obtain a saturated culture.
3. Inoculate 20 mL of o/n saturated culture into 1 L LB (for a 1/50 dilution) supplemented with the appropriate antibiotics.
4. When the culture has reached exponential phase growth (evaluated by reaching a measured OD_{600} of ~0.6), induce protein expression by addition of IPTG (1 mL of 1 M stock solution). Leave for 3 h at 37 °C with gentle agitation.
5. Harvest cells by centrifugation at 5000 × *g* for 15 min in a refrigerated centrifuge. Discard the supernatant and flash-freeze the pellets in liquid N_2 and store at −80 °C.

3.1.5 Homo/Heterodimer Purification

Carry out all protein purification steps at 4 °C.

1. Resuspend the *E. coli* cell pellets in cold lysis buffer (~40–50 mL for the cell pellet coming from a 1 L culture). Make sure that the suspension is completely homogenized with no clumps (*see* **Note 11**).
2. Disrupt the cells by sonication (or using alternative methods such as a French press). Centrifuge for 30 min at 16,000 × *g* to remove cell debris. Collect the supernatant containing the overexpressed proteins.
3. Load the clarified supernatant onto a 1 mL affinity column (the selection of which is dependent on the tag fused to the protein) which has been equilibrated previously with buffer A.
4. In the case of a His_6-tag, wash the column with 8% of buffer B (diluted in buffer A) to remove unspecific binding. Elute bound proteins by gradient elution, increasing buffer B to 70%. If another tag is used, elute the protein following the

affinity column recommendations. As an example, affinity chromatography using a 5 mL His-Trap column was performed to purify a heterodimer containing one His-tag in the L subunit. A 20-column volume gradient from 0 to 30% of buffer B was used in order to separate individual L or S constructs from the heterodimer (Fig. 4b).

5. If necessary, gel filtration chromatography can be performed as a final step to clean up the sample.

3.1.6 Autophosphorylation Assay

Radioactive samples and waste should be handled behind Perspex screens. When handling radioactive samples, it is imperative that you work behind a Perspex screen, wear suitable protective clothing and use a personal radiation monitor (as mandated by local institutional regulations). A portable radiation detector should be on hand to monitor potential contamination. All radioactive waste should be disposed of as mandated by local regulations.

1. Prepare 10 μL of the phosphorylation mixture with 4 μM of the homo or heterodimer HK and 1 μL of autophosphorylation buffer. Adjust to 10 μL with water.
2. Start the reaction by addition of 1 μL of the 10× ATP cocktail. Leave for between 5 and 30 min at RT, depending on the autophosphorylation capacity of the analyzed HK (e.g., 10 min for HK853 and EnvZ).
3. Stop the reactions by adding 2 volumes (20 μL) of 2× SDS loading buffer. Maintain the samples at room temperature for 15 min before performing SDS-PAGE (*see* **Note 12**).
4. Load 8 μL of each sample onto a 15% SDS polyacrylamide gel and separate the proteins by electrophoresis at 200 V until the bromophenol blue indicator marker reaches the bottom of the gel.
5. Extract the gel from the electrophoresis system. Remove the radioactive dye from and dispose of it in a suitable container. Transfer the gel to 3MM blotting paper and dry the gel.
6. Visualize the phosphorylated proteins by autoradiography and analyze the relative phosphorylation of the dimers using the appropriate software (*see* **Notes 13** and **14**).

3.2 Determining the Directionality of the Autophosphorylation Reaction

To visualize the molecular basis of the autophosphorylation reaction, crystallization studies can be performed (Fig. 5). The structure of a homodimeric HK performing the autophosphorylation reaction not only provides clues as to the directionality of this reaction but also points toward reaction mechanism that can be further confirmed by mutagenesis. For this purpose, the structures of different HKs have been solved by X-ray crystallography [10, 12].

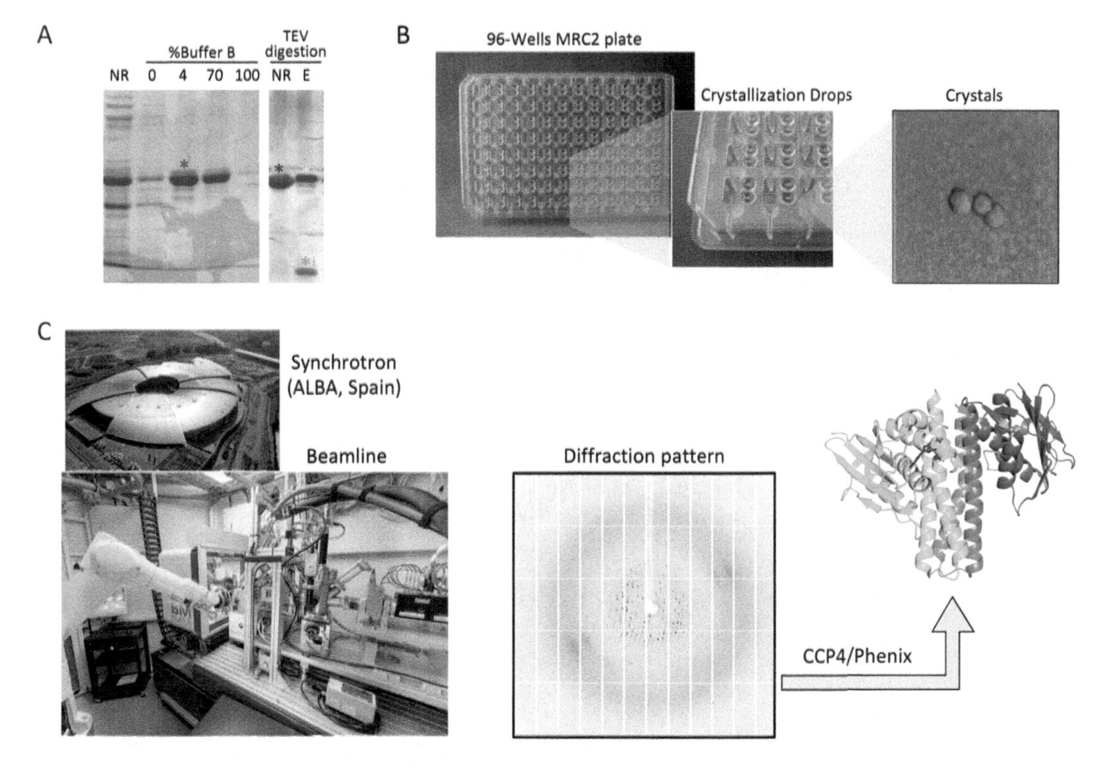

Fig. 5 Scheme of the steps to determine the 3D structure of a HK by X-ray crystallography. (**a**) SDS-PAGE showing the purification steps of the HK before and after removal of the tag. Red and blue asterisks indicate the band corresponding to the HK with and without tag, respectively. Yellow and orange asterisks indicate the bands corresponding to the TEV protease and the tag respectively that elute with 100% of buffer B. (**b**) MRC2 plates for protein crystal growth in an appropriate screening. (**c**) Pictures of synchrotron (ALBA, Spain) and X-ray beamline (XALOC in ALBA) used for protein crystal diffraction. Images containing diffraction patterns are processed with computer suites such as CCP4 or Phenix to obtain the final 3D structure of the HK

3.2.1 HK Expression and Purification for Crystallography

1. Express and purify the HK proteins as described in Subheadings 3.1.4–3.1.5.
2. To increase the probability of crystallization, we recommend that the HK purification tag be removed by proteolytic cleavage, using a protein:protease ratio typically between 50:1 and 20:1 (*see* **Note 15**). Incubate at 4 °C overnight and dialysis against buffer A (Fig. 5a).
3. Purify the cleaved protein by repeating the affinity chromatography as described in Subheading 3.1.5 (*see* **Note 16**). Collect the unbound (protease-cleaved) protein and concentrate to <2 mL using centrifugal filters.
4. Remove protein aggregates by separating the protein by gel filtration chromatography using a Superdex 200 column equilibrated with gel filtration buffer.

5. Pool the peak fractions containing monomeric HK protein and concentrate to 10–20 mg/mL using centrifugal filtration columns as described by the manufacturer.

3.2.2 Protein Crystallization, Data Collection and 3D Structure Determination

The vapor diffusion method using sitting drops is the most commonly used method to identify conditions suitable for protein crystallization [16]. We recommend using a liquid handling system (e.g., Mosquito by ttplabtech) that allows low volume drops (0.1–0.2 μL) to be used, in order to carry out more crystallization assays with the minimal volume of sample.

1. Set up crystallization assays by depositing equal amounts (e.g., 0.2 μL) of purified HK at 10–20 mg/mL and mother liquor in 96-well plates. Add 50–75 μL of the required mother liquor crystallization condition to the reservoir. Seal the plate with a transparent film and incubate at an appropriate temperature, usually 21 or 4 °C (Fig. 5b).
2. Monitor the growth of the crystal daily over the first week. Once crystals are observed, monitor once a week to define the optimal crystallization condition for further evaluation (*see* **Note 17**).
3. Once the conditions producing single crystals are found, scale up the volume of the crystallization assay (0.5–2 μL for the drop, 0.5–1.0 mL for the reservoir) to produce larger crystals.
4. Test the cryoprotectant capacity of the crystallization conditions by flash-freezing the solution in liquid nitrogen and taking some X-ray diffraction snapshots (*see* **Note 18**).
5. Collect complete X-ray diffraction data sets and solve the structure by molecular replacement using as a template one of the HK structures available in the PDB (Fig. 5c). Pipelines for automated structure solution using molecular replacement are available in the CCP4 Suite [17] (such as Balbes, MrBump, and MoRDa) or the PHENIX Suite [18], as well as in the Auto-Rickshaw web server [19].

3.2.3 Functional Characterization of Autophosphorylation Reaction

In the light of the structural data obtained following the autophosphorylation reaction, several potential catalytic residues for this reaction can be identified (Fig. 6a). To assess the effect of these residues in HK autophosphorylation, a mutagenesis study together with in vitro phosphorylation assays can be performed (*see* **Note 19**).The results of these assays can be compared with the mutagenic characterization of other HKs showing *cis* or *trans* autophosphorylation (e.g., HK853 showing *cis* and EnvZ showing *trans* as models).

1. Identify potential catalytic residues in the active site of the 3D structure by analysing the contacts of the substrate ATP (or a

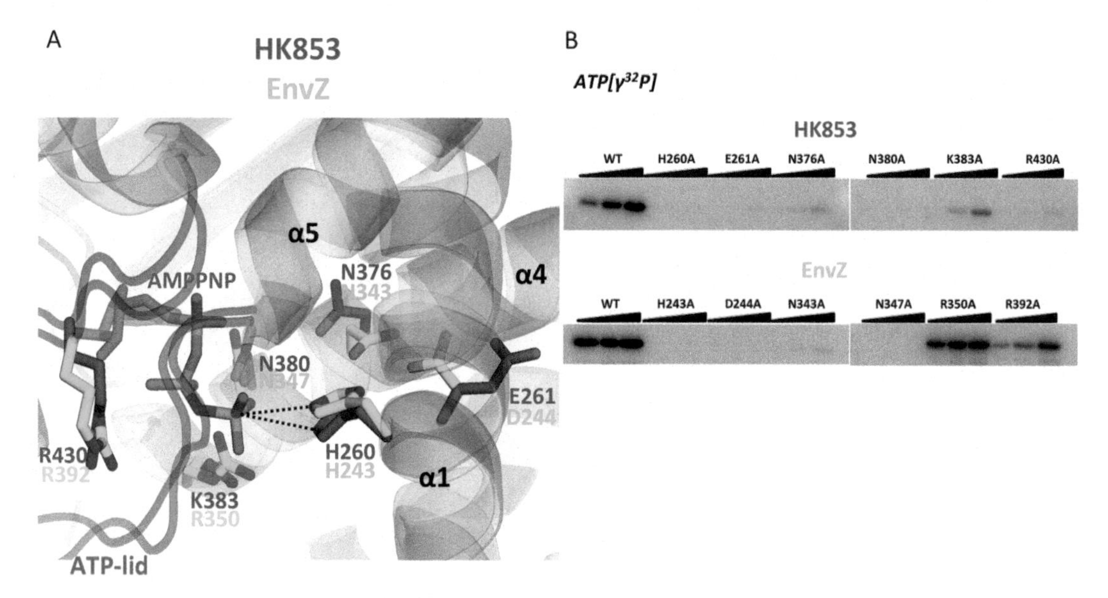

Fig. 6 Residues involved in the catalytic mechanism of the autophosphorylation reaction. (**a**) Detail of the active centre in the 3D structure of $EnvZ^{chim}$ (green) showing the close distance between AMP-PNP (in sticks with carbon and phosphate atoms in grey and orange, respectively) and the phosphoacceptor His. Superimposed DHp and CA domains of HK853 over $EnvZ^{chim}$ structure are shown in pink. Catalytically relevant residues are drawn as sticks and labelled. (**b**) Autophosphorylation activity assays of HK proteins where the relevant catalytic residues of HK853 and EnvZ have been mutated to Ala

nonhydrolyzable analog) and the phosphoacceptor His, as well ions or water molecules to which they are coordinated.

2. Compare the active centre organization with the active centre of other HKs previously characterized. Introducing mutations at the phosphoacceptor His as well as at the conserved acidic residue (e.g. HK853 E261 or EnvZ D244) after the His (if it is present) is also useful (Fig. 6a). Likewise, introducing a mutation at the conserved polar residue in α4 (e.g. HK853 N376 and EnvZ N343) which stabilizes the interaction between the two previous residues (the phosphoacceptor His and the acidic residue), may also reveal useful information.
3. Generate site-directed mutants of potential catalytic residues as described in Subheading 3.1.2.
4. Express and purify the mutant HK proteins as described in Subheadings 3.1.4–3.1.5.
5. Carry out autophosphorylation assays as described in Subheading 3.1.6. Quantify the effect of the individual mutations on HK activity compared with wild-type protein.
6. Compare the effect of these mutations with similar mutations reported in other HKs showing *cis* and *trans* autophosphorylation such as HK853 and EnvZ (Fig. 6a). For HK853 and

EnvZ, mutations at the conserved acidic residue completely abolish phosphorylation, highlighting the relevance of this residue in the autophosphorylation mechanism (Fig. 6b). Mutations at the conserved polar residue in α4 has a deleterious impact on autophosphorylation, demonstrating its role in stabilizing the interaction between the phosphoacceptor His and the acidic residue (Fig. 6b). Additional mutations at other specific residues, such as the conserved Asn in α5 which chelates the Mg^{2+} cation, completely abolishes phosphorylation due to impaired binding of ATP (Fig. 6b). Mutation of the conserved basic residues that contact the β and γ phosphates of ATP decreases activity due to a reduction in ATP binding and/or destabilization of the negative charge produced in the ADP product (Fig. 6b). Similar results obtained for the HK under investigation would demonstrate a canonical autophosphorylation mechanism [10]. Differences in the effect of the mutations would suggest a different mechanism of autophosphorylation and a deeper characterization of the HK would be required in order to identify molecular peculiarities.

4 Notes

1. HK DNA can be obtained by PCR using as template the genomic DNA of the corresponding bacteria or from a synthetic gene of the HK. Synthetic genes can be designed using codon optimization to enhance protein expression in *E. coli*.
2. Primers should be designed such that they incorporate the overhang necessary to subclone the HK cDNA into the required plasmid using the preferred cloning strategy (i.e., LIC [15], In-Fusion HD Cloning Technology (Clontech) or traditional restriction).
3. Plasmids where protein expression is under the control of an inducible promoter such as T7/lacO promoter should be used. It is necessary to use two plasmids with identical or different origins of replication but that provide resistance to different antibiotics. These plasmids should ideally allow the fusion of the HK to two different tags in order to provide short and long versions of the protein. In our study, we used as plasmids pLIC-SGC1 and pNIC28-Bsa4 [15], containing both N-terminal His-tag to produce long versions of the HK, and pET21d and pET24a (Novagene) without tag to produce short versions of the HK.
4. T4 polymerase is required to create overhangs with complementarity between the plasmid and insert in the case of LIC cloning.

5. It is important to choose a suitable *E. coli* strain for HK expression. It will depend on the presence of rare codons in the HK, if any toxicity is caused to the bacteria during expression and to the level of expression needed. For example, the cell strain BL21-codonplus (DE3)-RIL uses T7 RNA polymerase under the IPTG-inducible lacUV5 promoter to produce high levels of expression. This strain contains a plasmid with chloramphenicol (Clo) resistance that provides extra copies of the tRNA genes that recognize the Arg codons AGA and AGG, the isoleucine codon AUA and the Leu codon CUA.
6. The pH of the buffer will need to be adjusted depending on the pI of the protein. The addition of 1 mM of phenylmethylsulfonyl fluoride (PMSF) to inhibit protease activity, and 1 mM of β-mercaptoethanol to avoid oxidation is recommended during cell lysis.
7. An example of mutated residues involved in ATP binding could be N380A in HK853 and N347A in EnvZ. This residue corresponds to a conserved Asn that comprises the N-box in the CA domain and is responsible for binding the Mg^{2+} ion that stabilizes ATP. Mutation of this conserved Asn to Ala impairs Mg^{2+} ion binding.
8. Residues other than the conserved Asn in the N-box can be mutated to prevent ATP binding and impair the kinase activity. However, the inability to bind ATP should be evaluated by performing ligand binding assays.
9. The cytoplasmic catalytic portion of some HKs may undergo a phenomenon called spontaneous subunit exchange based on a continuously forming and breaking protein dimer. This phenomenon could affect the stability of the heterodimer formation. This is the case of the HK EnvZ that show *trans* autophosphorylation where the heterodimer $L^{Ø}S$ shows both subunits phosphorylated due to the formation of wild-type S dimer. In this case, it is necessary to produce the heterodimer $L^{Ø}S^{Ø}$ to confirm which subunit has become phosphorylated.
10. To generate the heterodimers, the plasmids used to express the L and S versions of the protein must have resistance to different antibiotics. Thus, the inclusion of both antibiotics in the culture medium will force their maintenance in the *E. coli* strain. It is also advisable that the expression of both proteins is under the control of the same promoter and that both plasmids present a similar number of copies, thus ensuring a similar expression level of L and S versions.
11. Purification of lower amounts of protein (e.g., produced in 20 mL of culture), can be performed in batch using a Ni^{2+} High Density resin (ABT) and following the indications of the supplier.

12. Boiling the samples will lead to the loss of the phosphorylated radioactive phosphohistidine. Samples should therefore be maintained at room temperature.
13. *Cis* autophosphorylation is confirmed by observation of radioactive signal in the S subunit of the heterodimer $L^{\emptyset}/S$ and not in the $L^{\emptyset}/S^{\emptyset}$ heterodimer. Conversely, *trans* autophosphorylation can be determined by observation of $L^{\emptyset}$ phosphorylation in the $L^{\emptyset}/S^{\emptyset}$ heterodimer, since the reaction takes place between subunits. A scheme of the heterodimers' hypothetical pattern of autophosphorylation and the experimental results obtained with HK853 (*cis*) and EnvZ (*trans*) are shown in Fig. 4c.
14. *Cis* or *trans* autophosphorylation has been related to the connection between helices α1 and α2 in the DHp domain, a connection which is variable in sequence and length as observed by the 3D structures of this domain from different HKs (EnvZ, CpxA, VicK, and HK853) (Fig. 2). Thus, the interchange of this connection with a HK showing a distinct autophosphorylation pattern could be a useful strategy to provide clues on the interaction specificity for response regulator recognition as described by Casino et al. [10].
15. For certain proteases, such as Tobacco Etch Virus Protease (TEV) or Human Rhinovirus 3C Protease (3C/PreScission), addition of 0.5 mM EDTA and 5 mM β-ME is known to increase cleavage efficiency.
16. This second purification step removes protein that still contains the tag, as the tagged HK protein remains bound to the column.
17. Optimize the initial crystallization conditions by varying the pH, salt and precipitant concentration around the original hit conditions. Another option is the use of additive screenings.
18. The absence of concentric rings characteristics of ice formation in the diffraction image will indicate that the solution is a good cryoprotectant. If ice-rings are observed, include cryoprotectant in the crystallization conditions and soak the crystals in this solution from 10 s to 2–3 min, and then flash-freeze in liquid nitrogen. Glycerol, sucrose or ethylene glycol, as well as high concentrations (>30%) of polyethylene glycols (PEGs), can be good options as cryoprotectants depending on the crystallization conditions.
19. The role of the catalytic residues proposed in HK853 and EnvZ was determined by their autophosphorylation capacity, but this capacity is directly dependent on the ability to bind ATP. Thus, to be able to distinguish if the contribution of a catalytic residue was solely based on the enzymatic reaction or on the ATP binding it is advisable to check the contribution of the

mutation in their ATP binding capacity through Isothermal Calorimetry assays (ITC) or Surface Plasmon resonance (SPR) using AMPPNP, a nonhydrolyzable ATP analog.

Acknowledgments

This work was supported by Spanish Government (Ministry of Economy and Competitiveness) grants BIO2016-78571-P to A.M. and BFU2016-78606-P to P.C. P.C. is the recipient of a Ramón y Cajal contract, from the Ministry of Economy and Competitiveness. C.M.-M. is the recipient of a Ph.D. fellowship from the Progama de becas, Secretaría de Educación Superior, Ciencia, Tecnología e Innovación of Ecuador Government (2015-AR2Q9228).

References

1. Gao R, Stock AM (2009) Biological insights from structures of two-component proteins. Annu Rev Microbiol 63:133–154
2. Dutta R, Inouye M (2000) GHKL, an emergent ATPase/kinase superfamily. Trends Biochem Sci 25:24–28
3. Marina A, Mott C, Auyzenberg A, Hendrickson WA, Waldburger CD (2001) Structural and mutational analysis of the PhoQ histidine kinase catalytic domain. Insight into the reaction mechanism. J Biol Chem 276:41182–41190
4. Bilwes AM, Quezada CM, Croal LR, Crane BR, Simon MI (2001) Nucleotide binding by the histidine kinase CheA. Nat Struct Biol 8:353–360
5. Yang Y, Inouye M (1991) Intermolecular complementation between two defective mutant signal-transducing receptors of Escherichia coli. Proc Natl Acad Sci U S A 88:11057–11061
6. Ninfa EG, Atkinson MR, Kamberov ES, Ninfa AJ (1993) Mechanism of autophosphorylation of Escherichia coli nitrogen regulator II (NRII or NtrB): trans-phosphorylation between subunits. J Bacteriol 175:7024–7032
7. Casino P, Rubio V, Marina A (2009) Structural insight into partner specificity and phosphoryl transfer in two-component signal transduction. Cell 139:325–336
8. Casino P, Rubio V, Marina A (2010) The mechanism of signal transduction by two-component systems. Curr Opin Struct Biol 20:763–771
9. Ashenberg O, Keating AE, Laub MT (2013) Helix bundle loops determine whether histidine kinases autophosphorylate in cis or in trans. J Mol Biol 425:1198–1209
10. Casino P, Miguel-Romero L, Marina A (2014) Visualizing autophosphorylation in histidine kinases. Nat Commun 5:3258
11. Cai SJ, Inouye M (2003) Spontaneous subunit exchange and biochemical evidence for trans-autophosphorylation in a dimer of Escherichia coli histidine kinase (EnvZ). J Mol Biol 329:495–503
12. Mechaly AE, Sassoon N, Betton JM, Alzari PM (2014) Segmental helical motions and dynamical asymmetry modulate histidine kinase autophosphorylation. PLoS Biol 12:e1001776
13. Letunic I, Doerks T, Bork P (2015) SMART: recent updates, new developments and status in 2015. Nucleic Acids Res 43(D1):D257–D260
14. Finn RD, Coggill P, Eberhardt RY, Eddy SR, Mistry J, Mitchell AL, Potter SC, Punta M, Qureshi M, Sangrador-Vegas A, Salazar GA, Tate J, Bateman A (2016) The Pfam protein families database: towards a more sustainable future. Nucleic Acids Res 44(D1):D279–D285
15. Savitsky P, Bray J, Cooper CD, Marsden BD, Mahajan P, Burgess-Brown NA, Gileadi O (2010) High-throughput production of human proteins for crystallization: the SGC experience. J Struct Biol 172:3–13
16. McPherson A, Gavira JA (2014) Introduction to protein crystallization. Acta Crystallogr F Struct Biol Commun 70(Pt 1):2–20
17. Winn MD, Ballard CC, Cowtan KD, Dodson EJ, Emsley P, Evans PR, Keegan RM, Krissinel

EB, Leslie AG, McCoy A, McNicholas SJ, Murshudov GN, Pannu NS, Potterton EA, Powell HR, Read RJ, Vagin A, Wilson KS (2011) Overview of the CCP4 suite and current developments. Acta Crystallogr D Biol Crystallogr 67:235–242

18. Adams PD, Afonine PV, Bunkoczi G, Chen VB, Davis IW, Echols N, Headd JJ, Hung LW, Kapral GJ, Grosse-Kunstleve RW, McCoy AJ, Moriarty NW, Oeffner R, Read RJ, Richardson DC, Richardson JS, Terwilliger TC, Zwart PH (2010) PHENIX: a comprehensive Python-based system for macromolecular structure solution. Acta Crystallogr D Biol Crystallogr 66:213–221

19. Panjikar S, Parthasarathy V, Lamzin VS, Weiss MS, Tucker PA (2005) Auto-rickshaw: an automated crystal structure determination platform as an efficient tool for the validation of an X-ray diffraction experiment. Acta Crystallogr D Biol Crystallogr 61(Pt 4):449–457

20. Emsley P, Cowtan K (2004) Coot: model-building tools for molecular graphics. Acta Crystallogr D Biol Crystallogr 60:2126–2132

21. DeLano WL (2002) The PyMOL molecular graphics system, version 11, Schrodinger LLC. http://www.pymolorg24. Murshudov GN, Vagin

Chapter 10

Manipulation of Bacterial Signaling Using Engineered Histidine Kinases

Kimberly A. Kowallis, Samuel W. Duvall, Wei Zhao, and W. Seth Childers

Abstract

Two-component systems allow bacteria to respond to changes in environmental or cytosolic conditions through autophosphorylation of a histidine kinase (HK) and subsequent transfer of the phosphate group to its downstream cognate response regulator (RR). The RR then elicits a cellular response, commonly through regulation of transcription. Engineering two-component system signaling networks provides a strategy to study bacterial signaling mechanisms related to bacterial cell survival, symbiosis, and virulence, and to develop sensory devices in synthetic biology. Here we focus on the principles for engineering the HK to identify unknown signal inputs, test signal transmission mechanisms, design small molecule sensors, and rewire two-component signaling networks.

Key words Histidine kinase, Two-component system, Bacterial signaling, Synthetic biology, Rewiring, Chimeras, Protein engineering, Leucine zipper

1 Introduction

Two-component systems allow bacteria to respond to changes through signal recognition by the histidine kinase (HK) and transfer of the phosphate to its response regulator (RR) [1, 2]. The RR can then activate a response, which is often gene transcription [3]. HKs are typically homodimeric, embedded in the membrane, and have one or more sensory domains that regulate the kinase domain. The HK domain, while highly conserved, contains many customizable functions including optional phosphatase activity, alteration of phosphotransfer partner specificity, and regulation of kinase heterodimerization (Fig. 1). The variation of these functions in the kinase domains generates an array of different phenotypic responses (Fig. 2a). HKs exhibit the most remarkable diversity in the sensory domains with varied domain architectures and sensing capabilities [1]. As an example, homologs of *C. crescentus'* cell-cycle

Kimberly A. Kowallis and Samuel W. Duvall contributed equally to this work.

Claire E. Eyers (ed.), *Histidine Phosphorylation: Methods and Protocols*, Methods in Molecular Biology, vol. 2077, https://doi.org/10.1007/978-1-4939-9884-5_10, © Springer Science+Business Media, LLC, part of Springer Nature 2020

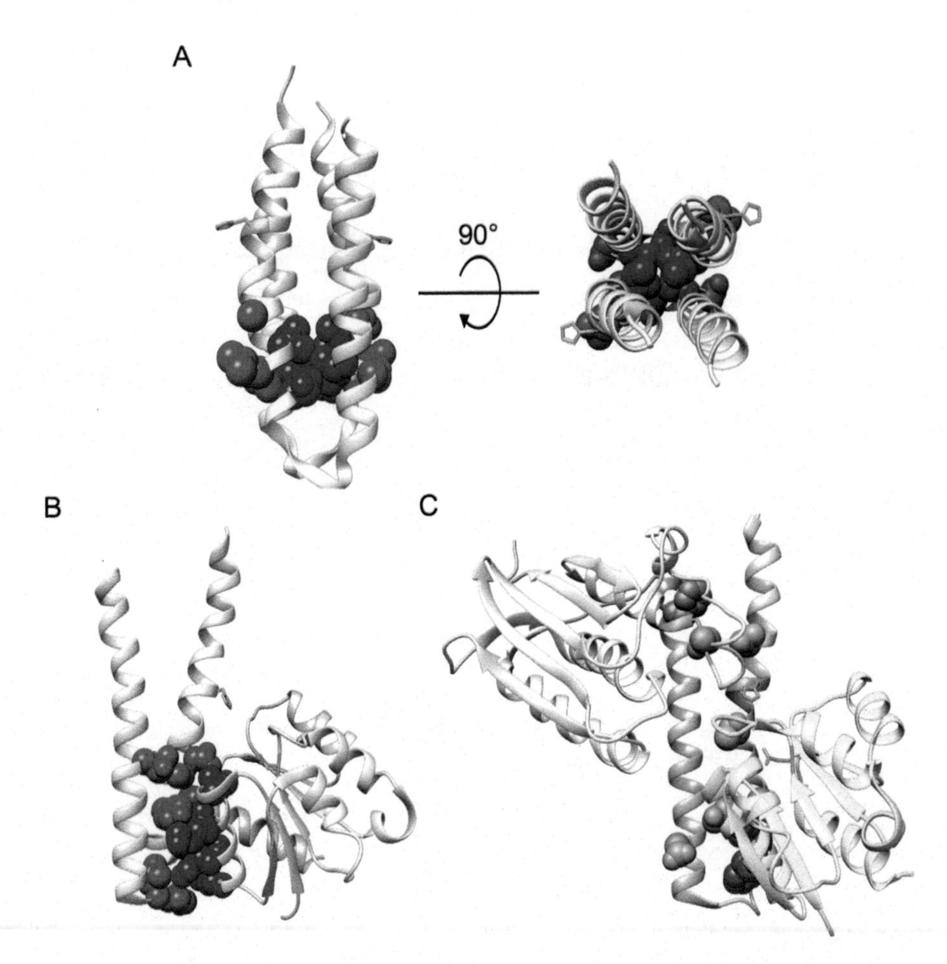

Fig. 1 Customizable residues for the DHp bundle of a histidine kinase (HK) and the regulatory domain of a response regulator (RR). Phosphorylatable histidine and aspartic acid residues are colored orange. (**a**) Residues involved in dimerization of the DHp are blue and buried in the bundle (PDB: 5B1N [88]). Residues in red are involved in RR specificity and are solvent exposed [23]. (**b**) The HK853-RR468 crystal structure (PDB: 3DGE [51]) highlights the residues involved in HK/RR specificity. Residues in red are located on the HK and blue are located on the RR. (**c**) HK853-RR468 crystal structure (PDB: 3DGE [51]) with residues in cyan denoting the DXXN phosphatase motif, residues in green denoting the GXGXG motif, with the second X (T in HK853) shown as spherical [51], blue showing the T267 and Y272 residues located on the HK capable of reducing phosphatase activity [51], and purple for the I17 and F20 residues located on the RR that can be mutated to prevent phosphate backtransfer [51]. All crystal structures images generated using Chimera [89]

kinase CckA vary in the number and identity of PAS sensory domains across the alphaproteobacteria (Fig. 3). This highlights the potential evolution of new connections that can occur between input signals and output responses through domain shuffling of sensory domains. In contrast to many eukaryotic sensory proteins that contain flexible linkers between domains, the linker region that connects the sensor to the kinase domain often contains one or more helical folds that allosterically transmit signals between domains [4–8].

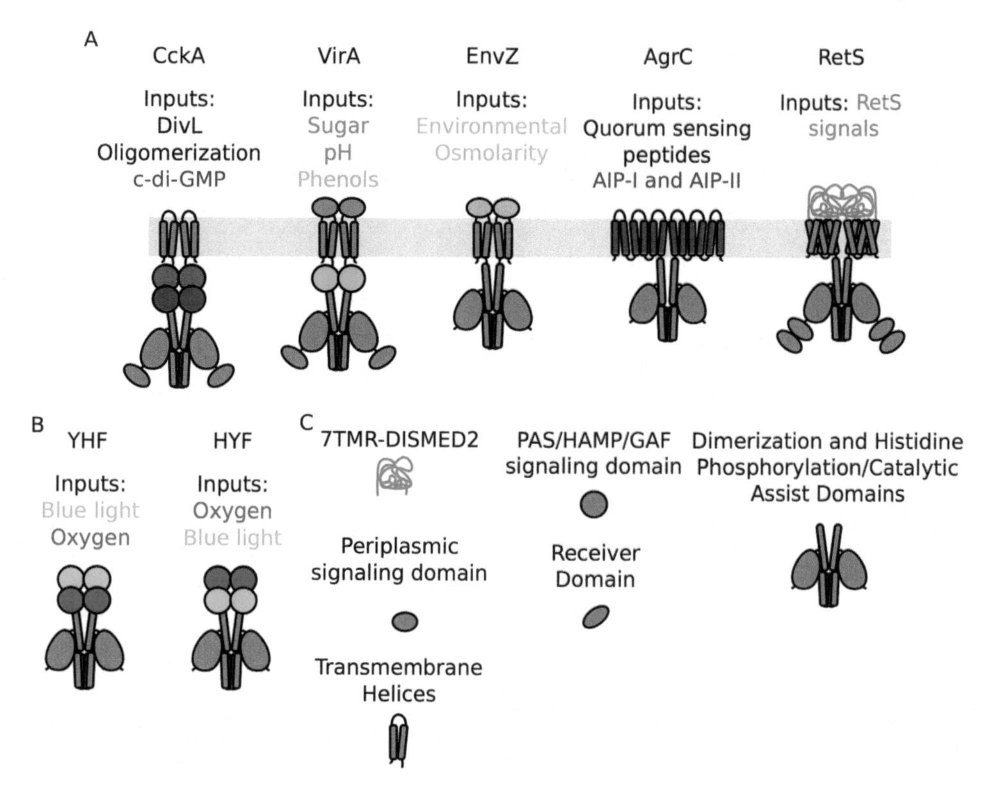

Fig. 2 (**a**) Domain architectures of five HKs with diverse sensing capabilities. Outputs include CckA activating transcription of cell-cycle regulated genes through the RR CtrA [90], VirA controlling virulence to plants and interkingdom gene transfer through the RR VirG [4], EnvZ controlling porin formation genes through the RR OmpR [91], AgrC activating virulence related genes [9], and RetS inhibiting GacS/GacA and PA1611 virulence pathways through heterodimer formation [92]. (**b**) Depiction of the synthetic chimeras of FixL and YtvA YHF and HYF HKs. (**c**) Legend of domain representations

Methods such as cysteine cross-linking [5, 8, 9] and mutagenesis have greatly contributed to the understanding of HK signal transmission mechanisms, but not without the unpredictable effects that point mutations can have on protein folding and activity. Crystallography [7, 10–14] has also been a powerful technique to analyze the conformational states of HKs, but due to many HKs having one or more transmembrane domains, it is often difficult to isolate soluble full-length HKs. Furthermore, the transmembrane region is not dispensable for activity in some cases [15, 16]. Additionally, the central dimerization and histidine phosphotransfer (DHp) domain must interact with distinct domains to carryout kinase versus phosphotransfer functions, which leads to HK conformational diversity. To address these challenges, HK engineering strategies have provided a hypothesis-driven approach to test bacterial HK signal transmission mechanisms.

In this review we discuss various protein engineering strategies that have been taken to understand the mechanisms of HKs and

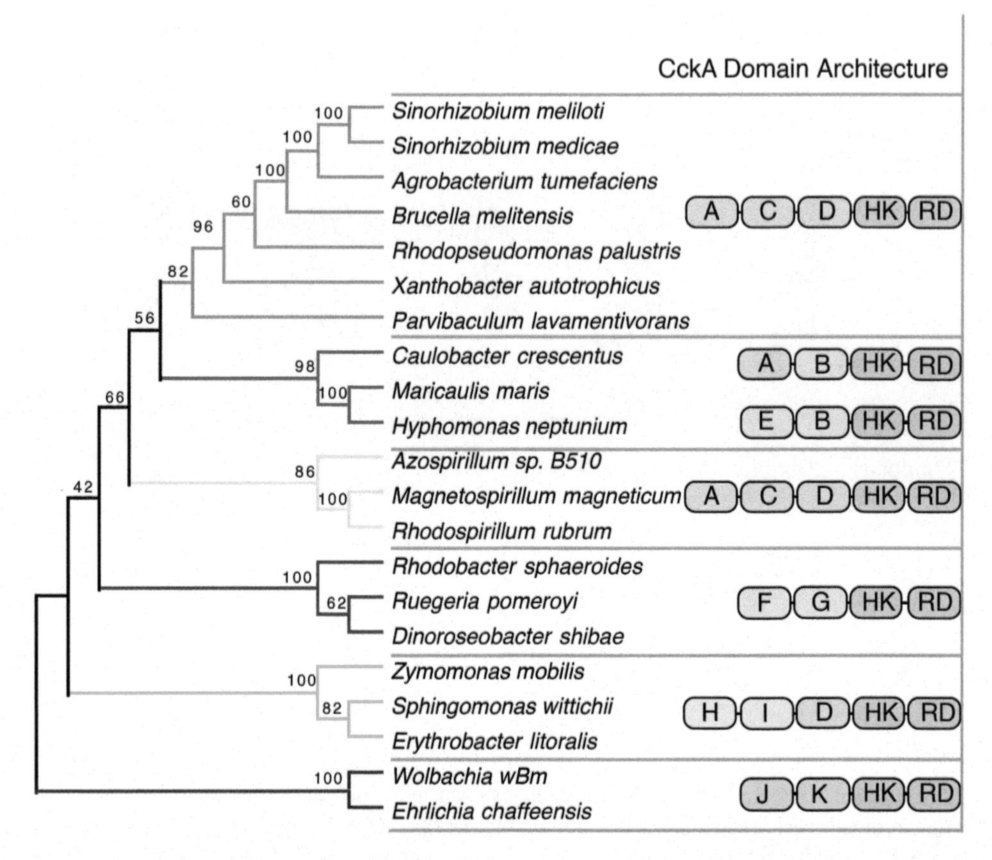

Fig. 3 Analysis of CckA homologs in alphaproteobacteria species highlights the diversity in signaling domains

two-component signaling networks. We review the design and characterization of chimeras that contain a sensory domain from one HK and the catalytic domain of another HK and examine how these strategies can help annotate unknown sensory functions [17, 18], discover phenotypes regulated by the signaling pathway [19, 20] and design sensors for small molecule detection [21, 22]. We also discuss a second strategy to test long-range HK allosteric signal transmission mechanisms that uses a leucine zipper to bias the conformational state to confer kinase activity, partial activity, or inactivity. Finally, we discuss strategies toward rewiring two-component system network through heterodimerization [23], reprogramming phosphotransfer RR specificity [24], and scaffold-dependent regulation [25].

2 First Generation Chimeric HKs Reveal the Importance of the Connecting Helical Linker in Signal Transmission

Early examples of engineered, chimeric histidine kinases were generated by fusing the periplasmic and transmembrane domains of the

E. coli aspartate sensing Tar chemoreceptor protein to the cytoplasmic HK domains of the osmolarity-sensing protein EnvZ of *E. coli* [17, 26]. Construction and design of chimeric HKs has revealed that the linker region acts as a "structural joint" for transmitting signals between domains [27], and that linker compositional changes can place the kinase in the "on" or "off" state [28].

It was further shown that the C-terminal region of many classes of sensory domains (e.g., HAMP, PAS, and GAF domains) exploit a C-terminal helix that can allosterically transmit a signaling event from the sensor binding site to the histidine phosphorylation site [29, 30]. The function of a chimeric HK is therefore dependent upon maintenance of the α-helix heptad periodicity. A second critical feature specific to PAS sensory domains is a structural motif that is positioned between the PAS domain and the C-terminal helix termed the DIT signal transmission motif [8, 18, 19, 31, 32]. The DIT motif is a set of residues at the C-terminus of PAS domains, and can present itself as D-(A/I/V)-(T/S) [29]. This motif forms hydrogen bonding and salt bridge interactions that couple the PAS fold to the C-terminal α-helix such that signal can alter the conformation of downstream domains [29]. Mutations to the conserved aspartic acid and threonine to aliphatic amino acids have been shown to abolish signal transduction through the α-helix [7]. The DIT motif mutations therefore present an approach to test if signals are transmitted from a given PAS domain to downstream effector domains through the helical linker.

3 Aromatic Tuning to Regulate HK Interactions at the Membrane-Cytosolic Interface

Transmission of a signal from the sensory domain in the periplasm to the cytoplasm is highly influenced by transmembrane helix interactions with the membrane. Signal transmission in this manner can present a unique challenge when heterologously expressing a kinase from another organism, as changes in membrane composition can cause conformational changes and repositioning of the transmembrane helix, altering kinase activity [11, 33–35]. The configuration of aromatic residues near the membrane interface has been shown to modulate HK activity in vivo [34, 36]. The repositioning of aromatic residues at the membranecytosolic interface to change the interactions between the transmembrane helix and the membrane has been termed aromatic tuning [37, 38]. As shown in Fig. 4, the tryptophan is repositioned by one residue in both the N-terminal and C-terminal directions and this impacts the helix tilt angle within the membrane. This aromatic tuning strategy can optimize the responsiveness of chimeric HKs by expanding the sensor's dynamic range, and in some cases, restoring kinase activity to an unresponsive chimera [36].

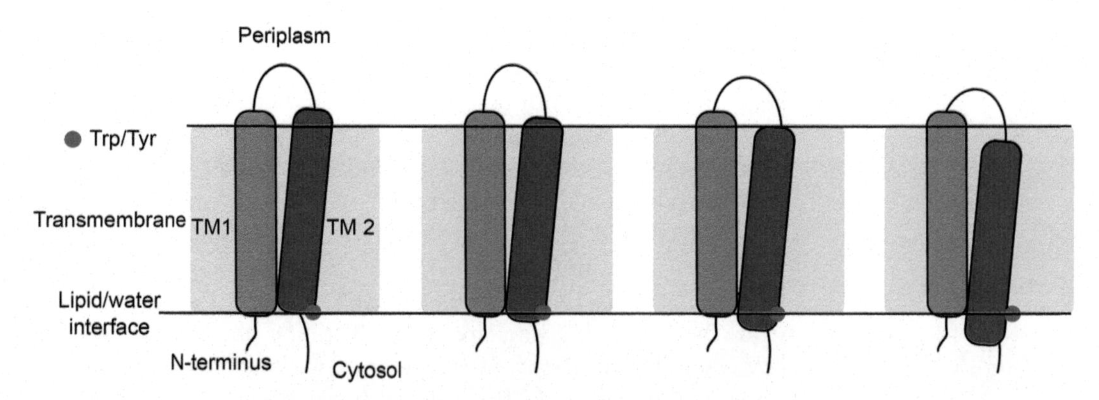

Fig. 4 The impact of repositioning the aromatic tryptophan or tyrosine located at the lipid–water interface of the transmembrane and cytosol upon HK signaling

4 Use of Chimeric HKs to Identify New Signaling Inputs

Due to the lack of conservation among sensory domains, it has been difficult to determine physiological signal inputs for the vast majority of HKs. One approach is engineering HK chimeras in which a sensory domain of unknown function is attached to well-characterized kinase domain. Combined with the cognate RR, a transcriptional gene reporter can be used to identify factors that activate or repress the signaling domain (Fig. 5a). This approach was used to characterize novel signals that regulate the sensory domain of the *Caulobacter crescentus* cell-cycle kinase CckA [19]. Prior to these studies, it was unclear what signals regulate CckA activity. In this study two design criteria were considered: first, selection of a well-understood kinase domain and downstream reporter, and second, selection of the fusion site between the CckA sensor and the kinase. For this assay, the kinase should be orthogonal to the existing two-component systems encoded in the bacteria to insulate the chimeric HK from native two-component signaling networks. In this example, Iniesta et al. selected a *C. crescentus* oxygen sensing FixL/FixJ β-galactosidase reporter used in strains lacking the native FixL/FixJ signaling pathway [39]. A panel of CckA-FixL chimeras was constructed and screened in order to identify the best fusion. The most responsive CckA-FixL chimera was coupled together with data from a transposon screen to identify protein signals that regulate CckA's activity and subcellular localization. This study revealed that a novel pseudokinase, DivL, promotes CckA activity through (direct or indirect) interaction with CckA's tandem sensory domain [19], highlighting the potential of coupling chimeric HKs with a fluorescent transcription reporter and high-throughput library screening to identify signal inputs that modulate HKs critical for development or pathogenesis.

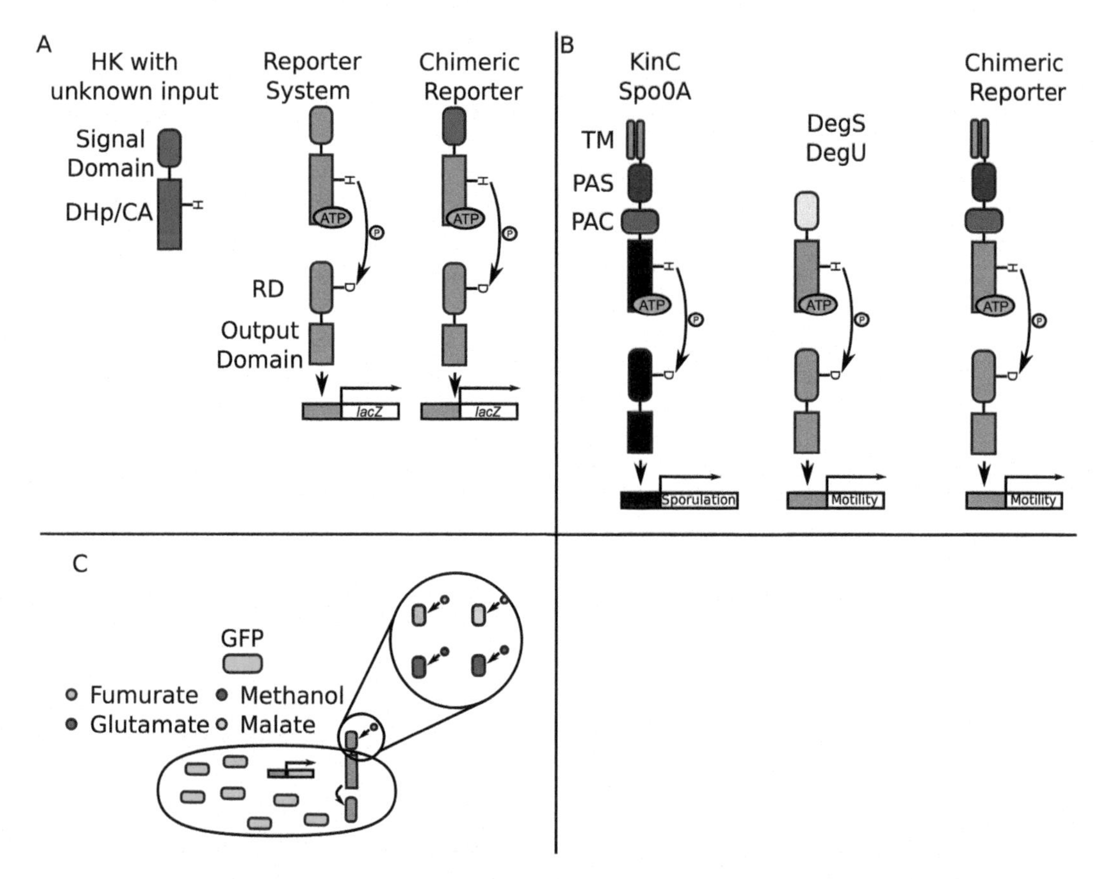

Fig. 5 (**a**) General design of a HK chimera to report on the activity of an unknown signaling domain via β-galactosidase expression. (**b**) A KinC/DegS reporter strain was generated such that membrane-damaging agent nystatin could be sensed via KinC and trigger motility regulation through the DegS pathway. (**c**) Small molecule reporters sensing fumurate, methanol, glutamate, and malate using the *E. coli* EnvZ/OmpR two-component system with GFP as a transcriptional readout

5 Rewired Developmental Response and Identification of Potassium Leakage Sensor Domain

B. subtilis is a model organism used to understand the process of bacterial differentiation into diverse cell types. The membrane potential sensing kinase KinC in *B. subtilis* has been implicated in the formation of biofilms in response to membrane damage due to exposure to surfactins and other antimicrobial agents [20]. Therefore it was proposed that KinC was a potassium leakage sensing kinase. To identify the domains of KinC involved in sensing the effects of potassium leakage, the sensor of KinC was attached to the well-characterized *B. subtilis* HK DegS, which controls genes that regulate swarming behaviour, in a DegS knockout strain. Through a swarm size assay with the KinC-DegS chimera, it was demonstrated that the KinC sensory domain acted to mediate sensing to potassium leakage [20]. Wild-type cells typically respond to potassium leakage by upregulation of KinC biofilm genes, while the KinC-DegS chimera rewires this response to potassium leakage to upregulate DegS-DegU motility genes (Fig. 5b). This case illustrates the potential for synthetic biology strategies to use HK chimeras to rewire the behavioral logic of a bacterium.

6 Chimeric HKs as Small Molecule Reporters

HKs can also be repurposed as sensors for metabolic engineering [21, 22, 40, 41], optogenetic applications [42] or sensing disease-related signals [43]. High-throughput screening methods using chimeric HKs have been designed to identify microorganisms that generate large quantities of chemicals for industrial processes [44, 45]. These chimeras use the EnvZ/OmpR two-component system that regulates GFP expression under *ompC* output. Unique chimeras were constructed using the sensor domains that detect malate (MalK), methanol (FlhS), glutamate (DegS), and fumurate (DcuS) (Fig. 5c) [21, 22, 40, 41]. These small molecule fluorescent biosensors will now enable future screens of bacterial libraries for isolation of unique chassis for production of feedstock and small molecules.

7 Tuning the Dynamic Range of HK Reporters by Varying the Degree of Phosphatase Function

An inherent challenge to implementing HKs as cell-based reporters is that each two component system has a defined signal detection range, which may not correspond to the application's required detection ranges [46]. Landry et al. utilized a computational model [47] to revealed that decreasing phosphatase activity lowered the detection threshold without negatively affecting the dynamic range [48].

The catalytic phosphatase residues are found in a conserved DxxxQ/H motif in the HisKA_3 family or E/DxxT/N motif in the HisKA family just downstream of the phosphorylatable histidine. The proposed mechanism for the HisKA_3 family asserts that the glutamine or histidine align a water molecule for the hydrolysis of the phosphate group, and that the aspartic acid forms a hydrogen bond with a conserved lysine of the response regulator [49]. Mutations to the glutamine/histidine or aspartic acid abolish or decrease phosphatase activity. The mechanism of the HisKA family appears to follow a similar mechanism as mutations to the threonine/asparagine result in loss of phosphatase activity [49, 50].

Other phosphatase activity mutants have been mapped to regions outside of the catalytic motif, and are thought to affect long-range conformational changes, interactions with the response regulator, the G2 box conformation, or nucleotide binding [50–52] (Fig. 1c). Landry et al. successfully applied the conserved G2 box motif mutation as a sensor tuning strategy by engineering NarX kinase as a detector of nitrate from fertilizer [48]. The engineered system sensed a larger range of nitrate conditions in the native soil environment than the wild-type NarX kinase. The authors suggest that this sensor can be coupled with nitrogen-fixing systems for agricultural and environmental safety applications [53, 54]. Furthermore, this presents an attractive general strategy that can be used to fine tune detection thresholds for histidine kinase biosensors in synthetic biology.

8 The Leucine Zipper Fusion Strategy to Test the Signal Transmission Mechanism of HKs

Studies of the long-range signal transmission from the sensory domain to the HK domain have led to a model of how these helical linkers propagate signal from the sensor to the HK domain. The model proposes that the helices form a coiled-coil interface that changes in response to signal binding. This conformational change is proposed to affect the position of the phosphorylatable histidine in relation to the ATP binding domain, resulting in either a kinase-active conformation, or switching to a phosphotransfer or a phosphatase-active conformation (Fig. 6) [7, 11, 13, 55, 56]. Other studies have suggested an electrostatic interaction between the signal transmission coiled-coil and the HK domain, which would similarly bias HK domain conformation [4, 5].

The ability to manipulate the conformation of the coiled-coil linkers has become a useful strategy for testing the HK signal transmission mechanism. The leucine zipper of the *S. cerevisiae* transcriptional activator GCN4 is a coiled-coil with a strong bias for one interface, with a K_d of dimerization in the 10 nM range [57]. Since the K_d of dimerization of the HK EnvZ is in the 0.4–10 μM range [23, 58], it is proposed that covalent attachment of the leucine zipper can stabilize the weaker HK coiled-coils in a single conformation (Fig. 6). This was first demonstrated to be an effective method to study signal transmission in a chemotaxis receptor [59] and later applied to the HKs VirA [4, 60, 61] and AgrC [9, 62]. The leucine zipper technique has been used to lock the catalytic domain of VirA in predicted kinase-on, kinase-off, and partial-kinase states and contributed to a three-state ratchet model for activation. The leucine zipper fusion strategy was applied to the AgrC kinase domain and

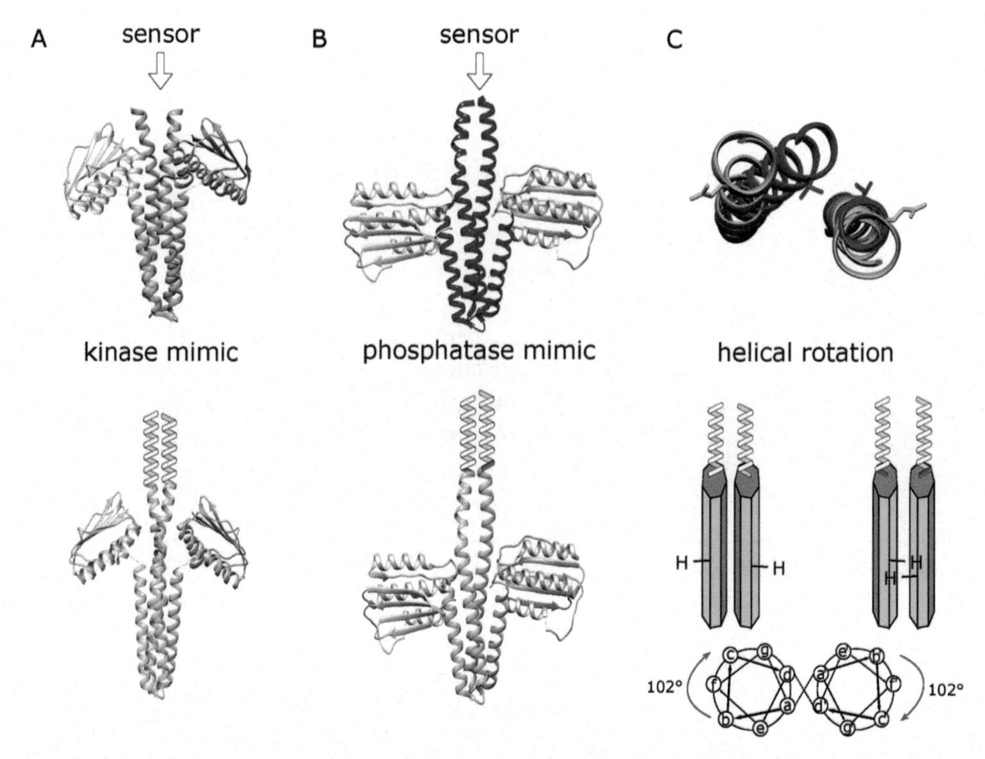

Fig. 6 Crystal structures of the (**a**, top) kinase mimic (H188E, PDB: 3GIE [11]) and (**b**, top) phosphatase mimic (H188V, PDB: 3EHJ [11]) of DesK show bending and helical rotation, apparent in (**c**, top) an overlay of the DHp domains from above. (**a–c**, bottom) The leucine zipper fusion strategy is utilized to lock the kinase domain into distinct conformational states by addition or deletion of residues from the helical coil. All crystal structure images generated using UCSF Chimera [89]

coupled with cysteine cross-linking to demonstrate the periodicity of the activation of AgrC [9, 62] in a mode similar to VirA kinase activation.

9 Leucine Zipper Fusions Reveal a Helical Bundle Ratchet Model in VirA Signal Transmission

VirA is a transmembrane hybrid HK that is expressed from a tumor-inducing plasmid in the plant-associated pathogen *A. tumefaciens*. VirA responds to at least four signals, including pH, temperature, and plant-derived phenolics and monosaccharides, integrating these signals to phosphorylate the RR VirG, which activates virulence factors [63, 64]. The leucine zipper approach was applied to investigate the mechanism by which VirA can integrate multiple signals with a goal of understanding plant host–bacterial communication (Fig. 7) [4, 60, 61].

The leucine zipper was fused directly to the N-terminus of either α1 or α4 within the GAF sensory domain. At the fusion site

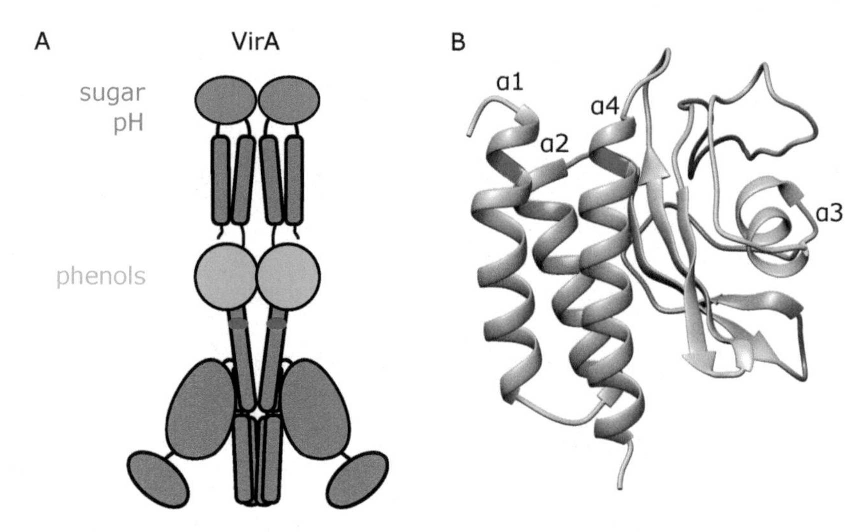

Fig. 7 (**a**) The domain architecture of VirA and (**b**) the homology model of the GAF domain of VirA compared to *Neisseria meningitidis* fRMsr [4, 93]. Leucine zipper fusion sites are denoted in red. All crystal structures images generated using UCSF Chimera [89]

an addition of three or four amino acids was incorporated to bias the coiled coil into a position corresponding to a 51° counterclockwise rotation, or 51° clockwise rotation from the neutral position, respectively. It is important to note that the kinase domain in each leucine zipper fusion contained the mutation G665E, which exhibits higher basal kinase activity but is still responsive to phenolic concentration [4, 60, 61, 65]. This hybrid protein allowed for detection of a negative impact of the leucine zipper fusion conformation upon kinase activity. The forced conformation of each fusion resulted in a distinct kinase activity measured by an in vivo β-galactosidase activity assay. These data contributed to theorizing of a ratchet model of activation in which phenolic signal binding causes rotation within the helical bundle, positioning the kinase domain in an active conformation [1, 4, 61].

It was further suggested that the rotation of the helical bundle places the phosphorylatable histidine into proximity with the ATP-binding domain, allowing phosphate transfer. To test this model, the leucine zipper fusion strategy was applied, fusing just 24 amino acids N-terminal to the phosphorylatable histidine, and the position of the histidine in each fusion was predicted. The panel of chimeras was extended to include seven fusion sites for a full heptad rotation, and the variable kinase activity in a β-galactosidase activity assay matched the prediction of the ratchet model. By comparing VirA to a protein with a solved x-ray crystal structure HK0853 [10], it was suggested that the active leucine zipper fusions correspond to a conformation in which the histidine would indeed be brought into proximity with the ATP-binding domain [4].

10 The Leucine Zipper Fusion Strategy Was Used to Demonstrate the Periodicity of AgrC

AgrC is a homodimeric transmembrane quorum sensing HK that regulates virulence upon sensing homologous or heterologous auto-inducing peptides in *S. aureus*. The linker between the sensor and kinase domain is predicted to have a helical fold [9]. A challenge in studying AgrC signal transmission in vivo is that the AgrC signaling cascade involves a positive feedback loop which can complicate transcriptional readouts [66]. Wang et al. used the leucine zipper fusion strategy to manipulate the conformation of this linker to study the AgrC signal transmission mechanism in vitro using nanometer-scale lipid bilayer discs [9, 62].

A panel of 15 leucine zipper fusions, corresponding to two heptad rotations, was constructed by fusing the GCN4 leucine zipper to the N-terminus of the AgrC linker, truncating one amino acid from the AgrC linker at a time. Each truncation results in a predicted 102° rotation of the helical linker per residue. An in vitro protein phosphorylation assay revealed that the kinase activity could be fit to a sine wave function, as a function of linker length, with a periodicity of 3.6 amino acids corresponding to one turn of the helix. The activities of each leucine zipper fusion were compared to the basal, induced, and inhibited activities of full-length AgrC and equivalent activities were observed. These results indicated that AgrC transmits signals through conformational twisting of the sensor-kinase helical linker.

A cysteine cross-linking experiment of full-length AgrC resulted in characterization of the conformation of the linker in each signal-binding state. This data, along with a study of AgrC homologs [62], allowed the degree of rotation achieved by each state to be defined and led to a rheostat model in which each signal binding event results in a rotation in the linker and the attached kinase domain, dictating activity. This model is similar to the "ratchet model" of VirA, in that it revealed a periodic trend in which insertion of residues modulates activity between kinase active, partial kinase activity and inactivity.

This fine-tuning of the dynamic range of kinase response may be an evolutionary tactic to regulate the amplitude of HK response. More broadly we envision that the leucine zipper strategy could be used to test other modes of signal transmission such as reverse signaling [67, 68], phosphatase activity, and phosphotransfer activity. This strategy could also be applied to kinases with diverse sensory domain architectures in which helical regions are proposed to be involved with signal transmission [29]. Furthermore, designed leucine zipper fusions could be used to confer constitutive, intermediate, or null kinase activity to manipulate downstream responses.

11 The Signal Processing Capabilities of Tandem Sensor Kinases

Several unannotated HKs contain two or more distinct sensory domains, raising the intriguing possibility that these HKs use Boolean-like logic to interrogate multiple environmental signals. As an example, Fig. 8 indicates that several HKs encoded within the genome of C. crescentus contain two or more sensory domains in tandem. Indeed, the VirA HK processes phenolic signals and sugars in an AND gate manner that confirms the presence of two signals prior to triggering activation the VirA-VirG virulence regulon (Fig. 2a) [4, 69]. However, a significant challenge remains in that we have poor knowledge of the identity of most HK sensory domains. The lack of knowledge about input signals prevents studying how most multisensor kinases processes signals.

To understand the large family of multisensor HKs, engineered synthetic tandem-sensor kinases composed of the oxygen sensing PAS domain of FixL and the blue-light sensing LOV domain of YtvA were developed (Fig. 2b) [70]. Their designs indicated that maintenance of α-helical secondary structure in their fusions was critical for kinase activity. This synthetic tandem-sensor kinase demonstrated that independent sensory domains could regulate kinase in an additive manner. Furthermore, these studies revealed that sensory domains more proximal to the HK domain had a larger impact on activity than the distal sensory domain. The examples of VirA and the synthetic tandem sensors reveal the capability to program advanced signal processing within the HK sensory domain. This could be commonly used in nature to ensure

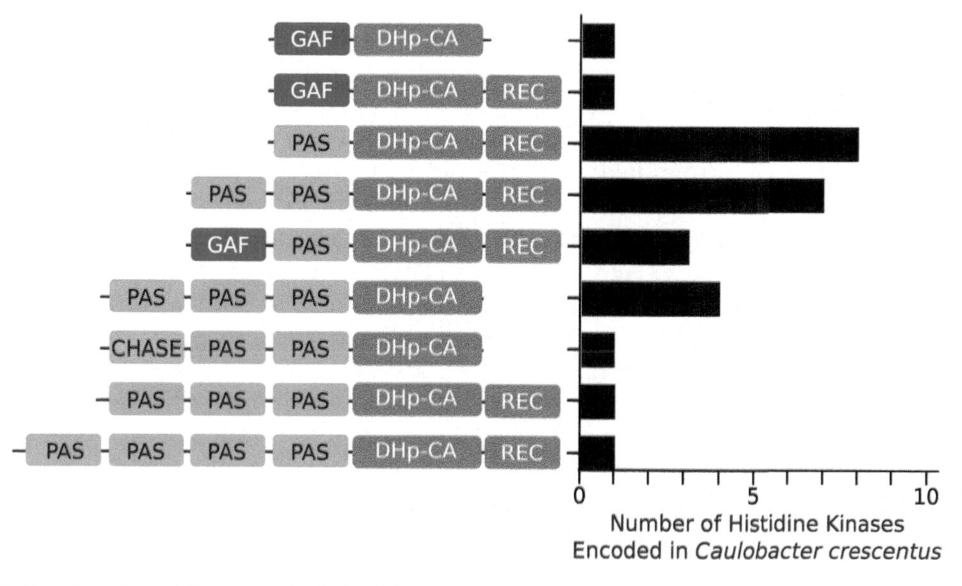

Fig. 8 The domain architectures of all *Caulobacter crescentus* histidine kinases that contain a PAS or GAF sensory domain

stringency of signaling response as highlighted in the VirA example [61], and also provides avenues to integrate and process signals for applications in synthetic biology.

12 Regulation of Multisensor Kinases Through Oligomerization

It has been proposed that a subset of multisensor domains control HK activity by toggling between dimer and tetramer states [71, 72]. The *B. subtilis* HK KinA regulates the sporulation pathways and utilizes a 4-PAS sensory domain. The input signal of KinA was not well understood but was predicted to be related to tetramerization [71, 73]. To tackle this problem, the diguanylate cyclase protein YdaM N-terminal PAS domains were fused to KinA kinase domain to promote tetramerization allowing investigation of the mechanisms whereby the sensory domain regulates KinA signaling activity [71]. Using cross-linking assays the authors observed that the YdaM fusion placed KinA into a tetramer conformation that was constitutively active [71, 72]. This example illustrates that multiple sensor kinases may have the capacity to regulate oligomeric changes as a mechanism of HK activity control. Future work is needed to understand how broadly this oligomerization mechanism is used and if it is a common feature of multisensor HK as suggested by studies of KinA [71] and CckA [72] multisensor kinases.

13 Strategies to Engineer HK Heterodimerization

Compared to eukaryotic signaling pathways, bacteria appear to exhibit minimal cross talk between signaling pathways in vivo and in vitro even though the DHp/CA domains of HKs are conserved across bacteria [74]. Bacterial cell fitness is reliant on reducing unwanted cross talk [74]. However, heterodimerization of HKs can generate more complex regulatory modes as highlighted by studies of the *P. aeruginosa* GacS and RetS HK heterodimer complex (Fig. 9) [75, 76]. These results highlight the importance of understanding how to engineer desired HK dimerization modes in order to wire HK heterodimerization to productively integrate information from two pathways. Comparison of EnvZ and its closest HK homolog RstB revealed residues that are required for these HKs to remain homodimers or function as heterodimers. Using covariation analysis, the base of the DHp bundle was identified as the region most likely contributing to homodimerization. To test this, residues from RstB were swapped into EnvZ. After two residue mutations, the engineered EnvZ variant was no longer able to heterodimerize with wild-type EnvZ as determined by a fluorescence resonance energy transfer (FRET) competition assay but

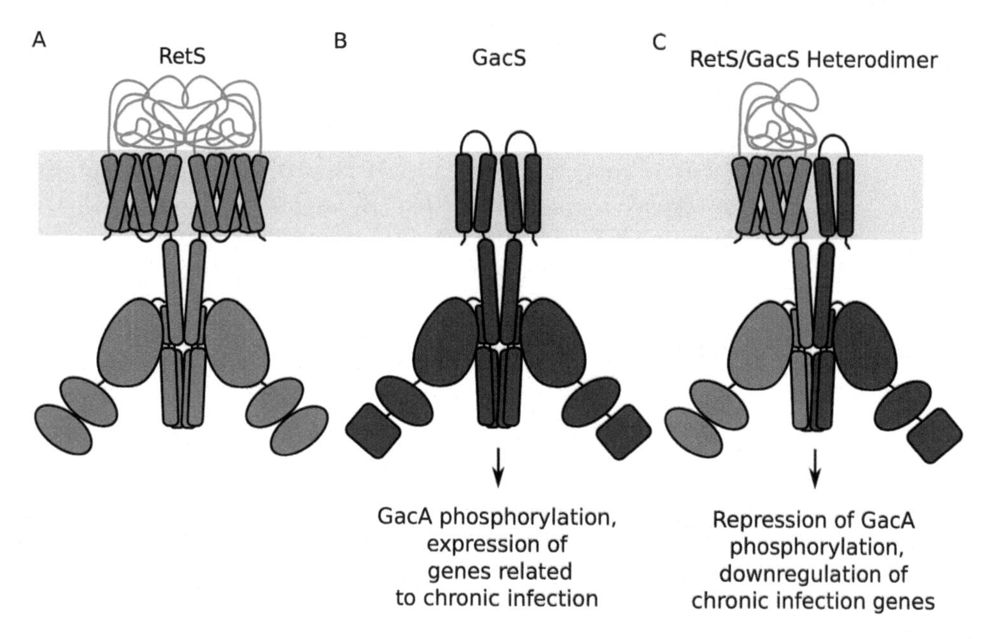

Fig. 9 Modeled structures of (**a**) RetS homodimer, (**b**) GacS homodimer and a (**c**) RetS-GacS heterodimer. When GacS is a homodimer and receives input signals, GacA phosphorylation occurs and expression of genes related to chronic infection are transcribed. When RetS and GacS form a heterodimer, repression of GacA phosphorylation and downregulation of chronic infection genes occur

maintained homodimerization ability [23]. A set of residues buried at the base of four-helix bundle was identified that proved to be sufficient for regulating dimerization specificity, while leaving RR phosphotransfer specificity unperturbed (Fig. 1a) [23].

The engineering of HK heterodimers may offer the potential to sense and integrate signals from each distinct sensory domain; however, this remains to be explored. In addition, future work will be needed to determine the degree of HK heterodimerization specificity and the implication of this level of regulation in microbial development.

14 Strategies to Rewire HK and RR Phosphotransfer

HKs and RRs have coevolved to permit specificity in phosphorylation activity in order to reduce HK-RR phosphotransfer cross talk [77]. By rewiring HK-RR protein interaction interfaces, signaling flow from one HK to a different RR output can be altered, a single HK can be engineered to phosphorylate several RR outputs, or engineered to allow integration of multiple signals through many HKs that all phosphotransfer to a single RR. The key interactions that modulate the phosphotransfer specificity are a suite of residues located on the solvent exposed surface of the DHp domain of the HK and the receiver domain of the cognate RR. The key residues that determine the specificity for the DHp tend to be solvent exposed and located at the base of the four-helix bundle, as well as in the loop region (Fig. 1a, b) [23, 24]. Skerker et al. identified residues within EnvZ that could be mutated to other HK residues (RstB, CpxA, PhoR, AtoS, and PhoQ) to confer phosphotransfer specificity to their RRs [24]. By identifying residues that covary between HKs and RRs [78, 79], they were able to demonstrate that this model is consistent with the interaction interface observed with several HK-RR cocrystal structures [51, 80, 81]. The entire loop regions of each kinase were substituted into EnvZ, resulting in EnvZ variants that recognize and phosphorylate the corresponding RR targets [24].

RR specificity residues have been identified through covariation analysis and can be applied to specific HKs through secondary structure alignment of DHp domains [24]. By applying these HK-RR interface mutations or switching the entire DHp loop region, multiple phosphorylation partners could be under the control of a single HK. This rewiring could also generate new connectivity between signaling input and outputs, providing an alternative approach to the sensor-HK chimeras that would allow much of the native HK function to be retained, while altering only those residues required to direct phosphotransfer to a desired RR.

15 Engineering Assembly and Activation of HK Signaling

Whitaker et al. took inspiration from eukaryotic signaling systems, which rely on scaffolds to colocalize and activate kinases, to design a modular prokaryotic system for synthetic biology [25, 82]. The downstream signaling of a chimeric HK [83] was rewired to activate noncognate RRs [84] in a scaffold-dependent manner (Fig. 10). It has been demonstrated that HKs will transfer phosphate to noncognate RRs with an increased effective concentration through direct tethering [85, 86]. Whitaker et al. sought to increase the local concentration of the HK and its downstream RRs by designing scaffolds with two protein–protein interaction domains, the SH3 peptide and leucine zipper, which recruited the SH3-tagged HK and the leucine zipper-tagged RR, respectively. The advantage of scaffold-dependent colocalization is that the RR specificity of the HK can be successfully adjusted by inducing scaffold protein expression, as determined by a fluorescent gene expression reporter assay. However, the system was sensitive to the expression level of each component.

The system was further engineered so that the scaffold is necessary not only for colocalization of the HK and RR but also for HK activation. The kinase domain was engineered with an SH3 domain and an internal SH3 peptide such that the SH3 domain would block phosphotransfer to the RR until the scaffold SH3 peptide

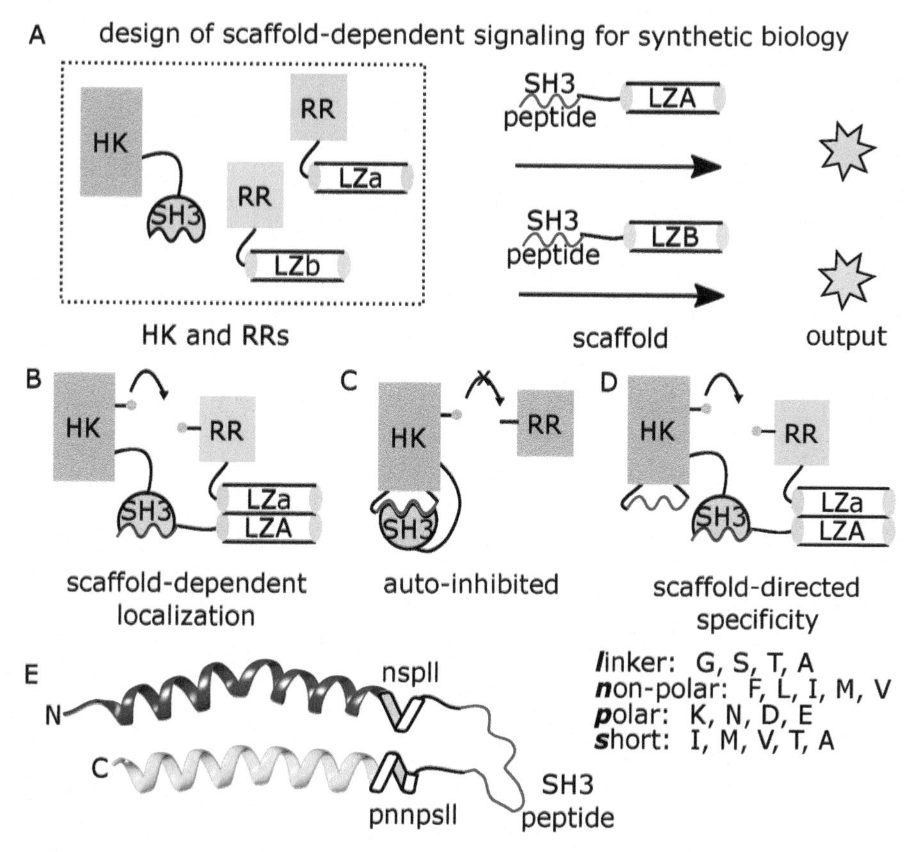

Fig. 10 (**a**) The design of scaffold-dependent signaling for HK synthetic biology. The SH3-bound HK and LZ-bound RRs are expressed in the cell. Induction of scaffold dictates the downstream activity of the HK. (**b**) Scaffold-dependent localization increases the effective concentration of the RR near the HK. (**c**) The HK can be engineered with an autoinhibitory interaction. (**d**) Induction of the scaffold colocalizes the components and relieves the autoinhibitory interaction to activate the signaling pathway. (**e**) The design of the autoinhibitory interaction in the DHp of the chimeric HK TAZ1 (PDB: 5B1N [88]). All crystal structure images generated using Chimera [89]

would compete for SH3 binding and activate the HK (Fig. 10). Using the same fluorescent gene expression reporter, it was demonstrated that the scaffold was effective in colocalizing and activating two-component system signaling.

Insertion of the SH3 peptide into the kinase structure, without affecting activity, presented an interesting protein kinase engineering challenge. The turn between the helices in the DHp was chosen as the site for insertion because of its proximity to the RR binding site. However, direct insertion, or the use of flexible linkers into this region, did not yield an active kinase. A library of linkers, which mimicked the seven amino acids at the end of the DHp bundle, was created to extend the helical structure (Fig. 10e). Additionally, two flexible amino acids were placed between the helical linker and the SH3 peptide at either end. Variants from the library were first screened for their ability to phosphorylate RR, and then for their conditional activation only in the presence of scaffold. The most promising library candidate was tested to confirm autoinhibition and scaffold-dependent activation. The modular signaling system designed here couples the roles of phosphotransfer and RR specificity and can therefore be extended to other two-component systems.

16 Conclusions and Outlook

The engineering strategies discussed here have contributed to an understanding of how bacteria use two-component systems to respond to their environment. However, several challenges remain in our capability to predictably engineer HK function. Most notable is poor access to diverse well-characterized sensory domains and the current inability to predict sensory domain function based on primary sequence. Further characterization of sensory domain structure and function is thus needed to understand the general design rules for how signals are perceived and distinguished, including investigation of how multisensor domains integrate multiple signals to regulate HK activity. A second major HK engineering opportunity is the engineering of allosteric regulation through kinase-kinase interactions. Indeed, natural histidine kinase pairs suggest this as a mode of histidine kinase regulation, such as the *P. aeruginosa* RetS–GacS pair [76] and the *C. crescentus* DivL–CckA pair [19, 67, 87]. Further application of these HK engineering strategies will enhance our understanding of bacterial impact upon human health, agriculture, and the environment.

References

1. Gao R, Stock AM (2009) Biological insights from structures of two-component proteins. Annu Rev Microbiol 63:133–154. https://doi.org/10.1146/annurev.micro.091208.073214
2. Zschiedrich CP, Keidel V, Szurmant H (2016) Molecular mechanisms of two-component signal transduction. J Mol Biol 428(19):3752–3775. https://doi.org/10.1016/j.jmb.2016.08.003
3. Galperin MY (2006) Structural classification of bacterial response regulators: diversity of output domains and domain combinations. J Bacteriol 188(12):4169–4182. https://doi.org/10.1128/JB.01887-05
4. Lin YH, Pierce BD, Fang F, Wise A, Binns AN, Lynn DG (2014) Role of the VirA histidine autokinase of Agrobacterium tumefaciens in the initial steps of pathogenesis. Front Plant Sci 5:195. https://doi.org/10.3389/fpls.2014.00195
5. Matamouros S, Hager KR, Miller SI (2015) HAMP domain rotation and tilting movements associated with signal transduction in the PhoQ sensor kinase. MBio 6(3):e00616–e00615. https://doi.org/10.1128/mBio.00616-15
6. Schultz JE, Kanchan K, Ziegler M (2015) Intraprotein signal transduction by HAMP domains: a balancing act. Int J Med Microbiol 305(2):243–251. https://doi.org/10.1016/j.ijmm.2014.12.007
7. Diensthuber RP, Bommer M, Gleichmann T, Moglich A (2013) Full-length structure of a sensor histidine kinase pinpoints coaxial coiled coils as signal transducers and modulators. Structure 21(7):1127–1136. https://doi.org/10.1016/j.str.2013.04.024

8. Lesne E, Dupre E, Lensink MF, Locht C, Antoine R, Jacob-Dubuisson F (2018) Coiled-coil antagonism regulates activity of venus flytrap-domain-containing sensor kinases of the BvgS family. mBio 9(1): e02052–e02017. https://doi.org/10.1128/mBio.02052-17
9. Wang B, Zhao A, Novick RP, Muir TW (2014) Activation and inhibition of the receptor histidine kinase AgrC occurs through opposite helical transduction motions. Mol Cell 53 (6):929–940. https://doi.org/10.1016/j.molcel.2014.02.029
10. Marina A, Waldburger CD, Hendrickson WA (2005) Structure of the entire cytoplasmic portion of a sensor histidine-kinase protein. EMBO J 24(24):4247–4259. https://doi.org/10.1038/sj.emboj.7600886
11. Albanesi D, Martin M, Trajtenberg F, Mansilla MC, Haouz A, Alzari PM, de Mendoza D, Buschiazzo A (2009) Structural plasticity and catalysis regulation of a thermosensor histidine kinase. Proc Natl Acad Sci U S A 106 (38):16185–16190. https://doi.org/10.1073/pnas.0906699106
12. Ferris HU, Zeth K, Hulko M, Dunin-Horkawicz S, Lupas AN (2014) Axial helix rotation as a mechanism for signal regulation inferred from the crystallographic analysis of the *E. coli* serine chemoreceptor. J Struct Biol 186(3):349–356. https://doi.org/10.1016/j.jsb.2014.03.015
13. Wang C, Sang J, Wang J, Su M, Downey JS, Wu Q, Wang S, Cai Y, Xu X, Wu J, Senadheera DB, Cvitkovitch DG, Chen L, Goodman SD, Han A (2013) Mechanistic insights revealed by the crystal structure of a histidine kinase with signal transducer and sensor domains. PLoS Biol 11(2):e1001493. https://doi.org/10.1371/journal.pbio.1001493
14. Mechaly AE, Sassoon N, Betton JM, Alzari PM (2014) Segmental helical motions and dynamical asymmetry modulate histidine kinase autophosphorylation. PLoS Biol 12(1):e1001776. https://doi.org/10.1371/journal.pbio.1001776
15. Szurmant H, Bu L, Brooks CL 3rd, Hoch JA (2008) An essential sensor histidine kinase controlled by transmembrane helix interactions with its auxiliary proteins. Proc Natl Acad Sci U S A 105(15):5891–5896. https://doi.org/10.1073/pnas.0800247105
16. Goldberg SD, Clinthorne GD, Goulian M, DeGrado WF (2010) Transmembrane polar interactions are required for signaling in the Escherichia coli sensor kinase PhoQ. Proc Natl Acad Sci U S A 107(18):8141–8146. https://doi.org/10.1073/pnas.1003166107
17. Utsumi R, Brissette RE, Rampersaud A, Forst SA, Oosawa K, Inouye M (1989) Activation of bacterial porin gene-expression by a chimeric signal transducer in response to aspartate. Science 245(4923):1246–1249. https://doi.org/10.1126/Science.2476847
18. Moglich A, Ayers RA, Moffat K (2009) Design and signaling mechanism of light-regulated histidine kinases. J Mol Biol 385 (5):1433–1444. https://doi.org/10.1016/j.jmb.2008.12.017
19. Iniesta AA, Hillson NJ, Shapiro L (2010) Cell pole-specific activation of a critical bacterial cell cycle kinase. Proc Natl Acad Sci U S A 107 (15):7012–7017. https://doi.org/10.1073/pnas.1001767107
20. Lopez D, Fischbach MA, Chu F, Losick R, Kolter R (2009) Structurally diverse natural products that cause potassium leakage trigger multicellularity in Bacillus subtilis. Proc Natl Acad Sci U S A 106(1):280–285. https://doi.org/10.1073/pnas.0810940106
21. Ganesh I, Ravikumar S, Lee SH, Park SJ, Hong SH (2013) Engineered fumarate sensing Escherichia coli based on novel chimeric two-component system. J Biotechnol 168 (4):560–566. https://doi.org/10.1016/j.jbiotec.2013.09.003
22. Ganesh I, Ravikumar S, Yoo IK, Hong SH (2015) Construction of malate-sensing Escherichia coli by introduction of a novel chimeric two-component system. Bioprocess Biosyst Eng 38(4):797–804. https://doi.org/10.1007/s00449-014-1321-3
23. Ashenberg O, Rozen-Gagnon K, Laub MT, Keating AE (2011) Determinants of homodimerization specificity in histidine kinases. J Mol Biol 413(1):222–235. https://doi.org/10.1016/j.jmb.2011.08.011
24. Skerker JM, Perchuk BS, Siryaporn A, Lubin EA, Ashenberg O, Goulian M, Laub MT (2008) Rewiring the specificity of two-component signal transduction systems. Cell 133(6):1043–1054. https://doi.org/10.1016/j.cell.2008.04.040
25. Whitaker WR, Davis SA, Arkin AP, Dueber JE (2012) Engineering robust control of two-component system phosphotransfer using modular scaffolds. Proc Natl Acad Sci U S A 109(44):18090–18095. https://doi.org/10.1073/pnas.1209230109
26. Krikos A, Conley MP, Boyd A, Berg HC, Simon MI (1985) Chimeric chemosensory transducers of Escherichia coli. Proc Natl Acad Sci U S A 82(5):1326–1330
27. Jin T, Inouye M (1994) Mutational analysis of the cytoplasmic linker region of Taz1-1, a

Tar-Envz chimeric receptor in Escherichia coli. J Mol Biol 244(5):477–481. https://doi.org/10.1006/Jmbi.1994.1746

28. Zhu Y, Inouye M (2003) Analysis of the role of the EnvZ linker region in signal transduction using a chimeric Tar/EnvZ receptor protein, Tez1. J Biol Chem 278(25):22812–22819. https://doi.org/10.1074/jbc.M300916200
29. Moglich A, Ayers RA, Moffat K (2009) Structure and signaling mechanism of Per-ARNT-Sim domains. Structure 17(10):1282–1294. https://doi.org/10.1016/j.str.2009.08.011
30. Harper SM, Christie JM, Gardner KH (2004) Disruption of the LOV-Jalpha helix interaction activates phototropin kinase activity. Biochemistry 43(51):16184–16192. https://doi.org/10.1021/bi048092i
31. Crosson S, Rajagopal S, Moffat K (2003) The LOV domain family: photoresponsive signaling modules coupled to diverse output domains. Biochemistry 42(1):2–10. https://doi.org/10.1021/bi026978l
32. Bury A, Hellingwerf KJ (2018) Design, characterization and in vivo functioning of a light-dependent histidine protein kinase in the yeast Saccharomyces cerevisiae. AMB Express 8 (1):53. https://doi.org/10.1186/s13568-018-0582-7
33. Shivaji S, Prakash JS (2010) How do bacteria sense and respond to low temperature? Arch Microbiol 192(2):85–95. https://doi.org/10.1007/s00203-009-0539-y
34. Draheim RR, Bormans AF, Lai RZ, Manson MD (2006) Tuning a bacterial chemoreceptor with protein-membrane interactions. Biochemistry 45(49):14655–14664. https://doi.org/10.1021/bi061259i
35. Norholm MH, von Heijne G, Draheim RR (2015) Forcing the issue: aromatic tuning facilitates stimulus-independent modulation of a two-component signaling circuit. ACS Synth Biol 4(4):474–481. https://doi.org/10.1021/sb500261t
36. Lehning CE, Heidelberger JB, Reinhard J, Norholm MHH, Draheim RR (2017) A modular high-throughput in vivo screening platform based on chimeric bacterial receptors. ACS Synth Biol 6(7):1315–1326. https://doi.org/10.1021/acssynbio.6b00288
37. de Planque MR, Boots JW, Rijkers DT, Liskamp RM, Greathouse DV, Killian JA (2002) The effects of hydrophobic mismatch between phosphatidylcholine bilayers and transmembrane alpha-helical peptides depend on the nature of interfacially exposed aromatic and charged residues. Biochemistry 41 (26):8396–8404
38. de Planque MR, Goormaghtigh E, Greathouse DV, Koeppe RE 2nd, Kruijtzer JA, Liskamp RM, de Kruijff B, Killian JA (2001) Sensitivity of single membrane-spanning alpha-helical peptides to hydrophobic mismatch with a lipid bilayer: effects on backbone structure, orientation, and extent of membrane incorporation. Biochemistry 40(16):5000–5010
39. Crosson S, McGrath PT, Stephens C, McAdams HH, Shapiro L (2005) Conserved modular design of an oxygen sensory/signaling network with species-specific output. Proc Natl Acad Sci U S A 102(22):8018–8023. https://doi.org/10.1073/pnas.0503022102
40. Ravikumar S, David Y, Park SJ, Choi JI (2018) A chimeric two-component regulatory system-based *Escherichia coli* biosensor engineered to detect glutamate. Appl Biochem Biotechnol 186:335–349. https://doi.org/10.1007/s12010-018-2746-y
41. Selvamani V, Ganesh I, Maruthamuthu MK, Eom GT, Hong SH (2017) Engineering chimeric two-component system into *Escherichia coli* from *Paracoccus denitrificans* to sense methanol. Biotechnol Bioproc E 22 (3):225–230. https://doi.org/10.1007/s12257-016-0484-y
42. Tabor JJ, Levskaya A, Voigt CA (2011) Multichromatic control of gene expression in Escherichia coli. J Mol Biol 405(2):315–324. https://doi.org/10.1016/j.jmb.2010.10.038
43. Daeffler KN, Galley JD, Sheth RU, Ortiz-Velez LC, Bibb CO, Shroyer NF, Britton RA, Tabor JJ (2017) Engineering bacterial thiosulfate and tetrathionate sensors for detecting gut inflammation. Mol Syst Biol 13(4):923. https://doi.org/10.15252/msb.20167416
44. Schrader J, Schilling M, Holtmann D, Sell D, Filho MV, Marx A, Vorholt JA (2009) Methanol-based industrial biotechnology: current status and future perspectives of methylotrophic bacteria. Trends Biotechnol 27 (2):107–115. https://doi.org/10.1016/j.tibtech.2008.10.009
45. Ziolkowska JR (2014) Prospective technologies, feedstocks and market innovations for ethanol and biodiesel production in the US. Biotechnol Rep 4:94–98. https://doi.org/10.1016/j.btre.2014.09.001
46. Ang J, Harris E, Hussey BJ, Kil R, McMillen DR (2013) Tuning response curves for synthetic biology. ACS Synth Biol 2 (10):547–567. https://doi.org/10.1021/sb4000564
47. Batchelor E, Goulian M (2003) Robustness and the cycle of phosphorylation and dephosphorylation in a two-component regulatory system. Proc Natl Acad Sci U S A 100

(2):691–696. https://doi.org/10.1073/pnas.0234782100

48. Landry BP, Palanki R, Dyulgyarov N, Hartsough LA, Tabor JJ (2018) Phosphatase activity tunes two-component system sensor detection threshold. Nat Commun 9 (1):1433. https://doi.org/10.1038/s41467-018-03929-y
49. Huynh TN, Noriega CE, Stewart V (2010) Conserved mechanism for sensor phosphatase control of two-component signaling revealed in the nitrate sensor NarX. Proc Natl Acad Sci U S A 107(49):21140–21145. https://doi.org/10.1073/pnas.1013081107
50. Huynh TN, Noriega CE, Stewart V (2013) Missense substitutions reflecting regulatory control of transmitter phosphatase activity in two-component signalling. Mol Microbiol 88 (3):459–472. https://doi.org/10.1111/mmi.12195
51. Casino P, Rubio V, Marina A (2009) Structural insight into partner specificity and phosphoryl transfer in two-component signal transduction. Cell 139(2):325–336. https://doi.org/10.1016/j.cell.2009.08.032
52. Zhu Y, Inouye M (2002) The role of the G2 box, a conserved motif in the histidine kinase superfamily, in modulating the function of EnvZ. Mol Microbiol 45(3):653–663
53. Sutton MA, Howard CM, Erisman JW, Billen G, Bleeker A, Grennfelt P, Van Grinsven H, Grizzetti B (2011) The European nitrogen assessment: sources, effects and policy perspectives. Cambridge University Press, Leiden
54. Smanski MJ, Bhatia S, Zhao D, Park Y, Woodruff LBA, Giannoukos G, Ciulla D, Busby M, Calderon J, Nicol R, Gordon DB, Densmore D, Voigt CA (2014) Functional optimization of gene clusters by combinatorial design and assembly. Nat Biotechnol 32 (12):1241–1249. https://doi.org/10.1038/nbt.3063
55. Casino P, Rubio V, Marina A (2010) The mechanism of signal transduction by two-component systems. Curr Opin Struct Biol 20(6):763–771. https://doi.org/10.1016/j.sbi.2010.09.010
56. Ferris HU, Dunin-Horkawicz S, Hornig N, Hulko M, Martin J, Schultz JE, Zeth K, Lupas AN, Coles M (2012) Mechanism of regulation of receptor histidine kinases. Structure 20(1):56–66. https://doi.org/10.1016/j.str.2011.11.014
57. Zitzewitz JA, Ibarra-Molero B, Fishel DR, Terry KL, Matthews CR (2000) Preformed secondary structure drives the association reaction of GCN4-p1, a model coiled-coil system. J Mol Biol 296(4):1105–1116. https://doi.org/10.1006/jmbi.2000.3507
58. Hidaka Y, Park H, Inouye M (1997) Demonstration of dimer formation of the cytoplasmic domain of a transmembrane osmosensor protein, EnvZ, of Escherichia coli using Ni-histidine tag affinity chromatography. FEBS Lett 400(2):238–242
59. Cochran AG, Kim PS (1996) Imitation of Escherichia coli aspartate receptor signaling in engineered dimers of the cytoplasmic domain. Science 271(5252):1113–1116
60. Wang Y, Gao R, Lynn DG (2002) Ratcheting up vir gene expression in Agrobacterium tumefaciens: coiled coils in histidine kinase signal transduction. Chembiochem 3(4):311–317
61. Gao R, Lynn DG (2007) Integration of rotation and piston motions in coiled-coil signal transduction. J Bacteriol 189(16):6048–6056. https://doi.org/10.1128/JB.00459-07
62. Wang B, Zhao A, Xie Q, Olinares PD, Chait BT, Novick RP, Muir TW (2017) Functional plasticity of the AgrC receptor histidine kinase required for staphylococcal virulence. Cell Chem Biol 24(1):76–86. https://doi.org/10.1016/j.chembiol.2016.12.008
63. Winans SC (1992) Two-way chemical signaling in Agrobacterium-plant interactions. Microbiol Rev 56(1):12–31
64. Doty SL, Yu MC, Lundin JI, Heath JD, Nester EW (1996) Mutational analysis of the input domain of the VirA protein of Agrobacterium tumefaciens. J Bacteriol 178(4):961–970
65. McLean BG, Greene EA, Zambryski PC (1994) Mutants of Agrobacterium VirA that activate vir gene expression in the absence of the inducer acetosyringone. J Biol Chem 269 (4):2645–2651
66. Reyes D, Andrey DO, Monod A, Kelley WL, Zhang G, Cheung AL (2011) Coordinated regulation by AgrA, SarA, and SarR to control agr expression in Staphylococcus aureus. J Bacteriol 193(21):6020–6031. https://doi.org/10.1128/JB.05436-11
67. Childers WS, Xu Q, Mann TH, Mathews II, Blair JA, Deacon AM, Shapiro L (2014) Cell fate regulation governed by a repurposed bacterial histidine kinase. PLoS Biol 12(10): e1001979. https://doi.org/10.1371/journal.pbio.1001979
68. Ozaki S, Schalch-Moser A, Zumthor L, Manfredi P, Ebbensgaard A, Schirmer T, Jenal U (2014) Activation and polar sequestration of PopA, a c-di-GMP effector protein involved in Caulobacter crescentus cell cycle control. Mol

Microbiol 94(3):580–594. https://doi.org/10.1111/mmi.12777

69. Gao R, Lynn DG (2005) Environmental pH sensing: resolving the VirA/VirG two-component system inputs for Agrobacterium pathogenesis. J Bacteriol 187 (6):2182–2189. https://doi.org/10.1128/JB.187.6.2182-2189.2005
70. Moglich A, Ayers RA, Moffat K (2010) Addition at the molecular level: signal integration in designed Per-ARNT-Sim receptor proteins. J Mol Biol 400(3):477–486. https://doi.org/10.1016/j.jmb.2010.05.019
71. Eswaramoorthy P, Dravis A, Devi SN, Vishnoi M, Dao HA, Fujita M (2011) Expression level of a chimeric kinase governs entry into sporulation in Bacillus subtilis. J Bacteriol 193(22):6113–6122. https://doi.org/10.1128/JB.05920-11
72. Mann TH, Shapiro L (2018) Integration of cell cycle signals by multi-PAS domain kinases. bioRxiv. https://doi.org/10.1101/323444
73. Eswaramoorthy P, Guo T, Fujita M (2009) In vivo domain-based functional analysis of the major sporulation sensor kinase, KinA, in Bacillus subtilis. J Bacteriol 191(17):5358–5368. https://doi.org/10.1128/JB.00503-09
74. Rowland MA, Deeds EJ (2014) Crosstalk and the evolution of specificity in two-component signaling. Proc Natl Acad Sci U S A 111 (15):5550–5555. https://doi.org/10.1073/pnas.1317178111
75. Chambonnier G, Roux L, Redelberger D, Fadel F, Filloux A, Sivaneson M, de Bentzmann S, Bordi C (2016) The hybrid histidine kinase LadS forms a multicomponent signal transduction system with the GacS/GacA two-component system in Pseudomonas aeruginosa. PLoS Genet 12(5):e1006032. https://doi.org/10.1371/journal.pgen.1006032
76. Goodman AL, Merighi M, Hyodo M, Ventre I, Filloux A, Lory S (2009) Direct interaction between sensor kinase proteins mediates acute and chronic disease phenotypes in a bacterial pathogen. Genes Dev 23(2):249–259. https://doi.org/10.1101/gad.1739009
77. Capra EJ, Laub MT (2012) Evolution of two-component signal transduction systems. Annu Rev Microbiol 66:325–347. https://doi.org/10.1146/annurev-micro-092611-150039
78. Atchley WR, Wollenberg KR, Fitch WM, Terhalle W, Dress AW (2000) Correlations among amino acid sites in bHLH protein domains: an information theoretic analysis. Mol Biol Evol 17(1):164–178. https://doi.org/10.1093/oxfordjournals.molbev.a026229
79. Fodor AA, Aldrich RW (2004) Influence of conservation on calculations of amino acid covariance in multiple sequence alignments. Proteins 56(2):211–221. https://doi.org/10.1002/prot.20098
80. Varughese KI, Tsigelny I, Zhao H (2006) The crystal structure of beryllofluoride Spo0F in complex with the phosphotransferase Spo0B represents a phosphotransfer pretransition state. J Bacteriol 188(13):4970–4977. https://doi.org/10.1128/JB.00160-06
81. Podgornaia AI, Casino P, Marina A, Laub MT (2013) Structural basis of a rationally rewired protein-protein interface critical to bacterial signaling. Structure 21(9):1636–1647. https://doi.org/10.1016/j.str.2013.07.005
82. Bhattacharyya RP, Reményi A, Yeh BJ, Lim WA (2006) Domains, motifs, and scaffolds: the role of modular interactions in the evolution and wiring of cell signaling circuits. Annu Rev Biochem 75(1):655–680. https://doi.org/10.1146/annurev.biochem.75.103004.142710
83. Jin T, Inouye M (1993) Ligand binding to the receptor domain regulates the ratio of kinase to phosphatase activities of the signaling domain of the hybrid Escherichia coli transmembrane receptor, Taz1. J Mol Biol 232(2):484–492. https://doi.org/10.1006/jmbi.1993.1404
84. Skerker JM, Prasol MS, Perchuk BS, Biondi EG, Laub MT (2005) Two-component signal transduction pathways regulating growth and cell cycle progression in a bacterium: a system-level analysis. PLoS Biol 3(10):e334. https://doi.org/10.1371/journal.pbio.0030334
85. Townsend GE, Raghavan V, Zwir I, Groisman EA (2013) Intramolecular arrangement of sensor and regulator overcomes relaxed specificity in hybrid two-component systems. Proc Natl Acad Sci 110(2):E161–E169. https://doi.org/10.1073/pnas.1212102110
86. Capra EJ, Perchuk BS, Ashenberg O, Seid CA, Snow HR, Skerker JM, Laub MT (2012) Spatial tethering of kinases to their substrates relaxes evolutionary constraints on specificity. Mol Microbiol 86(6):1393–1403. https://doi.org/10.1111/mmi.12064
87. Tsokos CG, Perchuk BS, Laub MT (2011) A dynamic complex of signaling proteins uses polar localization to regulate cell-fate asymmetry in Caulobacter crescentus. Dev Cell 20 (3):329–341. https://doi.org/10.1016/j.devcel.2011.01.007
88. Eguchi Y, Okajima T, Tochio N, Inukai Y, Shimizu R, Ueda S, Shinya S, Kigawa T, Fukamizo T, Igarashi M, Utsumi R (2017)

Angucycline antibiotic waldiomycin recognizes common structural motif conserved in bacterial histidine kinases. J Antibiot 70 (3):251–258. https://doi.org/10.1038/ja.2016.151

89. Pettersen EF, Goddard TD, Huang CC, Couch GS, Greenblatt DM, Meng EC, Ferrin TE (2004) UCSF chimera—a visualization system for exploratory research and analysis. J Comput Chem 25(13):1605–1612. https://doi.org/10.1002/jcc.20084
90. Jacobs C, Ausmees N, Cordwell SJ, Shapiro L, Laub MT (2003) Functions of the CckA histidine kinase in Caulobacter cell cycle control. Mol Microbiol 47(5):1279–1290
91. Cai SJ, Inouye M (2002) EnvZ-OmpR interaction and osmoregulation in Escherichia coli. J Biol Chem 277(27):24155–24161. https://doi.org/10.1074/jbc.M110715200
92. Gooderham WJ, Hancock RE (2009) Regulation of virulence and antibiotic resistance by two-component regulatory systems in Pseudomonas aeruginosa. FEMS Microbiol Rev 33 (2):279–294. https://doi.org/10.1111/j.1574-6976.2008.00135.x
93. Zimmermann L, Stephens A, Nam SZ, Rau D, Kubler J, Lozajic M, Gabler F, Soding J, Lupas AN, Alva V (2017) A completely reimplemented MPI bioinformatics toolkit with a new HHpred server at its core. J Mol Biol 430:2237–2243. https://doi.org/10.1016/j.jmb.2017.12.007

Chapter 11

Development of a Light-Dependent Protein Histidine Kinase

Aleksandra E. Bury and Klaas J. Hellingwerf

Abstract

Phosphorylation plays a critical role in facilitating signal transduction in prokaryotic and eukaryotic organisms. Our study introduces a tool for investigation of signal diffusion in a biochemical regulation network through the design and characterization of a light-stimulated histidine kinase that consists of the LOV domain from YtvA from *Bacillus subtilis* and the histidine kinase domain Sln1 from *Saccharomyces cerevisiae*. We show that blue light can be used as a trigger for modulation of the phosphorylation events in this engineered two-component signal transduction pathway in a eukaryotic cell. At the same time, we demonstrate the robustness of LOV domains and their utility for designing fusion proteins for signal transduction that can be triggered with (blue) light, providing a ready toolkit to design blue light dependent two-component signalling pathways.

Key words Light-dependent histidine kinase, YtvA, Sln1, Phosphorylation, Two-component regulation system, Nuclear shuttling

1 Introduction

Phosphorylation plays a critical role in facilitating signal transduction in all organisms. There is a growing body of literature that describes the mechanism of phosphorylation of histidine (protein) kinases and their cognate response regulator(s) in prokaryotic and certain eukaryotic systems. Many phosphorylation steps are integral to the two-component signal transduction pathways [1]. Spatially and temporally controlled triggering of such phosphorylation events is crucial for sensing of the environmental changes of a living cell, and the translation of these signals into a physiological response [2].

Light is optimally suited for high-resolution spatiotemporal triggering of signal transduction networks. About 10 years ago, this notion led to the emergence of the field of optogenetics (for review *see*, e.g., [3]). In optogenetics one uses light to activate or deactivate a signal-transduction process in a cell that is otherwise not responsive to light, via engineering of hybrid photoreceptor proteins. These are proteins that sense light via a specialized light-

Claire E. Eyers (ed.), *Histidine Phosphorylation: Methods and Protocols*, Methods in Molecular Biology, vol. 2077, https://doi.org/10.1007/978-1-4939-9884-5_11,

sensitive domain, which upon light stimulation undergoes a conformational change, and transfers this signal via a (covalently) linked output domain [4]. A considerable amount of literature has been published on the design of molecular devices for light-controlled regulation of gene expression [5], protein localization [6], signal transduction [7] and protein-protein interactions [8].

In this chapter we describe the design and functional characterization of a light-dependent histidine-kinase fusion protein, achieved by fusion of the (light-sensitive) LOV domain from YtvA from *Bacillus subtilis* with the histidine kinase domain of the Sln1 histidine kinase from *Saccharomyces cerevisiae*. The choice of these two protein domains was guided by the application we have in mind for follow-up studies of this fusion protein, that is, subcellular light activation of stress responses in yeast; the Sln1 protein is responsible for the osmotic stress response in yeast cells. To construct this fusion protein, the plasma membrane associated signal-input domain of the Sln1 kinase was replaced by the LOV domain of YtvA. It was shown earlier that the *N*-terminus of the conserved and phosphorylatable histidine H576 of Sln1, a coiled-coil element of secondary structure, is identifiable from the sequence of this protein [9]. Following the ideas developed by Möglich et al. [10] we aligned the coiled-coil structure of the Jα helix that connects the LOV domain with the STAS domain of YtvA, with this coiled-coil structure of the Sln1 kinase. Based on this alignment, we designed a number of fusion proteins that would be expected to show light-modulation of kinase activity (C1–C11; *see* Table 1) [11]. Aligning the two coiled-coil regions was straightforward because the hepta-helical sequence feature, typical for this element of secondary protein structure, was straightforwardly identifiable in both proteins. We anticipate that a similar strategy could be employed to facilitate the design of other similar light-activated sensors.

2 Materials

Prepare all solutions using ultrapure water (prepared by purifying deionized water, to attain a sensitivity of 18 MΩ cm at 25 °C) and analytical grade reagents. Prepare and store all reagents at room temperature (unless indicated otherwise). Diligently follow all waste disposal regulations when disposing of waste materials.

2.1 Expression of the Fusion Protein

1. Production Broth medium: add 700 mL of water to 1 L graduated cylinder or beaker. Weigh 20 g of tryptone, 10 g of yeast extract, 5 g of glucose, 5 g of NaCl, 8.7 g of K_2HPO_4 and transfer to the cylinder. Stir with the magnetic stirrer until all components are dissolved. Make up to 1 L with water (*see* **Note 1**). Prep 1000× stock of ampicillin by dissolving 100 mg of ampicillin in 1 mL of water. Sterilize by pushing with the

Table 1
Details of the design of a series of LOV::Sln1 histidine kinase fusion proteins [7]

Construct name	Numbers of amino acid from YtvA	Numbers of amino acid from Sln1	Initial rate of phosphorylation in the dark	Initial rate of phosphorylation in the light	Ratio
C1	YtvA (1–146)	Sln1HKR1 (567–1221)	0.18	0.08	2.25
C2	YtvA (1–127)	Sln1HKR1 (538–1221)	0.08	0.08	1
C5	YtvA (1–127)	Sln1HKR1 (536–1221)	0.03	0.03	1
C6	YtvA (1–127)	Sln1HKR1 (536–1221) ↓549H	0.08	0.09	0.9
C8	YtvA (1–132)	Sln1HKR1 (540–1221)	0.01	0.01	1
C9	YtvA (1–127) FixL (259–281)[a]	Sln1HKR1 (567–1221)	n.d.	n.d.	n.d.
C10	YtvA (1–129)	Sln1HKR1 (553–1221)	0.20	0.16	1.3
C11	YtvA (1–129)	Sln1HKR1 (554–1221)	0.11	0.21	0.5

Note: C1 contains the LOV domain and the Jα linker from YtvA from *B. subtilis* and the histidine kinase domain from Sln1 from *S. cerevisiae*, while the other one, C2, comprises of only LOV part from YtvA connected to coiled coil sequence preceding the histidine kinase from Sln1. The key DIT motif of YtvA is positioned from 125 to 127 in the Jα linker. The phosphorylatable histidine of the protein kinase is located immediately downstream of this linker helix. In variant C6 an extra amino acid was inserted in the 549 position (↓549H)

n.d. not determined

[a]This sequence from the hepta-helical signature domain of the coiled-coil structure of FixL was inserted in-between the sequences of LOV domain of YtvA and the kinase domain of Sln1p [7]

syringe through the sterile 0.45 μm filter. Supplement medium with 1 mL of the ampicillin stock. Prep 500× stock of kanamycin by dissolving 25 mg in 2 mL of water. Sterilize by pushing with the syringe through the sterile 0.45 μm filter. Supplement medium with 1 mL of the ampicillin stock solution and 2 mL of the kanamycin stock solution (final concentration of the ampicillin—100 μg/mL, kanamycin—25 μg/mL).

2. Isopropyl β-D-thiogalactopyranoside (IPTG): prepare 1000× stock solution at 0.1 M in water. Sterilize by pushing with the syringe through the sterile 0.45 μm filter. Store in −20 °C up to 6 months.
3. Lysis buffer: 50 mM Tris–HCl pH 8.0, 10 mM NaCl, 10% (v/v) glycerol, 25 μg/mL DNase (dissolve in demi water and add immediately prior to use), 25 μg/mL RNase (dissolve in demi water and add immediately prior to use), 1 mg/mL lysozyme (dissolve in demi water and add immediately prior to use) (*see* **Note 2**). An EDTA-free protease inhibitor cocktail (complete, EDTA-free, Roche) should be added immediately prior to use. Chill to 4 °C.

2.2 Purification of the Fusion Protein

1. Water, filtered, 1 L.
2. 20% ethanol (v/v) in water, filtered, 1 L.
3. Binding buffer A: 50 mM Tris–HCl, pH 8.0, 50 mM NaCl, 20 mM imidazole.
4. Elution buffer B: 50 mM Tris–HCl, pH 8.0, 50 mM NaCl, 500 mM imidazole.
5. HisTrap FF column (GE Healthcare, 5 mL column).
6. Protein storage buffer: 20 mM Tris–HCl pH 7.5.
7. Anion exchange ResQ column (GE Healthcare, 6 mL column volume).
8. Binding buffer C: 50 mM Tris–HCl, pH 8.0, 50 mM NaCl.
9. Elution buffer D: 50 mM Tris–HCl, pH 8.0, 150 mM NaCl.
10. Dialysis tubing or dialysis cassettes with a suitable MW cut-off.

2.3 SDS-PAGE

1. 0.5 M Tris–HCl pH 8.8: dissolve 18.171 g Trizma Base in 80 mL of water, adjust the pH with 6 M HCl, fill to 100 mL.
2. 0.5 M Tris–HCl pH 6.5: dissolve 6.057 g Trizma Base in 80 mL of water, adjust the pH with 6 M HCl, fill to 100 mL.
3. 10% SDS: dissolve 10 g sodium dodecyl sulfate in 100 mL of water.
4. 10% APS: dissolve 0.2 g ammonium persulfate in 2 mL water. Store the APS in aliquots of 100 μL at −20 °C, avoid freezing thaw cycles.

5. 10× TGS electrophoresis buffer: weight 30.3 g Trizma Base, 144.0 g glycine, 10.0 g SDS. Transfer these compounds to 800 mL water in the cylinder or beaker, stir until all components are dissolved. Fill to 1 L with water to give a solution with an approximate pH of 8.3.
6. 30% acrylamide–bis solution.
7. 10% (w/v) SDS-PAGE gel:

Component	5.6% stacking gel, mL	10% running gel, mL
H_2O	2.4	4.8
30% acrylamide/bis	0.8	4.0
1.5 M Tris–HCl pH 8.8	–	3.0
0.5 M Tris–HCl pH 6.5	1	–
10% SDS	0.04	0.12
10% APS[a]	0.04	0.04
TEMED[a]	0.01	0.01

[a]Add these components after mixing and just before pouring the gel

8. 4× sample buffer: 200 mM Tris–HCl pH 6.8, 8% SDS, 0.4% bromophenol blue, 40% glycerol. Store at −20 °C (*see* **Note 3**).
9. Coomassie Brilliant Blue staining solution: 200 mL methanol (40%), 50 mL acetic acid (10%), 250 mL water, add 0.125 g Coomassie Brilliant Blue (CBB).
10. De-staining solution: 400 mL methanol (40%), 100 mL acetic acid (10%), 500 mL water.

2.4 In Vitro Kinase Assay

1. ATP: Prepare a master mix of ATP by spiking 1 mM of cold ATP (50 μL ATP from a 10 mM stock solution) with 30 Ci/mmol [γ-^{32}P]-ATP (4.5 μL [γ-^{32}P]-ATP from 3300 Ci/mmol stock) (*see* **Note 4**).
2. Reaction buffer: 50 mM Tris–HCl, pH 8.0, 100 mM KCl, 15 mM $MgCl_2$, 2 mM DTT (*see* **Note 5**) and 20% (v/v) glycerol, ATP (1–5 mM), 3300 Ci/mmol of [γ-^{32}P]-ATP.
3. Stop buffer: 0.25 M Tris–HCl, pH 8.0, 8% (w/v) SDS, 40% (v/v) glycerol, 40 mM EDTA, 0.008% (w/v) bromophenol blue, 4 mM β-mercaptoethanol.
4. Source of blue *L*ight *E*mitting *D*iode (LEDs with $\nabla\lambda^{max}$ = 464 nm) with an incident light intensity of 200 μEinstein·m^2/s.
5. Phosphor screen.

2.5 In Vivo Kinase Assay

1. Minimal selection medium SCM: add 700 mL of water to 1 L graduated cylinder or beaker. Weigh 20 g of glucose, 2 g of the amino acid dropout mix and transfer to the cylinder. Stir with the magnetic stirrer until all components are dissolved. Fill up to 900 mL, autoclave 20 min, 110 °C (*see* **Note 6**).
2. Yeast nitrogen base with ammonium sulfate: prepare 10× stock by dissolving 67 g of yeast nitrogen base with ammonium sulfate in 1 L of water. Sterilized by filtration.
3. Yeast nitrogen base with SCM: add 100 mL of yeast nitrogen base with ammonium sulfate to cooled SCM medium.
4. *Para*-formaldehyde: add 0.37 mL of p-formaldehyde to 99.63 mL of water (*see* **Note** 7).
5. 4 M NaCl stock (10× stock solution) in SCM, autoclave 20 min, 120 °C.
6. Phosphate-buffered saline pH 7.0 (PBS).
7. 4′,6′-diamidino-phenylindole (DAPI).
8. Enclosed space with red light source.
9. Source of blue *L*ight *E*mitting *D*iode (LEDs with λ^{max} = 464 nm) with an incident light intensity of 200 μEinstein·m^2/s.

3 Methods

Perform all experiments at room temperature. Experiments requiring darkness should be performed in under red light.

3.1 Design of the LOV Fusion Proteins

1. Identify the architecture of the relevant proteins and their constituent domains with information from the UniProt (https://www.uniprot.org/) database.
2. Define the coiled-coil structures in the two domains that are going to be connected. Within the coiled-coil structure define the hydrophobic and the hydrophilic amino acids.
3. Align the coiled-coil structures of the proteins of interest based on the hepta-helical motif of hydrophobic and hydrophilic amino acids identifiable in both proteins.
4. Based on the identified hydrophobic residues, and previous literature [10], select several points of connecting heterologous protein domains. Table 1 gives further detail about the specific fusion proteins designed in our study.

3.2 Construction of the Expression Vectors

1. Amplify the genes of interest from the chromosomal DNA of the organism that contains the protein of interest by PCR.
2. Design the primers for the PCR reaction with a 50 bp overlap extension to make the gene splicing possible. For the gene splicing procedure *see* [12].

3. Introduce the restriction sites in the PCR primers to clone the fused gene into the bacterial expression vector.
4. Use your favourite bacterial expression system. Here we use the pQE30 system that allows introduction of the 6×His tag on the protein N-terminus.

3.3 Heterologous Protein Expression in *E. coli*

1. Inoculate 20 mL of production broth medium with a single colony of the hybrid expression construct in a suitable bacterial expression strain from a fresh agar plate made of the same medium. Grow overnight at 37 °C with shaking (250 rpm).
2. Dilute 20 mL of the overnight start culture into 500 mL of fresh production broth medium containing the required antibiotics at 37 °C. Allow to grow for 1.5–2 h in an Erlenmeyer's flask with vigorous shaking until OD_{600} reaches 0.6–0.8.
3. Induce expression of the heterologous protein product by sterile addition of isopropyl β-D-thiogalactopyranoside (IPTG) to a final concentration of 0.1 mM.
4. Decrease the temperature of the incubator to 25 °C and allow growth to continue for approximately 16 h in the dark with vigorous shaking (*see* **Note 8**).
5. Harvest the cells by centrifugation, 20 min, 4000 × *g*, 4 °C, remove supernatant and store at −20 °C (*see* **Note 9**).
6. Thaw the pellet on ice and resuspend in 30 mL lysis buffer (*see* **Note 10**).
7. Sonicate the cell suspension for 4 min at a duty cycle of 50%. Leave on ice for 2 min and repeat sonication (*see* **Note 11**).
8. Centrifuge the sonicated cell suspension for 45 min at 10,000 × *g* at 4 °C.
9. Collect the supernatant (cell lysate) and filter through a 0.45 μm filter. Store at −20 °C if necessary, prior to purification (*see* **Note 12**).

3.3.1 Construction of the Yeast Sln1 Knock Out Strain

1. Knockout endogenous Sln1 and Hog1 by making use of homologous recombination. Design the deletion cassette [13] with homologues flanking regions of 50 bp. Knockout of Sln1 is lethal for yeast. Cells can be rescued from this by overexpression of the Ptp2 phosphatase, (*see* **Note 13**) driven by a constitutive promotor, and expressed from a separate plasmid [14].

3.3.2 Construction of the Yeast Expression Vectors

1. Clone the fused gene into the multiple cloning site of the yeast expression vector (here we use pRS shuttled vectors) (*see* **Note 14**).
2. Transform the knockout strain with the constructed vector containing the hybrid LOV-kinase.

3.4 Purification of the Recombinant Protein

Purify the His6-tagged recombinant proteins from the cell-free extracts in a two-step procedure using an AKTA pure protein purification system (GE Healthcare) or similar.

3.4.1 HisTrap FF Column

1. Perform a System Wash with 20% ethanol.
2. Wash the system with water (pump wash/system wash).
3. Assemble the system. Set up the HisTrap FF column and an appropriate loading loop. For 500 mL cell culture a 50 mL loading loop is most suitable given the sample volume of 30 mL described here (*see* **Note 15**).
4. Wash the column with the column volumes of water at 3 mL/min to remove residual traces of ethanol.
5. Insert the hose for pump A in the bottle of buffer A and the hose for pump B in the bottle of buffer B. Perform a pump wash.
6. Wash the column with 2 column volumes (CV) of buffer A to equilibrate.
7. Configure the loading loop and set the injection valve to "inject" until the piston reaches the bottom of a large loading loop. Then set the valve to "load."
8. Wash the column with no less than 10 CV of buffer A to remove all non-His-tagged protein.
9. Elute the protein with a gradient of 5 CV, from 0% to 100% of buffer B. Do not forget to start the fraction collector (2 mL) to collect the eluted protein.
10. Optionally wash with 10 CV of buffer B to remove all protein from the column.
11. Wash the column with 10 CV of buffer A.
12. Wash the column with 5 CV of nanopure water.
13. Wash the column with 5 CV of ethanol and store at 4 °C.
14. Dialysis your protein overnight against 20 mM Tris–HCl pH 7.5 and store at −20 °C if required.

3.4.2 Anion Exchange ResQ Column

1. Perform a System Wash with 20% ethanol.
2. Wash the system with Millipore water (pump wash/system wash).
3. Assemble the system. Configure the ResQ column and an appropriate loading loop.
4. Wash the column with the 2 column volumes of nanopure water at 3 mL/min to remove residual traces of ethanol.
5. Insert the hose for pump A in the bottle of buffer C and the hose for pump B in the bottle of buffer D. Perform a pump wash.

6. Wash the column with 2 CV of buffer C to equilibrate.
7. Configure the loading loop and set the injection valve to "inject" until the piston reaches the bottom of a large loading loop. Then set the valve to "load."
8. Wash the column with no less than 10 CV of buffer C.
9. Elute the protein with a gradient of 5 CV, from 0 to 100% of buffer D. Do not forget to start the fraction collector (2 mL) to collect the eluted protein.
10. Optionally wash with 10 CV of buffer D to remove all protein from the column.
11. Wash the column with 10 CV of buffer C.
12. Wash the column with 5 CV of nanopure water.
13. Wash the column with 5 CV of ethanol and store it in 4 °C.
14. Dialysis the protein overnight against 20 mM Tris–HCl pH 7.5 and store at −20 °C.
15. Check the purity of the proteins by SDS-PAGE and Coomassie staining.
16. Evaluate the quality of the purified protein by using a UV/VIS spectrophotometer to measure Abs_{450}.
17. Calculate the concentration of each protein based on the Lambert-Beer equation ($A = c \times l \times \varepsilon$ where, A—absorbance at 450 nm, c—concentration of the solution (mol/l), l—length of solution the light passes through (cm), ε—extinction coefficient). For all proteins containing the LOV domain, their concentration can be determined using the extinction coefficient of 14,000 M^{-1} cm^{-1} at 450 nm [15] (*see* **Note 16**).

3.5 In Vitro Activity of the Light Dependent Histidine Kinase

This method for protein kinase activity assays is modified slightly from an established procedure [16].

1. Add ~50 μL of the purified protein histidine kinase fusion to 395.5 μL of reaction buffer such that the final concentration of the histidine kinase in the reaction mixture is 30 μM. Keep the fusion kinase in the dark throughout (*see* **Note 17**).
2. Divide this mixture equally into two Eppendorf tubes.
3. Start the reaction by adding 27.25 μL ATP master mix to each reaction.
4. Keep one reaction tube in the dark and place the second reaction tube next to the source of the blue LED (*see* **Note 17**).
5. Remove 15 μL aliquots at time points $T = 0.5$, 1, 2, 5, 10, 20, 40, 60, and 120 min, and add immediately to 5 μL stop buffer.
6. For phosphoryl transfer experiments, perform **step 1** and leave for 60 min in the dark to enable histidine kinase

autophosphorylation. Add Ypd1 to the reaction mixture such that the molar ratio of kinase to phosphoryl transfer domain is 1:2.

7. Remove time-series samples and mix with stop buffer as described in **step 3**.
8. Analyse the reactions by SDS-PAGE using a 10% (w/v) acrylamide gel. Wrap the gel in plastic wrap and expose to a phosphor screen (*see* **Note 18**). Scan the screen to detect protein bands incorporating γ-^{32}P with a Typhoon Fla. 7000 system or similar. Save the resulting images as .tiff files (*see* **Note 19**). Figure 1 shows the results from the phosphorylation and phosphotransfer experiment using the example system.

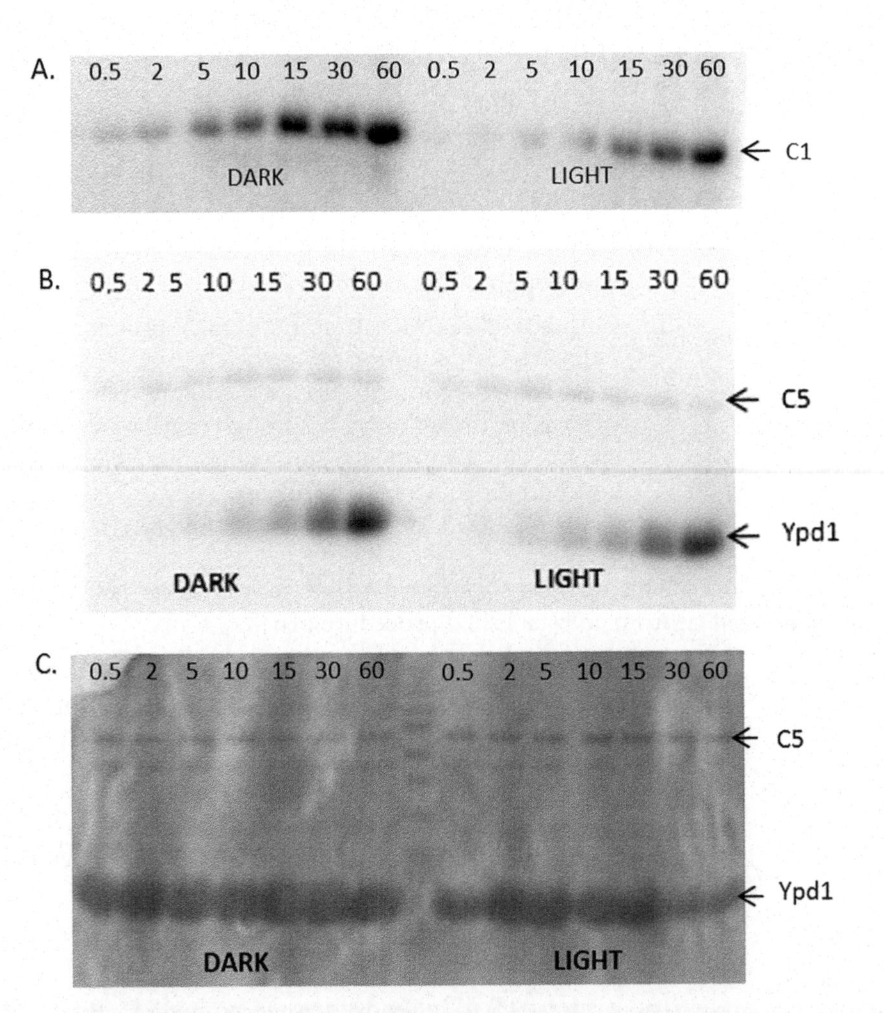

Fig. 1 (**a**) Autophosphorylation of the C1 fusion protein, Time series: 0.5, 2, 5, 10, 15, 30, 60 min, in the dark- and with blue-light illumination. The ATP concentration was 1 mM. (**b**) Autophosphorylation of the C5 fusion protein, followed by transfer of the phosphoryl group series: 0.5, 2, 5, 10, 15, 30, 60 min, under dark and light conditions. The molar ratio between the hybrid kinase and the Ypd1 phosphoryl transfer domain was 1:2. The ATP concentration was 1 mM. (**c**) Coomassie Blue stained gel

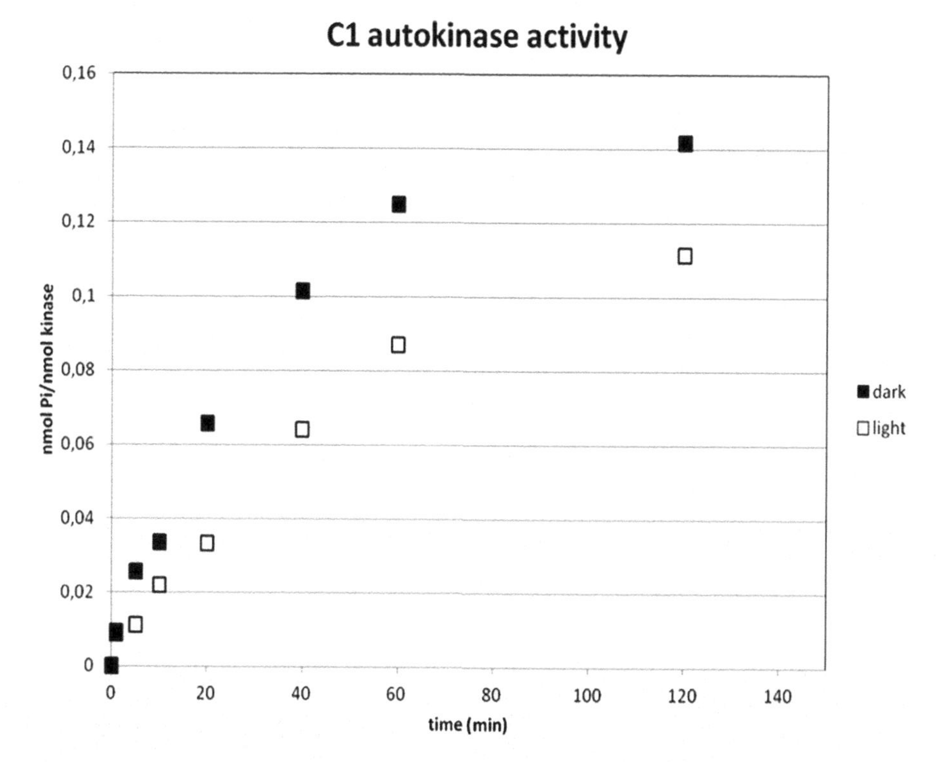

Fig. 2 Quantification of the autophosphorylation data of the C1 LOVSln1HKR1 fusion proteins in the presence of 1 mM of ATP

9. Quantify the relative intensity of the γ-^{32}P in the different protein bands using Image Quant software. Figure 2 shows example quantification data.

3.6 In Vivo Activity of the Light Dependent Histidine Kinase

1. Inoculate 5 mL of the SCM with the single colony of recombinant strains of *Saccharomyces cerevisiae*. Grow overnight in the dark with shaking (200 rpm) (*see* **Note 20**).
2. Dilute the overnight cultures to $OD_{600} = 0.05$ in 16 mL minimal selective medium and cultivate for 6 h in the dark (*see* **Note 21**).
3. Take 0.5 mL of the overnight dark sample and mix with 0.5 mL of 0.37% (v/v) *p*-formaldehyde. Incubate for 1 h in room temperature, spin down cells briefly by centrifugation, remove the supernatant and snap freeze samples in liquid nitrogen. Store at −20 °C until further investigation, *see* **step 10**.
4. Divide the overnight culture into two Erlenmeyer flasks, one for light/dark experiment and one for the salt stress experiment.
5. Divide the sample for the light/dark experiment in two such that the final volume of each of the cultures is 4 mL. Keep one sample in the dark under red light (*see* **Note 22**).

6. Light-stimulate the hybrid LOV-kinase protein with blue Light Emitting Diodes (LEDs) with λ^{max} = 464 nm (and full-width at half maximum ~20 nm), with an incident light intensity of 200 μEinstein·m^2/s (*see* **Note 23**).
7. Remove 0.5 mL aliquots at time points T = 1, 5, 10, 20, 40, 60, and 120 min and mix with *p*-formaldehyde as described in **step 3**.
8. To apply osmotic stress, add NaCl to 4 mL cell suspension to a final concentration of 0.4 M. Add an equivalent volume of SCM to 4 mL of cell suspension as a control for NaCl-induced osmotic stress [16, 17] (*see* **Note 24**).
9. Remove 0.5 mL aliquots at time points T = 1, 5, 10, 20, 40, 60, and 120 min and mix with *p*-formaldehyde as described in **step 3**.

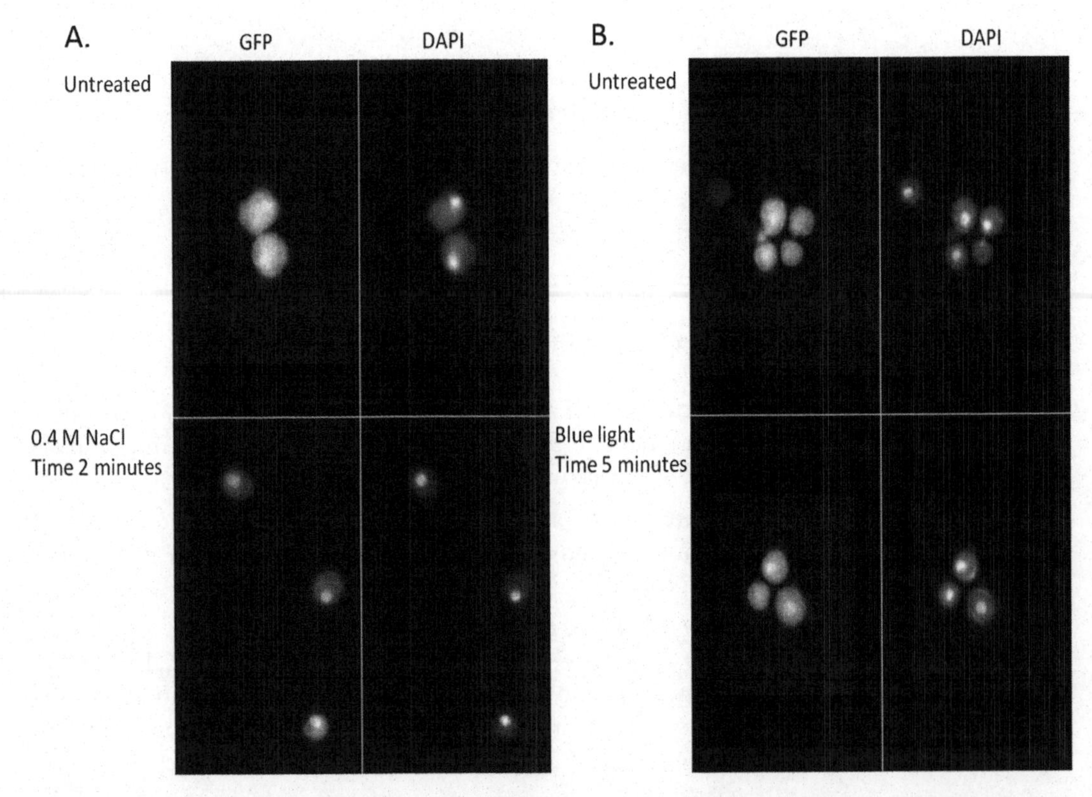

Fig. 3 (**a**) Subcellular localization of the HOG1::GFP fusion protein in response to changes in osmotic pressure and (**b**) after illumination of the cells with blue light. Activation of Sln1 signalling was initiated with: (**a**) a change in osmotic pressure, elicited by addition of 0.4 M NaCl (final concentration) to the cell suspension, and (**b**) illumination by exposure of the cells to blue light (200 μE incident intensity, 450 nm LED light). The strain used for (**a**) was: ΔHOG1, pRS416-HOG1::GFP; and for (**b**) ΔHOG1, pRS416-HOG1::GFP, ΔSln1, pRS325Act-C1LEU [7]

10. Wash the *p*-formaldehyde fixed cell with 1 mL of PBS and resuspend in 1 mL PBS. Stain the cells with 0.5 μg DAPI per mL culture to visualize the nuclei of the cells.
11. To observe the yeast cells, use an inverted microscope equipped with a 100× objective. To visualise GFP and DAPI, use a fluorescent light source and detect the emission at 470 nm and 395 nm, respectively. Capture the images using a digital camera (*see* **Note 25**).
12. Analyse the pictures using ImageJ software (or similar) without further manipulation (*see* **Note 26**). For quantitative analysis of the microscopy data, overlap the images of the DAPI-stained nuclei with those of the HOG1::GFP expressing yeast cells. Count those cells with nuclear- and cytoplasmic localization of the HOG1::GFP reporter protein, and calculate the percentage of cells with nuclear localization (here only cells with >1.5-fold nuclear accumulation were counted as positive for nuclear localization; by definition the other cells will have cytoplasmic localization). Figure 3 shows the accumulation of the HOG1::GFP in the nucleus.

4 Notes

1. Media should be sterilized for 20 min at 110 °C prior to adding antibiotics; the glucose in the medium will caramelize at higher temperatures.
2. If you want to use a Ni^{2+} column for affinity chromatography purification, for example, with a His6 purification tag, avoid using EDTA in the lysis buffer.
3. Use a face mask while weighing SDS. Using a 1 mL pipet tip which has been cut at the tip will improve the ease of pipetting glycerol accurately as this solution is very dense.
4. Follow the local rules how to work with radioactive compounds. Remember to protect yourself against the nuclear radiation. Always were a lab coat, glasses, and gloves, and work behind a suitable plastic protection shield.
5. The half-life of DTT is 30 min; always add fresh DTT before starting the reaction.
6. Prepare the amino acid mix lacking of the amino acids used for the positive selection of the expression vectors.
7. *p*-Formaldehyde is toxic: prepare the solution in a fume hood. Always use gloves.
8. As these are light-sensitive proteins, cover the Erlenmeyer flasks with a thick black fabric to keep the culture in the darkness, to improve stability of the overexpressed protein.

9. Freezing helps to lyse the cells. Keep the cells at −20 °C for at least 2 h. Do not omit this step. At this stage the protocol can be stopped and continued within 1–3 days. After that, you may see some degradation of the protein, which will lead to lower purification yields.
10. Use cold lysis buffer and a 25 mL pipette. Try not to make bubbles to avoid creating foam. Store the solution on ice.
11. Sonicate maximally 15 mL of the cell suspension at one time. While sonicating, cool the vessel with ice-water. After sonication let the cell suspension stand for 5 min to allow any foam to settle. This will ease pouring the suspension from the sonication vessel.
12. If freezing is required prior to purification, snap freeze the samples in liquid nitrogen.
13. Vector containing ptp2 phosphatase was a kind gift from J.S. Fassler [14].
14. The pHOG1::GFP vector was a kind gift from G. Smits (University of Amsterdam).
15. The columns can be reused several times for purification the same protein without regeneration. Test the loop for leaks before loading the sample. You can also use a pump (from old FPLC system) to load the sample. Do not forget to wash the pump and system before use.
16. The extinction coefficient for FMN is 14,000 M^{-1} cm^{-1} [15]. Prior to measuring the absorbance of the fusion kinases at the wavelength of 449 nm, the proteins need to be kept in the dark at room temperature for at least 6 h to bring the LOV domain back to the dark/ground state.
17. Kinase phosphorylation experiments and phosphoryl transfer experiments should be conducted either in the dark, with minimal red background light [18], or under constant illumination from blue Light Emitting Diodes (LEDs with λ^{max} = 464 nm) with an incident light intensity of 200 μEinstein·m^2/s.
18. Do not run the electrophoresis till the front reaches the end of the gel, to avoid contamination of the equipment with radioactivity. Carefully remove the gel below the dye front and dispose according to local safety guidelines; this region of the gel contains the free radioactive phosphate. Wrap the remaining gel in thin plastic wrap, to protect the screen and cassette from contamination with radioactivity.
19. Can also stain the gels with Coomassie brilliant blue to check if the radioactive bands observed correspond to the protein bands that you expect to be phosphorylated.

20. Do not use the YPD medium for experiments in which you want to use fluorescence microscopy. This medium shows high background fluorescence.
21. For experiments in the dark, wrap Erlenmeyer flasks with aluminium foil and secure this with tape. Prepare spare pieces of aluminium foil to cover the top of the flasks once you have made the proper dilutions.
22. Take the dark samples in dim red light [18].
23. Constantly illuminate with the blue LEDs using six LEDs fixed above a water bath incubator, in close distance (about 10–15 cm) to the water surface, to apply the appropriate light intensity.
24. Dilute the sterile 4 M stock solution of NaCl in SCM medium one in ten in the culture medium to achieve the require dilution of 0.4 M for osmotic stress treatment.
25. Acquire images using 100 ms exposure time for DAPI and 400 ms for GFP with the C11440, driven by the Nikon elements AR 4.50.001 software.
26. For analysis, export images as .tiff files and import into ImageJ software [19].

Acknowledgments

This work was supported by the Foundation for Fundamental Research on Matter (FOM) through program number 110 (Spatial design of biochemical regulation networks, SPAT), awarded to Prof. Dr. A. M. Dogterom (AMOLF, Amsterdam, The Netherlands).

References

1. Stock AM, Robinson VL, Goudreau PN (2000) Two-component signal transduction. Annu Rev Biochem 69:183–215
2. Engelberg D, Perlman R, Levitzki A (2014) Transmembrane signaling in Saccharomyces cerevisiae as a model for signaling in metazoans: state of the art after 25 years. Cell Signal 26:2865–2878
3. Deisseroth K (2011) Optogenetics. Nat Methods 8:26–29
4. van der Horst MA, Hellingwerf KJ (2004) Photoreceptor proteins, "star actors of modern times": a review of the functional dynamics in the structure of representative members of six different photoreceptor families. Acc Chem Res 37:13–20
5. Strickland D, Moffat K, Sosnick TR (2008) Light-activated DNA binding in a designed allosteric protein. Proc Natl Acad Sci U S A 105:10709–10714
6. Elgamoudi BA, Ketley JM (2018) Lighting up my life: a LOV-based fluorescent reporter for Campylobacter jejuni. Res Microbiol 169:108–114
7. Bury A, Hellingwerf KJ (2018) Design, characterization and in vivo functioning of a light-dependent histidine protein kinase in the yeast Saccharomyces cerevisiae. AMB Express 8:53
8. Yazawa M, Sadaghiani AM, Hsueh B, Dolmetsch RE (2009) Induction of protein-protein interactions in live cells using light. Nat Biotechnol 27:941–945
9. Tao W, Malone CL, Ault AD, Deschenes RJ, Fassler JS (2002) A cytoplasmic coiled-coil domain is required for histidine kinase activity

of the yeast osmosensor, SLN1. Mol Microbiol 43:459–473
10. Möglich A, Ayers RA, Moffat K (2009) Design and signaling mechanism of light-regulated histidine kinases. J Mol Biol 385:1433–1444
11. Möglich A, Moffat K (2007) Structural basis for light-dependent signaling in the dimeric LOV domain of the photosensor YtvA. J Mol Biol 373:112–126
12. Horton RM, Hunt HD, Ho SN, Pullen JK, Pease LR (1989) Engineering hybrid genes without the use of restriction enzymes: gene splicing by overlap extension. Gene 77:61–68
13. Goldstein AL, McCusker JH (1999) Three new dominant drug resistance cassettes for gene disruption in Saccharomyces cerevisiae. Yeast 15:1541–1553
14. Lu JMY, Deschenes RJ, Fassler JS (2003) Saccharomyces cerevisiae Histidine phosphotransferase Ypd1p shuttles between the nucleus and cytoplasm for SLN1-dependent phosphorylation of Ssk1p and Skn7p. Eukaryot Cell 2:1304–1314
15. Koziol J (1971) Fluorometric analyses of riboflavin and its coenzymes. Methods Enzymol 18:253–285
16. Fassler JS, West AH (2010) Genetic and biochemical analysis of the SLN1 pathway in Saccharomyces cerevisiae. Methods Enzymol 471:291–317
17. Kaserer AO, Andi B, Cook PF, West AH (2010) Kinetic studies of the yeast His-Asp phosphorelay signaling pathway. Methods Enzymol 471:59–75. https://doi.org/10.1016/S0076-6879(10)71004-1
18. Avila-Perez M, Vreede J, Tang Y, Bende O, Losi A, Gartner W, Hellingwerf K (2009) In vivo mutational analysis of YtvA from Bacillus subtilis: mechanism of light activation of the general stress response. J Biol Chem 284:24958–24964
19. Rost BR, Schneider-Warme F, Schmitz D, Hegemann P (2017) Optogenetic tools for subcellular applications in neuroscience. Neuron 96:572–603

Chapter 12

Empirical Evidence of Cellular Histidine Phosphorylation by Immunoblotting Using pHis mAbs

Rajasree Kalagiri, Kevin Adam, and Tony Hunter

Abstract

Immunoblotting is a ubiquitous immunological technique that aids in detecting and quantifying proteins (including those of lower abundance) and their posttranslational modifications such as phosphorylation, acetylation, ubiquitylation, and sumoylation. The technique involves electrophoretically separating proteins on an SDS-PAGE gel, transferring them onto a PVDF (or nitrocellulose) membrane and probing with specific antibodies. Here we describe an immunoblotting technique for detecting cellular phosphohistidine, a labile posttranslational modification, by optimizing experimental conditions such that the labile phosphohistidine signal is conserved throughout the experiment.

Key words Posttranslational modification, Histidine phosphorylation, pHis antibodies, Western blotting, Immunoblotting, Phosphoramidate bond

1 Introduction

Western blotting, also referred to as immunoblotting or protein blotting, is a technique that has been widely used to identify, characterize, and quantify proteins since its inception in the 1970s [1]. This versatile technique can be modified and adapted to a variety of sample preparations and experimental conditions. Western blotting is carried out in three steps: (1) SDS-PAGE, (2) blotting, and (3) immunostaining. In the first step, proteins are typically separated on an SDS-polyacrylamide gel based on their molecular weight. The separated proteins are then transferred electrophoretically onto a membrane (PVDF or nitrocellulose). The final part of the technique is protein detection by immunostaining, which involves use of protein or posttranslational modification (PTM)-specific antibodies. Antibodies against many stable PTMs such as O-linked phosphorylation (Ser, Thr, and Tyr), acetylation, ubiquitylation, and sumoylation [2–4] have been commercially available for a number of years, but only recently have these become available for labile N-linked phosphorylation. Consequently,

Claire E. Eyers (ed.), *Histidine Phosphorylation: Methods and Protocols*, Methods in Molecular Biology, vol. 2077, https://doi.org/10.1007/978-1-4939-9884-5_12,

proteins carrying N-linked phosphate are comparatively understudied [5]. Histidine undergoes N-linked phosphorylation on the imidazole ring forming a phosphoramidate (N-P) bond whose lability is conferred by the relatively high free energy of hydrolysis (−12 to −13 kcal/mol) [6]. Hydrolysis of the phosphoramidate bond is further accelerated by low pH and high temperature [7], common treatment conditions during biochemical analyses.

Recently, Fuhs et al. developed monoclonal antibodies (mAbs) against two stereoisomers of phosphohistidine (pHis; 1-pHis and 3-pHis) by incorporating the relatively stable pHis analogs, 1-pTza and 3-pTza (phosphoryltriazolylalanine), in degenerate peptides. These mAbs display the properties of pHis isoform stereospecificity, are sequence independent and exhibit submicromolar affinity to peptides and proteins which contain histidine phosphorylation [8]. Adapting standard immunoblotting techniques to preserve the phosphohistidine signal, in conjunction with the application of these newly developed pHis antibodies is helping unveil the mechanistic aspects of the pHis-containing proteins in normal and disease physiology [9–11].

In this chapter, we describe a modified immunoblotting technique which can be employed to detect histidine phosphorylated proteins in cell and/or tissue extracts using pHis mAbs. We demonstrate its utility using IMR-32 and HeLa cell lines established from neuroblastoma and cervical adenocarcinoma, respectively. This protocol can be further exploited to ascertain phosphohistidine levels in tissue samples and other cell lines.

2 Materials

All solutions should be prepared with deionized ultrapure water with resistivity of 18.2 MΩ cm and analytical grade reagents. All reagents should be handled and disposed of according to local safety guidelines.

2.1 Cell Culture and Lysate Preparation

1. Cell line: IMR-32 (human neuroblastoma); HeLa (cervical adenocarcinoma) cell lines.
2. Cell growth medium: RPMI-1640 (Rosewell Park Memorial Institute Medium) supplemented with 10% FBS (fetal bovine serum) for IMR-32 cell line; DMEM supplemented with 10% FBS for HeLa cell line and 100× Penicillin-Streptomycin solution (Purchase from Corning, Product number 30-002-cI; *see* **Note 1**).
3. Phosphate buffered saline (PBS; 1×): 0.8 mM sodium phosphate dibasic, 1.47 mM sodium phosphate monobasic, 137 mM NaCl, and 2.5 mM KCl.
4. Lysis buffer: 25 mM Tris–HCl pH 8.5, 140 mM NaCl, 0.1% (v/v) Tween 20.

5. Protease and phosphatase inhibitors: Protease inhibitor cocktail tablets (Roche), PhosSTOP tablets (Roche) and 1 mM PMSF (phenyl methyl sulfonyl fluoride) prepared in ethanol (*see* **Note 2**).
6. CO_2 incubator.
7. 10 cm^2 cell culture dishes
8. Probe sonicator.

2.2 SDS-Polyacrylamide Gel

1. 30% (w/v) acrylamide solution: Weigh 30 g of acrylamide and add to 60 mL of water. After the dissolution of the acrylamide, filter the solution through a 0.45 μM filter and store at 4 °C in a brown bottle (*see* **Note 3**).
2. 1% (w/v) bis-acrylamide solution: Weigh 1 g of bis-acrylamide and add to a cylinder with 100 mL of water. Filter through a 0.45 μM filter and store at 4 °C in a brown bottle.
3. Resolving and stacking gel buffer: 3 M Tris–HCl, pH 8.8. Weigh 363.4 g of Tris Base and add to 900 mL of water in a beaker. Adjust the pH to 8.8 with HCl. Make up the volume to 1 L before storing at 4 °C (*see* **Note 4**).
4. 20% (w/v) sodium dodecyl sulfate: Add 20 g of SDS to 80 mL of water and warm gently to dissolve SDS (*see* **Note 5**). Make up the volume to 100 mL after dissolution.
5. Ammonium persulfate: Prepare 10% (w/v) solution in water (*see* **Note 6**).
6. *N,N,N′,N′*-tetramethylethane-1,2-diamine (TEMED): Store at 4 °C.
7. SDS-PAGE running buffer (1×): 25 mM Tris–HCl pH 8.5, 0.1% SDS, and 192 mM glycine, precooled to 4 °C (*see* **Note 7**).
8. SDS-PAGE sample buffer (5×): 250 mM Tris–HCl pH 8.8, 10% SDS, 50 mM EDTA, 500 mM 1,4 dithiothreitol (DTT), 50% glycerol, and 0.02% bromophenol blue. Store the sample buffer at −20 °C (*see* **Note 8**).
9. Polyacrylamide gel electrophoresis equipment: Gel plates, electrophoresis chamber unit appropriate for gel plates, electric wires, and power supply with programmable voltage and current settings.

2.3 Immunoblotting

1. 0.45 μM PVDF membrane wetted in methanol for at least 2 min (*see* **Note 9**).
2. Transfer buffer (1×): 25 mM Tris–HCl pH 8.5, 0.1% SDS, 192 mM Glycine and 20% methanol (*see* **Note 10**).
3. Tris–NaCl buffer (10×): 500 mM Tris–HCl pH 8.5 and 1.5 M NaCl.

4. Blocking buffer: 0.2× Tris–NaCl buffer (pH 8.5), 0.1% casein (*see* **Note 11**).
5. Antibody dilution buffer: Blocking buffer with 0.1% Tween 20.
6. Wash buffer: 1× Tris–NaCl buffer with 0.1% Tween 20.
7. Wet transfer equipment: Wet transfer tank, two gel holder cassettes, filter paper, Whatman 3MM chromatography filter paper, foam pads, electric wires, and power supply with programmable voltage and current settings (*see* **Note 12**).

2.4 Immunostaining

1. Primary antibodies: Anti-N1-pHis (1-pHis) rabbit monoclonal antibodies purified from clone SC1-1 (Purchase from Millipore, product number MABS1330) and purified anti-N3-pHis (3-pHis) rabbit monoclonal antibodies from SC44-1 hybridoma cells [8] (*see* **Note 13**).
2. Antibody for loading control: anti-β-actin monoclonal antibody raised in mouse.
3. Secondary antibodies: goat-anti rabbit IgG with Alexa Fluor 680 conjugate, goat-anti mouse IgG with DyLight 800 conjugate.
4. Prestained protein molecular weight markers.

3 Methods

Detecting the pHis signal is pH and temperature dependent; hence, most of the solutions should be adjusted to a pH between 7 and 9 and stored at 4 °C, unless otherwise mentioned.

3.1 Whole Cell Lysate Preparation

Except for growing the cells, all steps should be carried out 4 °C. An alternative lysis strategy for tissue is provided in **Note 14**.

1. Grow the IMR-32 cells in RPMI medium and HeLa cells in DMEM medium, at 37 °C with 5% CO_2 in 10 cm^2 dishes until confluency reaches 70–90% (*see* **Note 15**).
2. Wash the cells twice with 5 mL of cold 1× PBS kept at 4 °C (*see* **Note 16**). Scrape the cells off the plate into 250 μL lysis buffer containing 1× protease and phosphatase inhibitor cocktail solution (*see* **Note 2**).
3. Mount the samples onto a probe sonicator for cell lysis. Keep the cells on ice during sonication to avoid sample heating. Program the sonicator to deliver 10 s bursts with a 10 s interval for three times at 40% amplitude (*see* **Note 17**).
4. Remove the cell debris by centrifugation at 14,000 × *g* for 10 min at 4 °C. Estimate the protein concentration in the clarified lysate by colorimetric assay and load around 30 μg of

protein onto the SDS-PAGE gel for analysis (*see* **Note 18**). Do not heat the samples before loading on the gel (*see* **Note 19**).

3.2 12% Sodium Dodecyl Sulfate-Polyacrylamide Gel Electrophoresis

It is imperative to carry out all the steps at 4 °C except for the polymerization of the gel.

1. Prepare the 12.5% resolving gel by mixing 2 mL of resolving gel buffer with 6.8 mL of 30% acrylamide and 1.66 mL of 1% bis-acrylamide. Make up the volume to 16 mL by adding 5.8 mL of water. To this mixture add 80 μL of 20% SDS, 60 μL of 10% ammonium persulfate (APS) and 10 μL of TEMED. Gently mix the solution before pouring into the gel cassette with 0.2 mm thickness. Fill two-thirds of the gel cassette with the mixture before layering isopropanol carefully on top (*see* **Note 20**). After the gel polymerizes remove the isopropanol and gently wash the gel with water to completely remove isopropanol.
2. Prepare the 4% stacking gel by adding 5.04 mL of water to 1.06 mL of 30% acrylamide, 0.8 mL of 1% bis-acrylamide and 1.0 mL of resolving buffer. To this mixture add 40 μL of 20% SDS, 30 μL of 10% APS and 5 μL of TEMED. Immediately insert the comb (10 well Teflon) without trapping air (*see* **Note 21**).
3. Mount the gel on the gel-running apparatus then fill it with SDS-PAGE running buffer precooled to 4 °C (*see* **Note 22**). To evaluate, the fidelity of the phosphohistidine signal, divide the samples into two tubes (Fig. 1). Keep one tube on ice at 4 °C. Add acid (0.01% acetic acid or formic acid) to the second tube till the pH drops to 4. Incubate for one minute before restoring the pH of the acidified sample to 8.5. After which boil the second tube for 15–30 min at 90 °C (*see* **Note 23**). Mix the protein samples with the 5× sample buffer. Load the protein samples (~30 μg of protein; *see* **Note 18**) into the wells along with a protein marker. Apply 100 V for 1–2 h for electrophoresis of the samples. This step should be carried out in a cold room or chamber at 4 °C.

3.3 Immunoblotting

All the buffers should be precooled and all steps performed at 4 °C, unless otherwise stated.

1. When the dye-front reaches bottom of the gel plates, switch off the current and open the gel plates using a spatula or similar (*see* **Note 24**). Carefully remove the stacking gel. Place the gel in the transfer buffer.
2. Cut the PVDF membrane and two Whatman 3MM filter papers to the size of the gel and make a small cut at the right top of the membrane to mark the orientation of the gel.

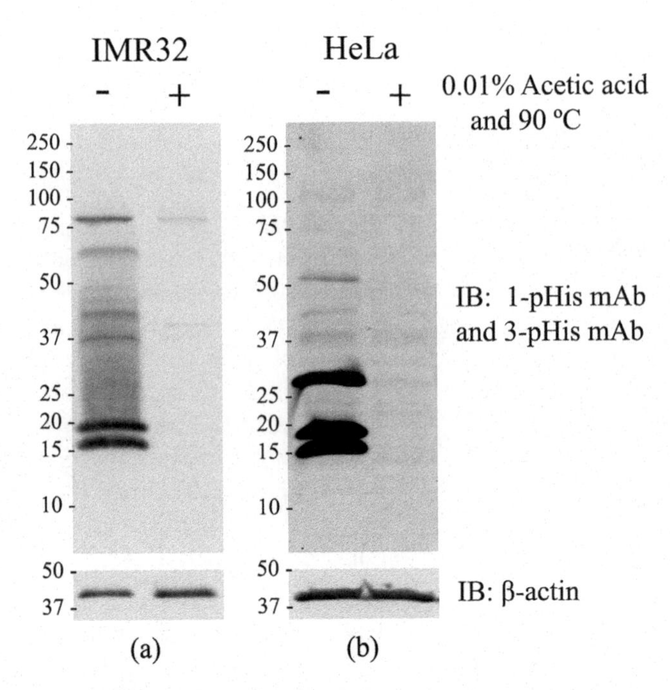

Fig. 1 Western blot detection of phosphohistidine proteins: Proteins containing 1- or 3-pHis isomers from either IMR-32 (**a**) and HeLa (**b**) cell lysates were detected using a combination of SC1-1 (1-pHis mAb) and SC44-1 (3-pHis mAb) monoclonal antibodies. Heat and acid treatment of the lysate prior to electrophoresis resulted in a considerable decrease in the phosphohistidine signal. β-actin is used as the loading control

Equilibrate the membrane in methanol for 20 s before equilibration in cold transfer buffer (*see* **Note 25**).

3. Soak the SDS-PAGE gel, foam pads and Whatman 3MM filter paper in transfer buffer along with the membrane. Place the membrane over the gel and sandwich with two Whatman 3MM filter papers followed by two foam pads on either side in a gel cassette (*see* **Note 26**).
4. Place the cassette in the tank so that the membrane is toward the anode (+) and the SDS-PAGE gel is toward the cathode (−). Fill the transfer buffer to the level indicated on the tank. Apply a voltage of 30 V for 12–18 h, or 75 V for 2 h at 4 °C (*see* **Note 27**).

3.4 Immunostaining

Precool all buffers and perform all the steps at 4 °C unless otherwise specified.

1. Immediately after the transfer, place the membrane in the blocking solution and incubate for 2 h at 4 °C or 45–60 min at room temperature on a shaking platform.

2. Dilute the primary antibody (1- or 3-pHis antibody) to a final concentration of 0.5 μg/mL in antibody dilution buffer. Immediately after removal of the blocking solution, incubate the membrane for 1–4 h at 4 °C or 1 h at room temperature (*see* **Note 28**). Wash the membrane 3× with 10 mL of wash buffer for 5 min each.
3. Dilute the secondary antibody (anti-rabbit IgG) 1:20,000 in antibody dilution buffer with 0.01% SDS. Incubate the membrane for 1–4 h at 4 °C, or 1 h at room temperature. Wash the membrane 3× with 10 mL of wash buffer for 10 min each.
4. Develop the membrane using a LI-COR Odyssey Infrared imaging system with 700 and 800 nm filters (Fig. 1) (*see* **Note 29**).
5. Repeat **steps 2–4** using primary and secondary antibodies against β-actin which is used as the protein loading control (*see* **Note 30**).

4 Notes

1. Filter FBS through 0.22 μM filter before adding to the RPMI or DMEM medium. The final concentration of penicillin-streptomycin antibiotic solution should be 1× in the media. Store the media at 4 °C after addition of FBS and antibiotics.
2. It is recommended to add protease and phosphatase inhibitors during lysis to prevent the degradation of proteins and loss of the phosphate signal. 10× solution of protease and phosphatase inhibitors can be prepared by dissolving one tablet each in 1 mL of lysis buffer separately and added to the sample immediately prior to sonication. The 10× stock solution can be stored at −20 °C for long term storage and 4 °C for short term storage. Protease and phosphatase inhibitor tablets can be replaced by making an in-house cocktail with appropriate concentrations of various inhibitors such as trypsin, chymotrypsin, thermolysin, papain, and pronase.

   PMSF is sparingly soluble and highly unstable in aqueous solutions hence it should be prepared immediately prior to use. Alternatively, it can be prepared in organic solvents such as ethanol or DMF where it is highly soluble and stable up to 2 years when stored at −20 °C.
3. Acrylamide is a potential neurotoxin. Hence wear gloves and mask while weighing and dissolving acrylamide.
4. Resolving and stacking gels have the same pH albeit the percentage acrylamide used will likely differ (depending on the size of the proteins to be resolved). The classic Laemmli's stacking gel pH is maintained at 6.8, which helps to focus the

proteins at the separating gel interface. However, the current protocol requires the stacking gel pH to be at 8.8. This modification prevents hydrolysis of the phosphohistidine signal that could occur at the more acidic pH of a conventional stacking gel.

5. SDS is an irritant. Wear gloves and mask while weighing. Prevent shaking or agitation which leads to inaccurate volume measurement due to frothing, instead dissolve SDS by keeping the buffer on hot plate with magnetic stirrer and slowly add SDS to it.
6. Ammonium persulfate solution should be filtered and stored at −20 °C as aliquots.
7. SDS-PAGE running buffer can be prepared as 10× native buffer without SDS and stored at 4 °C. Weigh 30 g of Tris base and 144 g of glycine and add 950 mL water before adjusting the pH to 8.5 using HCl. Make up the volume to 1000 mL. This 10× stock solution can be stored at 4 °C. To make a 1× solution, dilute 100 mL of the native 10× buffer into 890 mL of water kept 4 °C. Then add 10 mL of 10% (w/v) SDS solution to the buffer just before each experiment. SDS should be added at the end to prevent frothing and precipitation at 4 °C.
8. DTT is unstable in presence of metal ions, alkaline pH and higher temperatures. Addition of EDTA and storage at −20 °C improves the half-life of DTT in buffers. As SDS precipitates at lower temperatures, allow the sample buffer to reach room temperature before adding to the protein sample.
9. Nitrocellulose membrane can also be used for detecting pHis signals. 0.22 μM pore size membranes should be used for transferring smaller peptides and proteins below 15 kDa while 0.45 μM membranes can be used for proteins above 15 kDa.
10. The composition of the transfer buffer is similar to the SDS-PAGE running buffer except for the addition of 20% methanol. Prepare 1× buffer by diluting 100 mL of 10× native buffer into 690 mL of water kept at 4 °C. Add 10 mL of 10% (w/v) SDS solution along with 200 mL of methanol. The transfer buffer pH is not usually adjusted, but it is within the acceptable pH range for conserving pHis signal.
11. Casein can be replaced with serum free BSA as a blocking agent. Though casein is a phosphoprotein, no cross-reactivity with the pHis antibodies has been observed.
12. Semi-dry transfer is another popular western blotting technique used to transfer proteins from the gel to a membrane. Less heat convection occurs during the transfer, so semi-dry transfer is not used for immunoblotting pHis proteins.

13. There are other anti-pHis antibodies which also can be used for detection of phosphoproteins by immunoblotting which include SC50-3 (Purchase from Millipore, product number MABS1341) for 1-pHis detection and SC39–6 and SC56-2- (Purchase from Millipore, products numbers MABS1351 and MABS1352, respectively) for 3-pHis detection. Though the antibodies are sequence independent, SC44-1 appears to demonstrate a partial bias for 3-pHis on GHAGA sequences (Fig. 1) [8].
14. For the preparation of tissue lysates, homogenize tissue in a manner that minimizes freeze–thaw, such as grinding in liquid nitrogen, focusing on keeping mortar, pestle, and any apparatus cold by adding liquid nitrogen throughout the entire process. For instance, liver powder is resuspended in a lysis buffer containing 10 mM Tris base pH 8.8, 0.1% SDS, 1% sodium deoxycholate, 0.5 mM EDTA, and 150 mM NaCl, but other lysis buffers of pH 8–8.8 should be sufficient. Protease inhibitor tablets (Roche), PhosSTOP tablets (Roche) (*see* **Note 2**), and 1 mM PMSF are added fresh to lysis buffer and chilled prior to adding to tissue. Sonication must be performed on ice or maintained at 4 °C. Clear tissue lysates at 14,000 × g for 10 min. All other steps should follow the cell culture protocol outlined herein.
15. If the cells are retrieved for growth from liquid nitrogen, change the media after one day to remove DMSO.
16. Be gentle while washing adherent cells with the PBS otherwise cells will be washed off with the buffer.
17. Sonicator amplitude should be optimized for individual cell lines and tissue samples based on the volume and viscosity of the samples. As an alternative to sonication, the complete volume of lysate can be passed through a 23-G needle fitted to a syringe of sufficient volume. Repeat this process five times, keeping the lysate at a temperature of 4 °C (on ice or in a cold room).
18. DC assay (Detergent Compatible assay from Bio-Rad) can be used to estimate protein concentration in the lysate. Tween 20 present in the lysis buffer does not interfere with DC colorimetric assay. Serial dilutions of bovine serum albumin (BSA) can be used to generate a standard curve, based on which protein concentration in the lysate can be estimated.
19. It is imperative not to heat samples in sample buffer before loading on the gel to avoid pHis hydrolysis. If the samples are viscous, add small amounts of 8 M urea or 6 M guanidinium chloride (prepared in lysis buffer to maintain desirable pH) and vortex briefly before loading.

20. Layering isopropanol on top of the resolving gel speeds up the gel polymerization by preventing contact with oxygen. It can be replaced by isobutanol, 95% (v/v) ethanol or water.
21. Insert the comb into the stacking gel so that the depth of the well is not more than half of the stacking gel length. This ensures that there is sufficient room for the stacking of the proteins before entering the resolving gel. If air is trapped in the stacking gel while inserting the comb, add a small amount of stacking gel mixture to the gel cassette along the sides of the comb which will displace the trapped air.
22. After the gel solidifies keep it at 4 °C for at least 10 min before loading the protein.
23. Ensure that the pH of the acidified sample does not drop below pH 3 which may precipitate the proteins and make loading onto the SDS polyacrylamide gel difficult. Restore the pH of the acidified sample to pH 8.5 before boiling to avoid gel running issues. Heating should follow acidification for complete elimination of pHis signal.
24. Allow the dye front to just exit the gel, otherwise it stains the membrane.
25. Handle the PVDF membrane at the corners with tweezers only. If nitrocellulose membrane is used instead of PVDF for blotting, skip the methanol treatment step.
26. While placing the membrane over gel for transfer it is important to ensure that no air is trapped. Air bubbles can be removed by gently rolling a pipet over the membrane. Repeat this step after sandwiching with the Whatman paper and foam pads.
27. Electrophoretic transfer from the gel to the membrane at 30 V is highly recommended to prevent the buffer system from heating up and to conserve the pHis signals. If a higher voltage is required to reduce transfer time, the system should be maintained at 4 °C. To distribute the heat generated during electrophoretic transfer evenly across the tank and decrease localized effects of heat on the gel, place the gel tank on a magnetic stirrer with a magnetic bead inside the tank.
28. Membranes can be left in the primary antibody solution for up to 18 h at 4 °C.
29. Develop the membrane based on the secondary antibody conjugate. We observe consistent phosphohistidine signals across experiments with the LI-COR Odyssey Infrared imaging detection system.
30. Secondary antibodies that are used to detect phosphohistidine proteins and actin on the same blot should be tagged with two fluorophores with different emission wavelengths. **Steps 2–4**

(Subheading 3.4) can be carried out with a mixture of primary and secondary antibodies against phosphohistidine proteins and actin simultaneously, or can be done in tandem. However, if performed sequentially, the phosphohistidine blot should be developed first due to diminution of the signal.

References

1. Towbin H, Staehelin T, Gordon J (1979) Electrophoretic transfer of proteins from polyacrylamide gels to NC sheets: procedure and applications. Proc Natl Acad Sci U S A 76:4350–4354
2. Glenney JR, Zokas L, Kamps MP (1988) Monoclonal antibodies to phosphotyrosine. J Immunol Methods 109:277–285
3. Heffetz D, Fridkin M, Zick Y (1989) Antibodies directed against phosphothreonine residues as potent tools for studying protein phosphorylation. Eur J Biochem 182:343–348
4. Fujimuro M, Sawada H, Yokosawa H (1994) Production and characterization of monoclonal antibodies specific to multi-ubiquitin chains of polyubiquitinated proteins. FEBS Lett 349 (2):173–180
5. Kee JM, Muir TW (2012) Chasing phosphohistidine, an elusive sibling in the phosphoaminoacid family. ACS Chem Biol 7(1):44–51
6. Attwood PV, Piggott MJ, Zu XL, Besant PG (2007) Focus on phosphohistidine. Amino Acids 32(1):145–156
7. Wei YF, Matthews HR (1991) Identification of phosphohistidine in proteins and purification of protein-histidine kinases. Methods Enzymol 200:388–414
8. Fuhs SR, Meisenhelder J, Aslanian A, Ma L, Zagorska A, Stankova M et al (2015) Monoclonal 1- and 3-phosphohistidine antibodies: new tools to study histidine phosphorylation. Cell 162:198–210
9. Fuhs SR, Hunter T (2017) pHisphorylation: the emergence of histidine phosphorylation as a reversible regulatory modification. Curr Opin Cell Biol 45:8–16
10. Hindupur SK, Colombi M, Fuhs SR, Matter MS, Adam K (2018) The protein histidine phosphatase LHPP is a tumour suppressor. Nature 555(7698):678–682
11. Srivastava S, Panda S, Li Z, Fuhs SR, Hunter T, Thiele J (2016) Histidine phosphorylation relieves copper inhibition in the mammalian potassium channel KCa3.1. elife 5:pii:e16093

Chapter 13

Immunohistochemistry (IHC): Chromogenic Detection of 3-Phosphohistidine Proteins in Formaldehyde-Fixed, Frozen Mouse Liver Tissue Sections

Natalie Luhtala and Tony Hunter

Abstract

The development of antibodies that specifically detect histidine-phosphorylated proteins is a recent achievement and allows potential roles of histidine phosphorylated proteins in pathological and physiological conditions to be characterized. Immunohistochemical analyses enable the detection of proteins in tissues and can reveal alterations to the quantity and/or localization of these proteins through comparisons of normal and diseased specimens. However, the sensitivity of phosphohistidine modifications to phosphatases, acidic pH, and elevated temperatures poses unique challenges to the detection process and requires a protocol that bypasses traditional procedures utilizing paraffin-embedding and antigen-retrieval methods. Here, we detail a method for a brief fixation by 4% (v/v) paraformaldehyde on freshly collected tissues in the presence of PhosSTOP to block phosphatase activity, followed by a float on sucrose to protect the tissue prior to freezing. Specimens are then embedded in a cryopreservation medium in molds and frozen using an isoflurane, dry ice bath to best preserve the tissue morphology and phosphohistidine signal. We validate this technique in normal mouse liver using SC44-1, a monoclonal anti-3-pHis antibody used to uncover a role for a protein histidine phosphatase as a tumor suppressor in the liver. Furthermore, we demonstrate that the antibody signal can be eliminated by preincubating SC44-1 with a peptide treated with phosphoramidate to phosphorylate histidine residues. Thus, we present an IHC protocol suitable for specific detection of 3-phosphohistidine proteins in mouse liver tissue, and suggest that this can be used as a starting point for optimization of IHC using other phosphohistidine antibodies or in other tissue types, generating information that will enhance our understanding of phosphohistidine in models of disease.

Key words Immunohistochemistry, IHC, Phosphohistidine, pHis, Histidine phosphorylation, pHis antibodies, Histidine phosphorylation antibodies

1 Introduction

Given the recent, successful development of antibodies that specifically recognize protein histidine phosphorylation [1–4], and the reported roles for histidine phosphorylated (pHis) proteins in mammalian disease and physiology [5–11], it is of interest to adapt traditional immunohistochemistry (IHC) protocols to

Claire E. Eyers (ed.), *Histidine Phosphorylation: Methods and Protocols*, Methods in Molecular Biology, vol. 2077, https://doi.org/10.1007/978-1-4939-9884-5_13,

better understand where pHis modifications occur within tissues, and to determine whether there are disease-mediated phenotypic changes in pHis. Immunohistochemistry (IHC) is the process by which antibodies detecting specific proteins or protein modifications are applied to tissues to determine the localization of these factors amongst the varied cell types in a tissue environment. Although IHC can be used in combination with immunofluorescent (IF) detection methods, the workflow outlined here includes a secondary antibody conjugated to horseradish peroxidase (HRP) for the chromogenic detection of antigens through the conversion of 3,3′-diamino benzidine (DAB) to a brownish precipitate that remains permanently detectable on slides by light microscopy. Due to the longevity of the signal on slides, this protocol permits the comparison of tissues collected from multiple sources over time.

To our knowledge, there are no published studies using the recently developed pHis antibodies for IHC or for any microscopic analysis of tissue. However, Fuhs et al. used immunocytochemistry (ICC) with IF detection to study pHis protein localization in cultured cells [1]. For this study, cells were fixed by one of two methods, methanol or 4% (v/v) paraformaldehyde (PFA); boiling slides for 5 min in 100 mM citrate buffer served as a negative control, due to the acid and heat sensitivity of histidine phosphorylation. A detailed protocol for ICC analysis with IF detection is included in a separate chapter (*see* Chapter 14).

Consequently, in the development of an IHC protocol, we selected methanol and 4% (v/v) PFA as two methods for comparison in the fixation of fresh, frozen, mouse liver tissue. Both methods avoided the requirement for antigen retrieval which proved too harsh for pHis detection based on the Fuhs et al. analysis [1]. For both methods, pieces of liver from C57BL/6 mice, freshly isolated, were embedded in a cryopreservative medium in plastic molds and frozen on an isopentane–dry ice bath and stored at −80 °C prior to sectioning. For fixation by methanol, frozen sections (prepared from unfixed, frozen tissue) were fixed for 2 min after cryosectioning and processed for staining afterward. For the 4% (v/v) PFA fixation, pieces of liver, processed immediately after removing from dissected mice, were fixed on ice for 3.5 h in the presence of PhosSTOP phosphatase inhibitors, washed in TBS at 4 °C, then floated overnight at 4 °C on 30% (w/v) sucrose, and embedded in cryoprotectant and frozen in plastic molds in the morning. After developing slides in parallel for each technique, we only observed signal that could be blocked by coincubation of the monoclonal antibody (mAb) with histidine phosphorylated peptide for liver samples prepared by 4% (v/v) PFA fixation (in the presence of PhosSTOP), highlighting the importance of phosphatase inhibitors in preserving pHis signal in tissue. Thus, we present this protocol, which specifically detects

3-pHis using SC44-1 mAb, in formaldehyde-fixed, fresh, frozen mouse liver tissue, as a starting point for future studies of pHis protein localization in other tissues.

2 Materials

Use sterile, ultrapure water (MilliQ water is sufficient) and analytical grade reagents to prepare solutions. Equilibrate all solutions to the indicated temperatures prior to performing the experiment. Tissue, chemical and biohazard waste disposal must all be performed in accordance with local regulations. Specifically, any research involving animals or human subjects must be planned well in advance with protocols approved by the appropriate organizations according to your institution's rules.

2.1 Tissue Fixation and Cryopreservation in Sucrose

1. Tris-buffered saline (TBS; 10×): 1.5 M NaCl, 0.1 M Tris–HCl, pH 7.5. Dilute enough 10× solution to 1× in water for washing samples three times in 15 mL of TBS. Chill wash buffer to 4 °C.
2. Sucrose: 30% (w/v) in 1× TBS. Dissolve slowly with a stir bar in room temperature water. This can be prepared one day before the experiment; prepare at least 20 mL per sample. Chill to 4 °C.
3. Paraformaldehyde (PFA): 4% (v/v) PFA in TBS. Use ultrapure, electron microscopy grade PFA (such as 16% (w/v) formaldehyde, Polysciences, Inc., #18814). The day of fixation, in a chemical fume hood, dilute an entire 10 mL ampule of 16% (w/v) formaldehyde into a small glass bottle (with a cap) containing 4 mL 10× TBS and 26 mL of water (prechilled at 4 °C). Dissolve 1 tablet of PhosSTOP (Roche) for every 10 mL of solution (4 tablets for 40 mL) (*see* **Note 1**). Parafilm the top of the bottle or tube and store on ice in the chemical fume hood until use.
4. Labeled 15 mL conical tubes (Falcon) for each tissue sample that will be analyzed; prechill at 4 °C, then place on ice in the chemical fume hood.
5. 6-Well plates filled with TBS for washing individual tissue specimens.
6. Dissecting scissors and forceps.
7. Tissue specimen(s) (*see* **Note 2**).

2.2 Tissue Cryo Embedding

1. Disposable embedding molds, such as Polysciences, Inc., #18986. Label in advance for each sample.

2. Optimal cutting temperature (OCT) compound, such as TissueTek Optimal Cutting Temperature Compound (OCT), #4583.
3. Forceps: 4–5″ long for handling tissue; 10″ long if needed for securing molds while freezing.
4. P200 pipet tips to secure tissue in OCT.
5. Styrofoam container with dry ice for transporting cryo molds to storage in −80 °C.
6. Dry ice–isopentane (2-methylbutane, Fisher #O3551-4) bath (in Styrofoam container) for freezing embedded tissue in molds.
7. Ziploc bags sufficient to accommodate individual molds. Label in advance and freeze at −80 °C prior to making cryo molds.

2.3 Blocking of Antibody with pHis Peptide

1. Peptide to phosphorylate: 10 mg/mL in water, pH 7–9 (*see* **Note 3**).
2. Phosphoramidate: 81 mg/mL in water (*see* **Note 4**).
3. Diluent solution: 10 mM Tris–HCl, pH 8.8.
4. Nitrocellulose.
5. Dot blot blocking buffer: 0.2× Tris Buffered Saline (TBS) with 0.1% (w/v) casein.
6. TBST: 1× TBS containing 0.1% (v/v) Tween 20.
7. Anti-pHis antibodies: such as SC44-1 anti-3-pHis monoclonal rabbit antibody diluted to 0.5 μg/mL [1] (*see* **Note 5**).
8. Secondary antibody to detect the primary pHis antibody for dot blot (*see* **Note 6**).
9. TBST: 1× TBS containing 0.1% (v/v) Tween 20.
10. Blocking buffer: 5% (v/v) normal goat serum in TBST.
11. Rotator at room temperature that will accommodate 1.5 mL Eppendorf tubes.

2.4 Staining and Mounting of Coverslips to Cryosection Slides

1. Slide staining apparatus: staining racks and jars sufficient to process slides and complete washes.
2. Hydrophobic barrier pen for IHC.
3. Humid chamber.
4. TBST: 1× TBS containing 0.1% (v/v) Tween 20.
5. Blocking buffer: 5% (v/v) normal goat serum in TBST.
6. IHC detection reagent (HRP), such as SignalStain Boost, rabbit, CST #8114, a polymer-based HRP-conjugated antibody.
7. 3,3′-diaminobenzidine (DAB) peroxidase (HRP) substrate (e.g., ImmPACT (SK-4105, Vector Labs) prepared according

to manufacturer's instructions immediately prior to use (*see* **Note 7**).

8. Hematoxylin, such as Fisher #CA401-1D.
9. Slide staining jars that accommodate glass slide racks, prepared for serial processing of slides in the hood, and placed in this order: 50% (v/v) ethanol, 50% (v/v) ethanol, 75% (v/v) ethanol, 95% (v/v) ethanol, 100% (v/v) ethanol, 100% (v/v) ethanol, 100% (v/v) xylene, 100% (v/v) xylene, 100% (v/v) xylene. Ethanol reagents are diluted into ultrapure water (*see* **Note 8**).
10. Mounting media, such as Sigma DPX #06522 mounting media.
11. Coverslips of sufficient size to cover mounted tissue on slides (*see* **Note 9**).
12. 3 mL disposable plastic transfer pipettes, at least 2.
13. Cardboard of sufficient size to hold all slides for drying.

3 Methods

We recommend that immunoblots are performed on the tissue prior to IHC, using anti-pHis antibodies of interest (*see* Chapter 12). Immunoblots involve fewer steps from tissue isolation to lysis in phosphatase inhibitors, and results should be easier to interpret, clearly revealing the number and size of proteins detected, which will aid in interpreting IHC results. Immunoblotting can be used to probe a tissue model individually with multiple pHis antibodies to discern which, if any, produces a pattern of interest. If pHis proteins cannot be detected by immunoblot in the tissue of interest with the pHis antibody of choice, it is unlikely that IHC will be successful. Fig. 1 outlines the steps detailed below.

3.1 Tissue Fixation and Cryopreservation in Sucrose

1. Fill a sufficient number of 6-wells of a multiwell plate with TBS to rinse each piece of isolated tissue. Immediately after isolating fresh tissue, use dissecting forceps and scissors to cut a piece for processing (~3–5 mm thick, no larger than 1–2 cm across). Holding the tissue with forceps, immerse the piece in TBS to remove blood, fur, etc. from your sample.
2. Immediately place the rinsed tissue in a labeled, 15 mL conical tube containing 4% (v/v) PFA in TBS containing PhosSTOP. We use 5 mL of solution for pieces of liver tissue that are about 1 cm × 1 cm × 0.5 cm (0.5 cm^3) (*see* **Note 10**). Leave on ice in the chemical fume hood for 3–4 h.
3. When incubation is complete, discard PFA solution and, with forceps, transfer the pieces of tissue to a 15 mL conical tube containing 15 mL of prechilled TBS. Place on a rocker at 4 °C

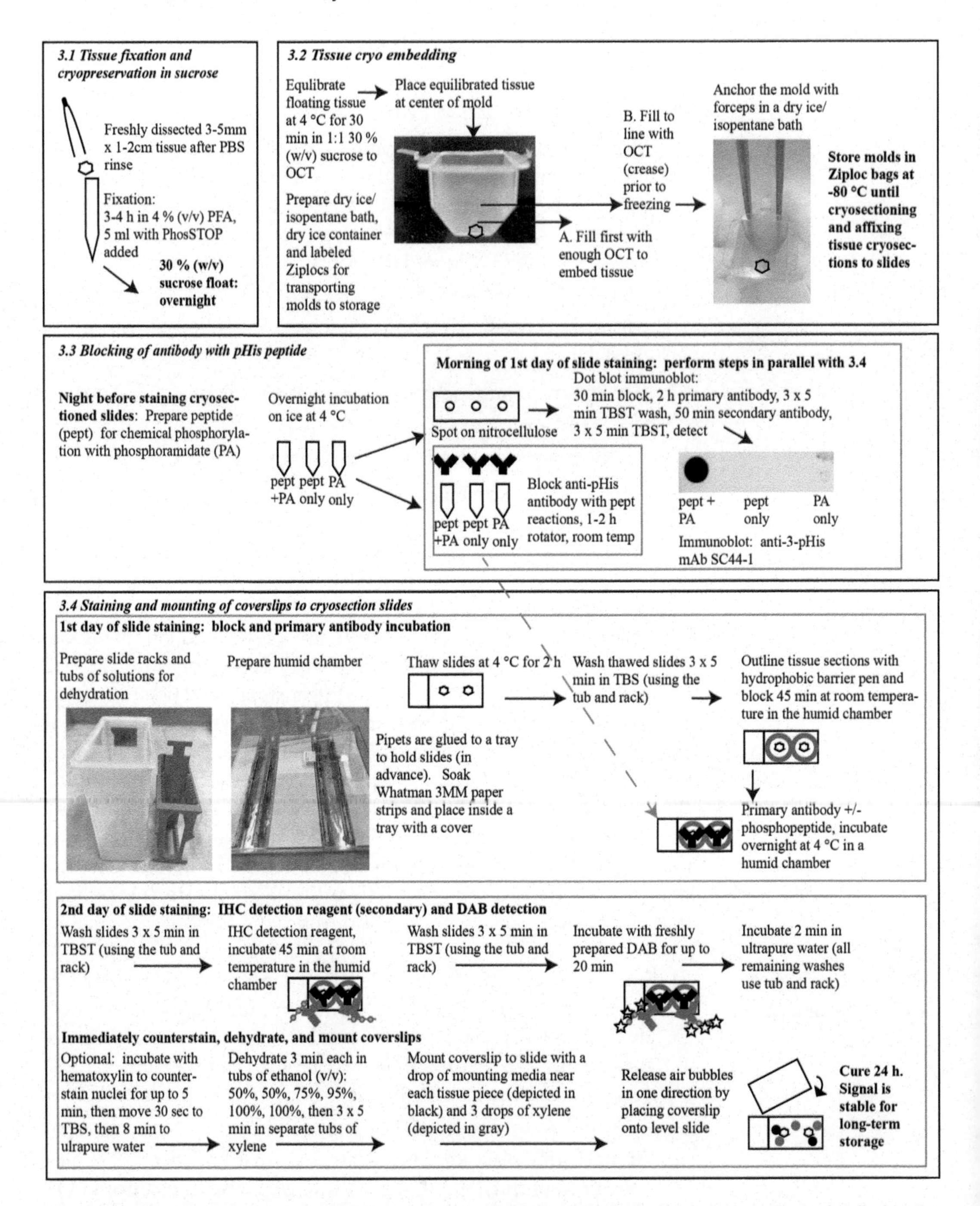

Fig. 1 Workflow for IHC. Steps and important timings are in bold; italic font refers to sections detailed in the text. Photographs and schematics are provided to better illustrate certain steps or materials

and wash for 15 min. Discard the buffer, and repeat the wash two more times.

4. After the final wash, discard the buffer, removing all residual buffer using a P1000 tip. Add 15 mL of prechilled 30% (w/v) sucrose in TBS. Invert the tube twice. The tissue will float on top. Cap the tube and place on ice at 4 °C overnight.
5. In the morning, invert the tube 2× again. The tissue should now have sunk to the bottom of the tube. If so, the sucrose has impregnated the tissue sufficiently to preserve it during the freezing process and you can proceed (*see* **Note 11**).

3.2 Tissue Cryo Embedding

1. Remove all but 1 mL of the sucrose solution from the 15 mL tube containing the tissue. Add 1 mL of OCT to the tube. Place on a rocker at 4 °C for 30 min.
2. While the tissue is equilibrating in OCT, prepare two Styrofoam containers with dry ice. One container must contain sufficient space to hold all of the cryo molds to transport these to the −80 °C freezer for storage. The other will contain dry ice for the isopentane bath, which can be used to freeze 2–3 molds at a time. Wearing safety glasses for protection, pour isopentane over the dry ice, wait until bubbling and vaporization of isopentane ceases, then repeat. Continue until there is enough volume of liquid isopentane to sufficiently cover the dry ice such that cryomolds can be balanced on top of the dry ice in the isopentane bath, with the liquid covering the base of the mold without inundating the sample in isopentane.
3. Once tissue pieces have equilibrated for 30 min in the OCT/-sucrose mixture, place a sufficient volume of OCT at the base of a labeled, disposable cryomold such that it can fully cover the tissue piece. Use a P200 pipette tip as a tool to lift the OCT media and wrap it in a circular motion around the piece of tissue. Take care to avoid the introduction of air bubbles. Gently press the tissue to the bottom, again avoiding bubbles (*see* **Note 12**). Carefully, squeeze out additional OCT from the bottle to slowly cover the tissue piece, filling the mold to the crease where the mold widens (*see* **Note 13**). The tissue should be entirely covered by OCT and without bubbles in proximity of the tissue.
4. Avoiding the introduction of bubbles, move the mold to the isopentane bath. Balance the mold carefully on the dry ice making sure that the bottom is level (*see* **Note 14**).
5. Continue processing samples and adding to the isopentane bath as space allows, but samples can be moved to the Styrofoam container containing dry ice once the OCT is opaque white.

6. Once all samples have been processed and placed on dry ice, store these at −80 °C in labeled Ziploc bags prior to cryosectioning and staining.
7. Prepare 5 μm sections on poly-L-lysine–coated glass slides using a cryostat (*see* **Note 15**).

3.3 Blocking of Antibody with pHis Peptide

1. One night before staining and mounting the cryosections onto slides, prepare a reaction to phosphorylate your chosen peptide overnight in phosphoramidate (*see* **Notes 16** and **17**). Label three 1.5 mL Eppendorf tubes: one will contain phosphoramidate (PA), one will serve as a control to contain unphosphorylated peptide and have water instead of PA, and one will contain PA and no peptide. Place these on ice.
2. Add 2.5 μL of 10 mg/mL peptide (or water for the PA only control) at pH 7–9 and 10 μL of 81 mg/mL PA (or water for the unphosphorylated peptide control) to the appropriate tubes on ice. This can be scaled up as needed for the experiment. The amount of PA may need to be increased to conserve the molar excess of PA if your peptide is much larger than 1 kDa (*see* **Note 18**).Incubate overnight on ice in the cold room to phosphorylate the peptide (*see* **Note 19**).
3. After incubation, remove 1 μL from each reaction and pipet as a spot onto a small piece of nitrocellulose labeled in pencil to indicate +PA, −PA, −peptide. Allow to dry on a flat, clean surface for 5 min. Holding the blot with forceps, blow compressed air on the nitrocellulose to complete the drying process, or allow the spots to dry at room temperature for longer.
4. Process the dot blot as an immunoblot using the anti-pHis antibody (*see* **Note 20**).
5. Incubate the dot blot for 30 min in dot blot blocking buffer.
6. Incubate the dot blot in primary antibody to detect phosphohistidine for 2 h.
7. Wash the dot blot for 3 × 5 min in TBST.
8. Incubate the dot blot for 50 min in secondary antibody conjugated with a substrate to detect the phosphohistidine antibody used (*see* **Note 21**).
9. Wash the dot blot for 3 × 5 min in TBST.
10. Detect the pHis signals. An example of a dot blot from a histidine phosphorylation experiment is presented in Fig. 1 (*see* also Chapter 12).
11. Dilute the remaining phosphorylated and unphosphorylated peptide reactions (to reduce the levels of PA present) by adding 189 μL of 10 mM Tris pH 8.8 to the remainder of the reaction (which should be about 11.5 μL).

12. Prepare three 1.5 mL Eppendorf tube containing anti-pHis antibody diluted to 2 μg/mL in 0.5 mL of blocking buffer and label for pHis peptide, unphosphorylated peptide control (should not block the antibody), and a no peptide control (to verify that the peptide itself does not interfere with the signal). Add 5 μL of the appropriate reactions (now diluted in 200 μL) to the pHis peptide and unphosphorylated peptide tubes. Reactions can be scaled up as needed for IHC staining.
13. Wrap Parafilm around the caps to seal, and place on a rotator at room temperature for 1–2 h. Remove Parafilm and place on ice (*see* **Note 22**).

3.4 Staining and Mounting of Coverslips to Cryosection Slides

If you do not already have solutions prepared to fill the tubs of xylene and ethanol needed in the dehydration and coverslip mounting steps, prepare these as indicated in Subheading 2.4.

1. Thaw cryosection slides at 4 °C for at least 2 h.
2. Prepare a humid chamber while slides thaw: soak Whatman 3MM paper in deionized water to cover the bottom of a tray (with a cover) sufficient to hold slides. Create an area for slides to sit that is not on the damp surface, either using pieces of Parafilm, or by place pieces of plastic pipettes vertically to sit at the left and right ends of slides and act as supports.
3. Wash slides at 4 °C with TBS for 3× 5 min each using a staining rack and tub.
4. Move slides to room temperature, leaving them on the rack in the final TBS wash. Remove slides one-by-one to avoid drying of the tissue on the slides. Tap the edges of the slide on a stack of dry paper towels to remove excess TBS, being cautious not to touch the pieces of tissue mounted on the front of slide.
5. Once residual liquid has been removed, outline the pieces of tissue (the shape of a circle or rectangle is usually best) with a hydrophobic barrier pen. Lay the slide in the humid chamber, either on a strip of Parafilm or on top of vertically placed pipets that hold the slides at the left and right sides. Add sufficient blocking buffer to cover the piece of tissue outlined by the hydrophobic barrier pen. This should form a bubble over the piece of tissue that will not dry during incubation (*see* **Note 23**).
6. Repeat **steps 4–5** for each individual slide until all slides have bubbles of blocking solution and are sitting in the humid chamber. Place a cover on the humid chamber and incubate for at least 45 min at room temperature, ensuring that it is not touched.
7. Add the primary antibody solutions that were prepared in Subheading 3.3: pHis peptide blocked, unphosphorylated

peptide control, and no peptide control. Working with slides one-by-one as in **step 4**, remove the blocking buffer by tapping the slide on a stack of paper towels, then place flat in the humid chamber. Add a sufficient volume to each tissue outline of slides to cover for an overnight incubation. Be sure to record which primary antibody solution is added to each tissue section. Incubate at 4 °C overnight with the humid chamber closed and ensure that it will not be touched or moved during the incubation.

8. In the morning, equilibrate to room temperature the IHC detection reagent (such as SignalStain Boost, rabbit) designed for the host species of your antibody. Move the humid chamber carefully to room temperature.
9. Using a tub that holds a slide staining rack, wash each slide in TBST for 3 × 5 min, using a sufficient volume of TBST to cover the slides placed in the tub. As in **step 4**, process slides one-by-one, tapping off the antibody and placing on the staining rack in the tub of TBST. Repeat this process until all slides have been moved to the rack immersed in the tub of TBST.
10. Leave slides on the rack in the tub with the final TBST wash. Process slides again as in **step 4**, one-by-one, to tap away residual liquid. Add sufficient IHC detection reagent to cover each tissue section (usually 100–300 μl), and place on the humid chamber as for the blocking buffer. Once all slides have been transferred to the humid chamber, cover and incubate for 45 min at room temperature, ensuring that the humid chamber is not touched during this incubation.
11. Wash the slides in TBST for 3 × 5 min each, as in **step 9**.
12. Remove residual TBST from slides by tapping the edges of each slide on a stack of paper towels, then place flat in the humid chamber. Process a few slides at a time, preferably those that are being directly compared, and add sufficient DAB peroxidase HRP substrate to cover the outlined tissue sections, working quickly to avoid drying of tissue sections. Using a timer, monitor the color change on the tissue sections as the substrate is converted to a brownish precipitate. This is usually observed by 10 min (*see* **Note 24**).
13. Once the color has developed to the required shade, place the slides on the slide staining rack in the tub of ultrapure water. Incubate for 2 min once all slides are on the rack.
14. Place the slide staining rack in a tub of hematoxylin and incubate for up to 5 min, to achieve the required shade, if you wish to counterstain nuclei.
15. Place the slide staining rack in TBS for 30 s then move to ultrapure water for 8 min.

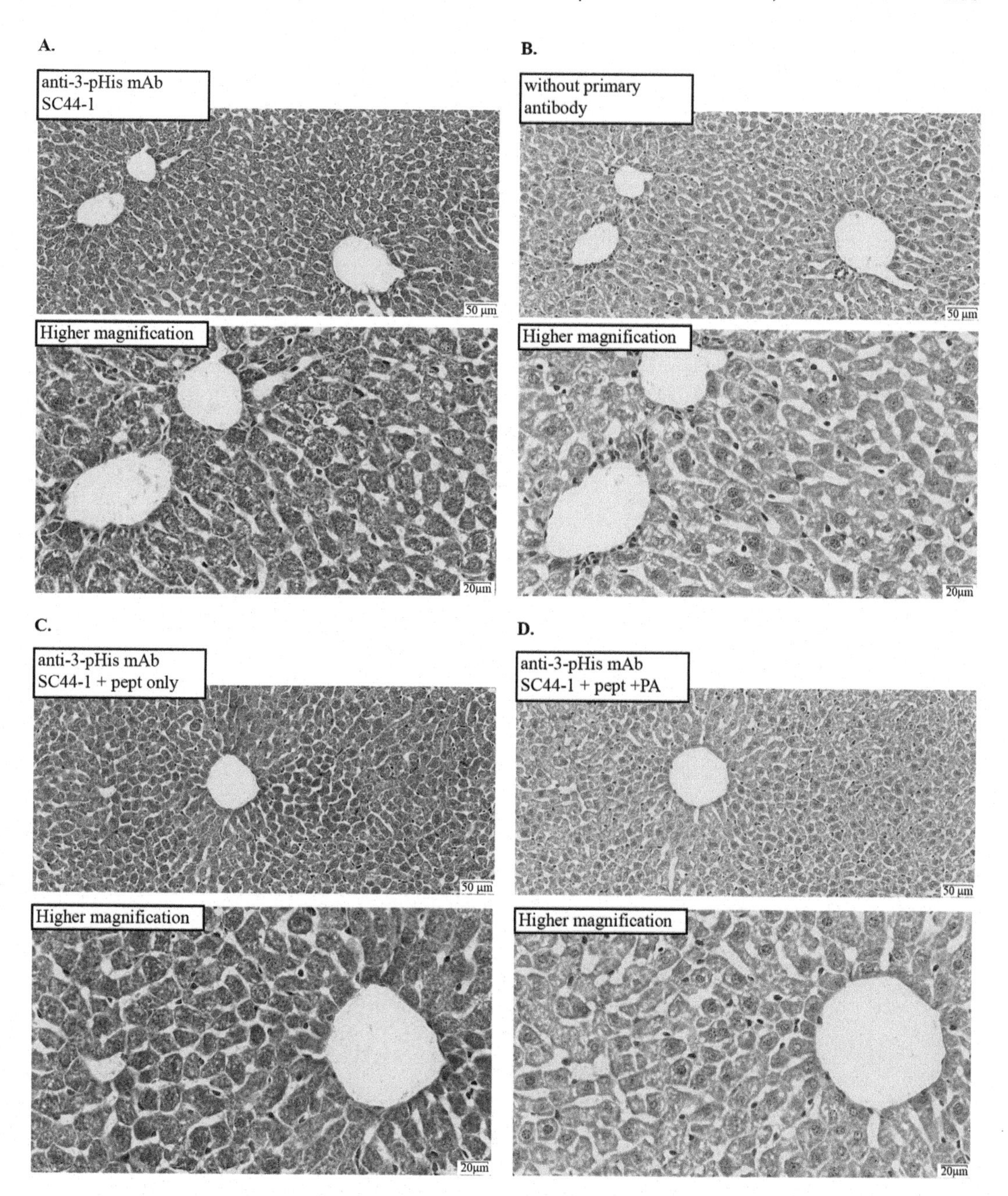

Fig. 2 Chromogenic detection of SC44-1 anti-3-phosphohistidine mAb signals in cryosections from formaldehyde-fixed, cryoprotected mouse liver. IHC images of slides for serial cryosections of C57BL/6 normal mouse liver. To best compare controls to experimental conditions, two separate slides were processed and analyzed: one was mounted with both (**a**) and (**b**) cryosections and another mounted with both (**c**) and (**d**) cryosections. Low magnification (top) and higher magnification (bottom) images with a scale bar are shown for each. Histidine phosphorylated peptides or controls were used in primary antibody incubations to probe phosphorylation-dependence of the IHC signal: (**a**) antibody (SC44-1) incubated without peptide (pept); (**b**) blocking buffer alone (no antibody) used in primary incubation; (**c**) antibody incubated with pept, untreated with phosphoramidate (PA); (**d**) antibody incubated with pept treated with PA (i.e., phosphorylated). Results support the phosphorylation dependence of IHC signal

16. Dehydrate slides by incubating the slide staining rack for 3 min each in tubs of ethanol of progressively higher percentages (v/v): 50%, 50%, 75%, 95%, 100%, 100%.
17. Continue dehydration using three tubs of xylene (working in a chemical fume hood), incubating the slide staining rack for 5 min in each. While washing, move a stack of paper towels, coverslips, mounting media, two 3 mL disposable transfer pipettes, and a piece of cardboard of suitable size for holding all of the mounted slides to the hood.
18. Leave the slide staining rack in the final xylene wash tub, and remove one slide at a time to mount coverslips. Tap the edges of the slide on a stack of paper towels. Holding the slide above the paper towels and level to a flat surface, add a couple of drops of mounting media to the slide (using the transfer pipette) on the face that contains the tissue section(s) (*see* **Note 25**). Add three drops of xylene from the final wash to the tissue section face of the slide while still holding it level. Mount the coverslip from one side to allow air bubbles to be released. Place the mounted slide on the cardboard in the hood, and leave for 24 h to cure.
19. View the cured slides by light microscopy. The slides can be stored in a slide storage box; the color is stable for long-term storage. Examples of results from our validation of SC44-1 anti-3-pHis signal in liver tissue are found in Fig. 2.

4 Notes

1. You can aliquot 10 mL of the 4% (v/v) PFA solution to a 15 mL conical tube (Falcon) placed on ice to reduce the number of PhosSTOP tablets used. PhosSTOP is needed to inhibit protein histidine phosphatase activity from enzymes such as PP1, PP2A, and PP2B [12], and possibly protein-tyrosine phosphatases.
2. Normal mouse C57BL/6 liver has been validated with this method, but other tissues can be optimized using this protocol as a starting point. Isolate fresh immediately prior to fixation.
3. We used the H_2N-Ala-Gly-Ala-Gly-His-Ala-Gly-Ala-Gly-NH_2 peptide used previously to validate pHis antibodies [1]. In selecting the sequence for a peptide to phosphorylate, it is important to note that phosphoramidate produces phosphoramidate (P-N) bonds on histidine, lysine and possibly on arginine residues [13]. For example, a study characterizing PHPT1 activity against lysine phosphoramidate bonds selected a histone H1 variant for incubation with phosphoramidate that did not contain histidine residues [14]. For this reason, in your

peptide sequence, you might choose to substitute lysine and possibly arginine residues within a sequence of interest containing phosphohistidine to uniquely analyze the contribution of the phosphohistidine.

4. This can be purchased or synthesized [15] in a chemical lab equipped and approved for this process.
5. This antibody was validated for IHC in normal mouse liver using the SC44-1 anti-3-pHis rabbit monoclonal antibody, which appears to demonstrate a partial bias for 3-pHis on GHAGA sequences [1].
6. Secondary antibody for the dot blot should be coupled to the appropriate substrate for the detection system desired, such as goat anti-rabbit IgG with Alexa 680 conjugate for Li-Cor Odyssey detection.
7. Prepare DAB HRP substrate during the final TBST wash after secondary antibody incubation (Subheading 3.4, **step 11**).
8. For 50% (v/v) ethanol, 95% (v/v) ethanol can be diluted into ultrapure water. For other ethanol dilutions, use absolute ethanol in the dilution. We reuse ethanol and xylene solutions for IHC staining, but do not use xylene from deparaffinization steps in other protocols in this protocol. Solutions used in tubs in the chemical fume hood during the protocol should be stored prior to and afterward in tightly capped containers to prevent evaporation and placed in an appropriate storage area for flammable liquids in accordance with regulations.
9. We used 22 mm × 50 mm No. 1 thickness cover glass from Fisher.
10. The volume of 4% (v/v) PFA needed for each piece of tissue will depend on its size. We recommend a volume of 10–20 times the tissue volume, but you might need to optimize this if your antibody fails to recognize overfixed antigens or if underfixation affects tissue morphology.
11. If the tissue does not sink, you can try a gradient of sucrose concentrations: 10% (w/v), 20% (w/v), and 30% (w/v), incubating the tissue at 4 °C until it sinks or for 24 h in each. However, there might not be any difference in the morphology even if the tissue does not sink, and it is unclear how longer incubation times might affect the pHis signal detected. We recommend analysing a portion of the tissue that does not sink in 30% (w/v) sucrose to decide whether the gradient is necessary. In the future, verify that your sucrose solution is freshly made at the appropriate concentration, use a larger volume of solution to tissue, and agitate periodically during the incubation to prevent the formation of a gradient.

12. The idea is to coat the surface of the tissue with OCT such that air bubbles are not trapped between the tissue and the OCT media. If it proves challenging to prevent the introduction of bubbles to the tissue, transfer the tissue from the 15 mL conical tube to a Petri dish containing enough OCT to cover the piece. Use a P200 tip to "wrap" the tissue in OCT and eliminate bubbles, then move the tissue to the bottom of the cryo mold.
13. The molds that we use are tapered at the bottom and broaden to the widest point, which is visible as a creased line around the mold (as shown in Fig. 1).
14. To balance the cryo molds in the isopentane–dry ice bath, you can use forceps that are longer than the depth of the bath. Insert these to stand as a support for the mold, arms down in the bath, placing the mold against the forceps' arms (as shown in Fig. 1). You will need more than one set of forceps if you have multiple molds to process and wish to progress rapidly. You can also simply hold the mold using forceps in the bath to a depth that immerses the portion of the mold containing the OCT-embedded tissue without allowing isopentane to touch the inside of the mold. This takes more time since only one sample is processed at a time but may better ensure that molds do not spill or become immersed in isopentane.
15. Due to the challenging nature of cryosectioning, we rely on the expertise of a core facility to prepare 5 μm sections on poly-L-lysine–coated glass slides using a cryostat. We typically have serial sections cut, with two sections mounted per slide. In this manner, we can make the following comparisons in parallel: anti-pHis antibody alone vs. no primary antibody and anti-pHis antibody blocked with peptide vs. anti-pHis antibody blocked with phosphorylated peptide.
16. If you are using anti-1-pHis antibodies in your study, rather than anti-3-pHis antibodies as we have in our optimization of the IHC method, you will likely need a different peptide control.
17. PA-phosphorylated peptides contain a mixture of the 1- and 3-pHis isomers. Due to the relatively greater thermodynamic instability of 1-pHis as compared to 3-pHis [1], this isomer is poorly retained in the final reaction that is used to block the anti-pHis antibodies and hence may not effectively block enough antibody to observe a decrease in IHC signal in validation experiments. Thus, if you are using anti-1-pHis antibodies for IHC, we recommend blocking the anti-1-pHis antibodies using peptides synthesized to incorporate a stable pHis mimetic (1-pTza) at the histidine residue rather than PA-phosphorylated peptides. This approach was utilized to

validate anti-pHis antibody staining (ICC/IF) in the Fuhs et al. publication [1].

18. These amounts represent a greater than 100-fold molar excess of phosphorylated peptide (≅700 Da, 0.6 μg added to blocking reaction) to immunoglobulin protein (IgG ≅ 150 kDa, 1 μg used in blocking reaction). It is important to conserve at least this ratio, particularly since peptides are not phosphorylated to 100%. If you use more or less antibody (i.e., if the 2 μg/mL concentration of antibody is not ideal for your samples), you can alter the amount of peptide that you prepare accordingly. This protocol utilizes a greater than 200-fold molar excess of PA to peptide. You can attempt to use less PA than this, but due to the instability of pHis, you might encounter difficulty in generating enough pHis peptide to sufficiently block the anti-pHis antibody in validation experiments.
19. You can also phosphorylate peptides using PA by incubating the reaction for 2 h at room temperature if you are unable to perform the reaction overnight.
20. It is best to process the dot blot rapidly in parallel with the remaining steps of the protocol and use the peptide reactions to block the anti-pHis antibody (**steps 11–13**, Subheading 3.3) prior to observing the dot blot results. It is possible to incubate the dot blot in primary antibody overnight while completing the protocol, skipping **steps 7–10** of Subheading 3.3, and completing these steps of the dot blot protocol the following day. However, the levels of pHis available to bind the anti-pHis antibody will likely be reduced and yield lower (possibly undetectable) signal on the dot blot. Nevertheless, even if the signal on your dot blot is weak, you may still have enough phosphorylated peptide to block the anti-pHis antibody.
21. Secondary antibodies conjugated to dye for Li-Cor Odyssey detection, or conjugated to HRP for chemiluminescent detection can be used.
22. If you observe background in peptide incubations with antibody, you can clear the antibody blocking reactions after the 1 h incubation by spinning at 10,000 × g for 5 min and using the supernatant.
23. It is useful to try the pen on a piece of paper towel before applying to the slide to ensure that the media is flowing well. Hold the slide with the tissue section facing you to apply the outline; placing it flat and outlining with the pen is likely to release too much media and smudge. Any discontinuity in the outline will allow your solutions to leak away from the tissue. It is worthwhile practicing ahead of time with the pen to prepare you for how to outline the tissue sections effectively.

24. To make appropriate comparisons, we examine the tissue treated with anti-pHis antibody alone first, incubating the other controls for at least as much time as this slide.
25. It is best to put the mounting media near but not on the tissue section(s), and to avoid introducing bubbles. Air bubbles on the tissue will pose problems in viewing the tissue by microscopy. This process is diagrammed in Fig. 1.

References

1. Fuhs SR, Meisenhelder J, Aslanian A, Ma L, Zagorska A, Stankova M, Binnie A, Al-Obeidi F, Mauger J, Lemke G, Yates JR 3rd, Hunter T (2015) Monoclonal 1- and 3-phosphohistidine antibodies: new tools to study histidine phosphorylation. Cell 162 (1):198–210. https://doi.org/10.1016/j.cell.2015.05.046
2. Kee JM, Oslund RC, Couvillon AD, Muir TW (2015) A second-generation phosphohistidine analog for production of phosphohistidine antibodies. Org Lett 17(2):187–189. https://doi.org/10.1021/ol503320p
3. Kee JM, Oslund RC, Perlman DH, Muir TW (2013) A pan-specific antibody for direct detection of protein histidine phosphorylation. Nat Chem Biol 9(7):416–421. https://doi.org/10.1038/nchembio.1259
4. Kee JM, Villani B, Carpenter LR, Muir TW (2010) Development of stable phosphohistidine analogues. J Am Chem Soc 132 (41):14327–14329. https://doi.org/10.1021/ja104393t
5. Hindupur SK, Colombi M, Fuhs SR, Matter MS, Guri Y, Adam K, Cornu M, Piscuoglio S, Ng CKY, Betz C, Liko D, Quagliata L, Moes S, Jenoe P, Terracciano LM, Heim MH, Hunter T, Hall MN (2018) The protein histidine phosphatase LHPP is a tumour suppressor. Nature 555(7698):678–682. https://doi.org/10.1038/nature26140
6. Cai X, Srivastava S, Surindran S, Li Z, Skolnik EY (2014) Regulation of the epithelial Ca(2)(+) channel TRPV5 by reversible histidine phosphorylation mediated by NDPK-B and PHPT1. Mol Biol Cell 25(8):1244–1250. https://doi.org/10.1091/mbc.E13-04-0180
7. Khan I, Steeg PS (2018) The relationship of NM23 (NME) metastasis suppressor histidine phosphorylation to its nucleoside diphosphate kinase, histidine protein kinase and motility suppression activities. Oncotarget 9 (12):10185–10202. https://doi.org/10.18632/oncotarget.23796
8. Panda S, Srivastava S, Li Z, Vaeth M, Fuhs SR, Hunter T, Skolnik EY (2016) Identification of PGAM5 as a mammalian protein histidine phosphatase that plays a central role to negatively regulate CD4(+) T cells. Mol Cell 63 (3):457–469. https://doi.org/10.1016/j.molcel.2016.06.021
9. Srivastava S, Li Z, Soomro I, Sun Y, Wang J, Bao L, Coetzee WA, Stanley CA, Li C, Skolnik EY (2018) Regulation of KATP channel trafficking in pancreatic beta-cells by protein histidine phosphorylation. Diabetes 67 (5):849–860. https://doi.org/10.2337/db17-1433
10. Srivastava S, Panda S, Li Z, Fuhs SR, Hunter T, Thiele DJ, Hubbard SR, Skolnik EY (2016) Histidine phosphorylation relieves copper inhibition in the mammalian potassium channel KCa3.1. eLife 5. https://doi.org/10.7554/eLife.16093
11. Srivastava S, Zhdanova O, Di L, Li Z, Albaqumi M, Wulff H, Skolnik EY (2008) Protein histidine phosphatase 1 negatively regulates CD4 T cells by inhibiting the K+ channel KCa3.1. Proc Natl Acad Sci U S A 105 (38):14442–14446. https://doi.org/10.1073/pnas.0803678105
12. Matthews HR, MacKintosh C (1995) Protein histidine phosphatase activity in rat liver and spinach leaves. FEBS Lett 364(1):51–54
13. Kowalewska K, Stefanowicz P, Ruman T, Fraczyk T, Rode W, Szewczuk Z (2010) Electron capture dissociation mass spectrometric analysis of lysine-phosphorylated peptides. Biosci Rep 30(6):433–443. https://doi.org/10.1042/BSR20090167
14. Ek P, Ek B, Zetterqvist O (2015) Phosphohistidine phosphatase 1 (PHPT1) also dephosphorylates phospholysine of chemically phosphorylated histone H1 and polylysine. Ups J Med Sci 120(1):20–27. https://doi.org/10.3109/03009734.2014.996720
15. Wei YF, Matthews HR (1991) Identification of phosphohistidine in proteins and purification of protein-histidine kinases. Methods Enzymol 200:388–414

Chapter 14

Subcellular Localization of Histidine Phosphorylated Proteins Through Indirect Immunofluorescence

Kevin Adam and Tony Hunter

Abstract

Immunofluorescence (IF) takes advantage of biological and physical mechanisms to identify proteins in cell or tissue samples, exploiting the specificity of antibodies and stimulated fluorescence light emission. Here, we describe an immunofluorescence staining method for the identification of histidine phosphorylated proteins that uses neutral/alkaline conditions and targeted reagents to overcome the chemical lability of histidine phosphorylation. This method describes how 1- and 3-phosphohistidine (pHis) monoclonal antibodies can be used to reveal the localization of proteins containing these elusive phosphoramidate bonds in cells. Standard procedures and materials for IF staining with adherent and nonadherent cells are described.

Key words Immunofluorescence, Fluorescence, Antibody, Phosphohistidine, Histidine phosphorylation, Immunostaining, Microscopy

1 Introduction

Immunofluorescence staining (IF) takes advantage of both the binding specificity of antibodies against specific epitopes and the selective visibility of different wavelengths through stimulated light emission (fluorescence) to localize the epitope-containing protein within cells or tissues. This powerful combination was described for the first time by Coons et al. almost 80 years ago [1]. An additional 20 years was required to develop this method practically, and demonstrate its utility for the detection and localisation of proteins in bacteria, for viral protein studies and for diagnosis of disease [2, 3]. Nowadays, IF is well-established as a standard procedure in clinical and research applications, to study and validate the localization of a wide range of proteins both outside and inside the cell [4, 5]. Critically, several methods for cell fixation and membrane permeabilization have been developed over the years to allow the labeling of proteins in a spatial context within cellular compartments, which is essential to visualize subcellular localization [6, 7].

Claire E. Eyers (ed.), *Histidine Phosphorylation: Methods and Protocols*, Methods in Molecular Biology, vol. 2077,
https://doi.org/10.1007/978-1-4939-9884-5_14,

Whereas immunohistochemistry (IHC) tissue staining methods generally exploit a color-producing reaction using an enzyme conjugated to antibodies, IF uses fluorophore-conjugated antibodies. This feature enables two different approaches to be taken for visualization—direct and indirect immunofluorescence, each with specific advantages and disadvantages. Direct IF is a "one-step" immunostaining procedure involving a fluorophore-conjugated antibody directly targeting its antigen. Commonly used in live cell flow cytometry, direct IF is a more simple procedure and requires a shorter incubation time. In contrast, indirect IF is a "two-step" procedure in which an unlabeled primary antibody targeting the antigen is detected by fluorophore-conjugated secondary antibodies that provide signal amplification and allow the use of different fluorophores.

A third combined IF approach can be envisaged, which uses both direct and indirect approaches simultaneously. This combined method can be used, for example, when an extracellular protein is labeled before permeabilization, and then an intracellular protein is labeled after permeabilization. However, the indirect approach remains the most commonly used; as well as permitting signal amplification, the use of different fluorophores increases flexibility so that costaining can be achieved using the same primary antibodies.

Irrespective of the approach employed, the existence of numerous fluorophores with different characteristics that can be conjugated to antibodies, using distinct excitation and emission wavelengths, is a huge advantage for colocalization of different targets/antigens [8]. Particularly because IF is compatible with imaging using confocal microscopy technologies, it permits detection of a defined fluorescent emission signal in a specific focal plane with higher-resolution than IHC, and also makes it possible to combine several successive focal planes to obtain a 3D localization [9].

Histidine phosphorylation has been known since 1962, when Boyer et al. observed for the first time the presence of a phosphorylated imidazole structure from bovine liver mitochondria [10]. Since then, it has become particularly well known due to its involvement in signal transduction in, for example, bacteria, due to autophosphorylation of transmembrane histidine kinases in two-component systems (TCSs) [11]. However, because of the acid lability and thermosensitivity of this posttranslational modification, it took over 50 years to obtain the first phosphohistidine-specific antibodies, which finally became possible through the use of stable pHis analogues [12, 13]. Here, we describe standard procedures for systematic IF studies [14], in adherent and nonadherent cells, adapted to be compatible with the detection of histidine phosphorylation using 1- and 3-pHis monoclonal antibodies for an indirect immunofluorescence approach [15, 16].

2 Materials

Prepare all solutions using ultrapure water (prepared by purifying deionized water, to attain a sensitivity of 18 MΩ cm at 25 °C). All the reagents and buffers are prepared and stored at room temperature (RT), unless otherwise specified in the method. Diligently follow all local waste disposal regulations when disposing of waste materials. We do not add sodium azide to antibody reagents.

2.1 Preparation of Coverslips

1. Coverslips, 15–16 mm diameter, ~170 μm thickness (*see* **Note 1**).
2. Sterilization solution: 70% MeOH, 1% (v/v) HCl.
3. Poly-L-lysine solution: Resuspend 1 mg of poly-L-lysine in 20 mL sterile tissue culture grade water to obtain a 50 μg/mL solution (*see* **Note 2**).
4. Two forceps or cotton pliers: one curved extremity, one flat extremity. Use the forceps to handle and manipulate coverslips (Fig. 1).
5. Tissue culture hood.
6. Twelve well plates.

2.2 Cell Culture and Preparation of Cells

1. Cell culture hood.
2. Cell incubator (*see* **Note 3**).
3. Cells undergoing analysis.
4. Cell culture medium appropriate for the cells being investigated.
5. Phosphate Buffered Saline solution stock (PBS 10×): 1.37 M sodium chloride (NaCl), 27 mM potassium chloride (KCl), 8 mM sodium phosphate dibasic (Na_2HPO_4), 14.7 mM sodium phosphate monobasic (NaH_2PO_4). PBS stock solution

a

b

Fig. 1 Manipulate the coverslips with forceps. (**a**) Hold the coverslips on one edge with the flat end forceps. (**b**) Remove the coverslips from the well by first lifting up with the curved end forceps in one hand, then pick up the coverslips with the flat end forceps in the other hand

is typically around pH 7.4. Dissolve 80 g NaCl, 2 g KCl, 1.14 g Na_2HPO_4, 1.76 g NaH_2PO_4 in 800 mL of H_2O and adjust pH to 7.4 with HCl at RT. Add water to 1 L. Autoclave and store at RT (*see* **Notes 4** and **5**).

6. PBS (pH 8.5): Add 50 mL of PBS 10× solution stock to 450 mL water. Mix and adjust pH to 8.5 with sodium hydroxide (*see* **Note 6**). Pre-cool and keep the solution at 4 °C.
7. Cyto-centrifuge (if required for certain types of suspension cells).

2.3 Cell Fixation, Permeabilization, and Blocking

1. PBS (pH 8.5): Add 50 mL of PBS 10× solution stock to 450 mL water. Mix and adjust pH to 8.5 with sodium hydroxide (*see* **Note 6**). Pre-cool and keep the solution at 4 °C.
2. Paraformaldehyde (PFA) 8% (v/v), pH 8.5: Prepare in a fume/chemical hood from pre-scored ampoules of 16% (v/v) PFA under inert gas, adding PBS volume to volume (*see* **Notes 7** and **8**). Check the pH and adjust to 8.5 if needed using NaOH (*see* **Note 6**). Aliquot before storing at −20 °C (*see* **Note 9**).
3. PBS (pH 4): Add 50 mL of PBS 10× solution stock to 450 mL water. Mix and adjust pH to 4 with formic acid (*see* **Note 6**). Keep at RT.
4. PBS (pH 8.5), 0.1% (v/v) Triton X-100: Add 50 μL Triton X-100 to 50 mL PBS, pH 8.5. Resuspend well by vortexing at RT (*see* **Note 10**).
5. Tris-buffered saline (TBS) (pH 8.5): 138 mM NaCl, 5 mM KCl, 0.7 mM Na_2HPO_4, 25 mM Trizma base, 0.9 mM $CaCl_2$, 0.5 mM $MgCl_2$. Alternatively, prepare a stock solution of 10× TBS and dilute in water.
6. TBS with 0.1% (v/v) Tween 20 (TBST) (pH 8.5): Add 1 mL Tween 20 to 1 L TBS 1× (*see* **Note 10**). Dissolve completely at RT and adjust the pH to 8.5 with ~250 μL NaOH (10 N).
7. TBST (pH 8.5), 10% (w/v) BSA: Resuspend 1 g of BSA in 10 mL of TBST. Vortex for a few minutes at RT, then keep at 4 °C.
8. TBST (pH 8.5), 1% BSA: Dilute 1 mL of TBST, 10% BSA in 9 mL of TBST. Vortex and keep at 4 °C.

2.4 Immunostaining

1. TBS with 0.1% (v/v) Tween 20 (TBST) (pH 8.5): Add 1 mL Tween 20 to 1 L TBS 1×. Dissolve completely at RT and adjust the pH to 8.5 with ~250 μL NaOH (10 N).
2. TBST (pH 8.5), 1% BSA: Dilute 1 mL of TBST, 10% BSA in 9 mL of TBST. Vortex and keep at 4 °C.
3. Purified anti-N1-pHis rabbit monoclonal antibody [15, 17], clone SC1-1 hybridoma purification (2.5 μg/mL final). A commercially available version of anti-N1-pHis rabbit

monoclonal antibodies, clone SC1-1 (#MABS1330, Millipore, 1:100 ~5 μg/mL) can also be used.

4. Purified anti-N3-pHis rabbit monoclonal antibody [15, 17], clone SC44-1 hybridoma purification (2.5 μg/mL final). Another commercially available clone, clone SC56-2 (#MABS1352, Millipore, 1:100 ~5 μg/mL) can also be used.
5. Rabbit IgG for negative control (5 mg/mL in PBS—1000×) (*see* **Note 11**).
6. Anti-rabbit Alexa Fluor 647 (2 mg/mL—1000×) (*see* **Note 12**).
7. Anti-rabbit Alexa Fluor 488 (2 mg/mL—1000×) (*see* **Note 12**).
8. Two forceps or cotton pliers: one curved extremity, one flat extremity.
9. Parafilm.
10. Benchtop centrifuge.
11. Kimwipes.
12. Pipette tip box (or something of a size suitable to cover the coverslips during incubation).
13. Twelve well plates.

2.5 Nuclei Staining, Mounting and Imaging

1. DAPI (4′,6-diamidino-2-phenylindole): Dissolve 10 mg DAPI in 1 mL water (~36 mM). Aliquot the stock solution and keep in the dark at −20 °C. We recommend preparing an intermediate dilution at 1 mg/mL, diluting 100 μL stock solution in 900 μL PBS (*see* **Note 13**). To obtain a solution of DAPI at 2.5 μg/mL, dissolve the 1 mg/mL solution 1:400 in TBST.
2. PBS (pH 8.5): Add 50 mL of PBS 10× solution stock to 450 mL water. Mix and adjust pH to 8.5 with sodium hydroxide. Precool and keep the solution at 4 °C.
3. Mounting media (*see* **Notes 14** and **15**).
4. Glass slides (*see* **Note 16**).
5. Microscope: We capture images from a super-resolution Zeiss LSM 880 rear port laser scanning confocal microscope with an Airyscan FAST module. If using other systems, a 40× objective is required as a minimum for a magnification of 400× (when used with a 10× eyepiece) essential for studying cells and cell structure. A light source with a power supply for laser excitation is required for immunofluorescence. Detectors and filters must be selected based on the emission spectra of the fluorochrome used. A computer and a microscope imaging software are necessary to acquire and analyze data.

3 Methods

It is worth noting that the concentration of the reagents and the incubation time may need to be adapted dependent on cell type and morphology.

3.1 Coating Coverslips with Poly-L-Lysine (Suspension Cells Only)

1. Sterilize the required number of coverslips ready for use by soaking each coverslip separately in sterilization solution (70% MeOH, 1% (v/v) HCl). Once soaked, they can be kept together in this solution until use (*see* **Note 17**). Alternatively, expose the coverslips to 30 min of UV light just before use or autoclave coverslips in a closed petri dish.
2. Allow the sterilization/storage solution to evaporate from the sterilized cover slides and/or cover glass (*see* **Note 17**).
3. In a tissue culture hood, coat the surface of the coverslips by immersion in 50 μg/mL of poly-L-lysine for 1 h (*see* **Note 18**).
4. Remove excess solution by aspiration and gently wash the surface of the coverslips once with sterile tissue culture grade water.
5. Allow the coverslips to dry for 2 h in the tissue culture hood before adding cells.

3.2 Tissue Culture and Preparation of Cells

Calculate the number of cells required for the experiment, to include two negative controls that can be used to assess nonspecific background staining and the pHis signal under degradative conditions (pH 4, 90 °C) (*see* **Note 19**). The decision tree in Fig. 2 will help you to define what cell preparation technique is the best dependent on cell type.

3.2.1 Adherent Cells

1. Plate cells in a tissue culture hood at a density of 10^4–10^5 cells/well directly on coverslips previously positioned in the well, as shown in Fig. 1a (*see* **Note 20**).
2. Grow cells in standard culture medium until about 50% confluency over 24 h.

3.2.2 Nonadherent Cells

1. Resuspend ~10^4–10^5 suspension cells in 100 μL cold PBS pH 8.5 in a tissue culture hood.
2. If cells are sensitive to turgescence (bigger roundish cells with important cytoplasmic compartment), they can be incubated on coated slides at 4 °C for 30–60 min. If suspension cells are flat with an important nuclear compartment, cytospin them onto poly-L-lysine–coated slides, using a cyto-centrifuge at 800 rpm for 2 min, then proceed immediately to cell fixation step.

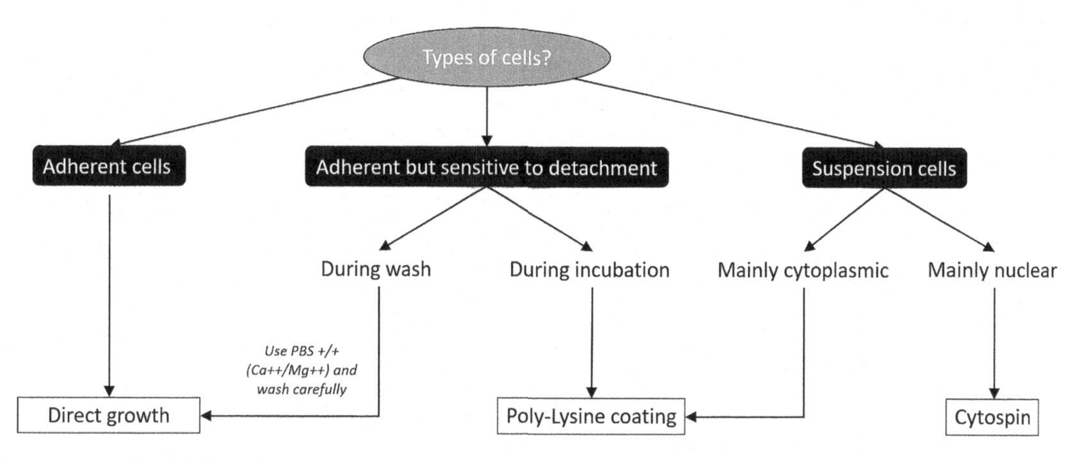

Fig. 2 Decision tree of cell preparation for IF. Adherence, sensitivity to detachment and morphology are used to define the optimal approach to mount cells for immunofluorescence

3.3 Cell Fixation

1. Place the plate with the cells on ice.
2. Remove the media from the cells by aspiration without directly touching the coverslips or cells. Do not let the cells dry out.
3. Wash the cells gently by addition of 1 mL cold sterile PBS, pH 8.5 and remove the solution by aspiration as **step 2**. Repeat once more.
4. Remove the plate from ice.
5. Cover the cells with PFA 8%, pH 8.5 for 15 min at RT (*see* **Note 21**). Check the pH of the new PFA solution before use as it decays to formic acid, which decreases the fixation efficiency (*see* **Note 22**). Do not exceed the 15 min incubation time as PFA is highly concentrated for "instant" fixation (*see* **Note 23**).

3.4 Cell Membrane Permeabilization and Blocking

1. Preincubate PBS pH 8.5 and PBS pH 4 to 4 and 90 °C respectively, to evaluate pHis-specific (acid- and heat-labile) signal. The recommended negative control without primary antibodies should be treated using conservative conditions (*see* **Note 12**).
2. After fixation, collect the PFA solution (*see* **Note 24**). Do not let the cells dry out.
3. Wash the cells 3× with 1 mL of the appropriate PBS solution, 5 min each time: cold PBS, pH 8.5 or warm PBS, pH 4 (*see* **Note 25**).
4. Permeabilize fixed cells by addition of 1 mL PBS (pH 8.5), 0.1% (v/v) Triton X-100 at RT for 15 min (*see* **Note 26**).
5. Wash 2× with 1 mL of the appropriate condition for PBS solution, 5 min each time.
6. Wash with 1 mL of TBST, 1% BSA at RT.
7. Block with 1 mL of TBST, 10% BSA at RT for 30 min (*see* **Note 27**).

3.5 Primary and Secondary Immunostaining

1. Dilute the required primary antibody (1- and 3-pHis mAbs; SC1-1 and SC44-1 respectively) in 0.1% TBST (pH 8.5) BSA 1% to 2.5 μg/mL (*see* **Note 28**). Just before use, centrifuge the diluted antibody for 5 min, 20,000 × *g* at 4 °C to remove any precipitate from the solution.
2. Cut a suitable size piece of Parafilm (~3 × 3 cm, depending on the number and diameter of the coverslips), place on a Kimwipes prewetted with water. Drop 10–25 μL of diluted antibody directly on the Parafilm surface for each coverslip.
3. Carefully transfer the coverslips to the Parafilm using both curved end and flat end forceps, one to easily separate the fragile slides and a second to lift the coverslip without breaking it (Fig. 1b). Be sure to turn the coverslips' top side (with cells on its surface) onto the droplet with the diluted antibody (*see* **Note 29**).
4. Cover the Parafilm and humid paper with a box (e.g., pipette tip box or bigger) to create a humid environment and limit solution evaporation. Incubate with the primary antibodies at RT for 60 min.
5. Using the curved end and flat end forceps, place the coverslips into a new plate well. Be careful to turn the coverslips to keep the cells on top. Wash immediately with 1 mL of TBST. Repeat the 1 mL wash three times, for 5 min each.
6. Dilute each secondary antibody 1/1000 in TBST, 1% (w/v) BSA to 1 μL/mL. Protect from light and centrifuge the diluted antibodies before use for 5 min at 20,000 × *g* at 4 °C to avoid any precipitate in the solution (*see* **Note 30**).
7. Add 1 mL of secondary antibody solution per coverslip and incubate at RT for 60 min in the dark.
8. Remove the solution and immediately wash cells with 1 mL of TBST at RT. Repeat the 1 mL wash three times, for 5 min each.

3.6 Nuclei Staining and Mounting

1. Just before use dilute DAPI to ~2.5 μg/mL (1:400) in 0.1% TBST (pH 8.5) (*see* **Note 31**).
2. Add solution to the cells on the coverslip and leave for 5 min.
3. Remove the staining solution and wash with PBS (pH 8.5) for 5 min. Repeat 2×.
4. Meanwhile, label the cover glass with relevant identifiable information (e.g., cell name, experimental condition, fluorochrome, date).
5. Drop 10 μL of mounting media on the cover glass for each coverslip.
6. Carefully transfer the coverslip on cover glass using the curved end and flat end forceps, as mentioned earlier (*see* **Note 29**).

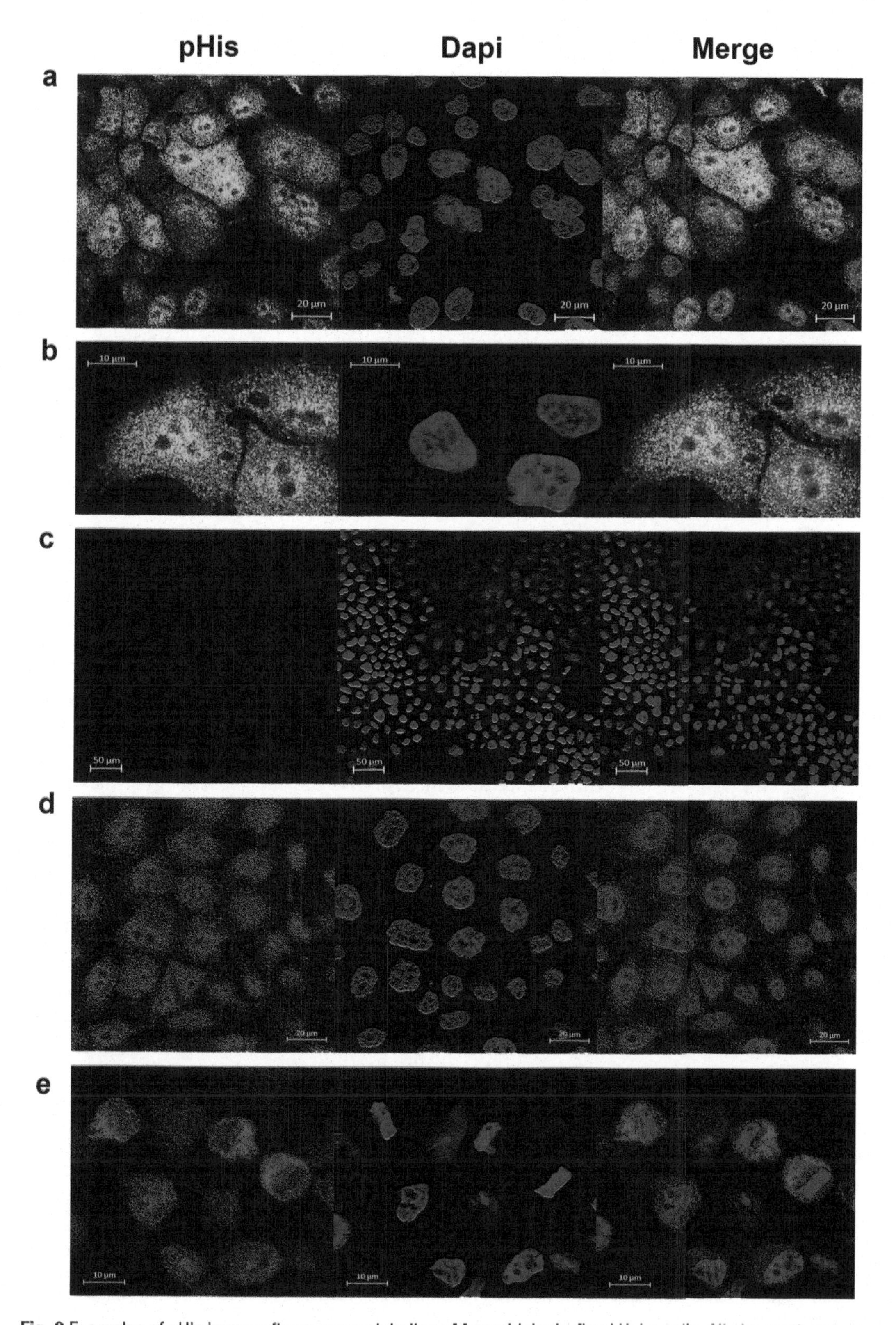

Fig. 3 Examples of pHis immunofluorescence labeling of formaldehyde-fixed HeLa cells. All observations were made using super-resolution microscopy on a Zeiss LSM 880 rear port laser scanning confocal microscope

Be careful to turn the coverslip's top side with cells onto the mounting media droplet avoiding any bubbles.

7. Let the coverslip air-dry in the dark for 5–10 min. Store in an opaque box on a flat surface at 4 °C overnight prior to use the confocal microscopy (*see* **Note 32**).

3.7 Imaging

1. Bring the slides to room temperature by leaving on a bench prior to scanning on a confocal microscope (Fig. 3).
2. Follow the microscope's instructions for more details about the use of a confocal system and its parameters.
3. Focus the cell image using the lower magnification (20×) with light exposure prior switching to the excitation laser (*see* **Note 33**).
4. Re-generate the cell image using a higher magnification (60/63×, oil immersion) to get a better resolution (*see* **Note 34**).

4 Notes

1. Round coverslips fit better in 12-well plates. A diameter of 12 mm is sufficient and small enough to limit the volume of antibody needed, whereas a 22 mm permits more cells to be mounted on the surface, increasing the choice of cellular area that can be observed. We would recommend 15–16 mm as a good compromise. The thickness of the coverslip can impact the quality and intensity of the image; 170 μm thickness is standardly compatible with all lens magnifications.
2. Poly-L-lysine is polycationic and interacts with anionic surfaces of cell membranes to facilitate adherence. It exists in either D or L chirality: poly-L-lysine (natural form) and poly-D-lysine (artificial form). Artificial poly-D-lysine is more resistant to enzymatic degradation compared to poly-L-lysine, but a preference can be cell dependent.
3. A standard cell culture incubator with 5–10% CO_2, 37 °C and sufficient humidity is fine for mammalian cells. Naturally, these parameters need to be adapted as a function of the cell type and experimental conditions.

Fig. 3 (continued) with an Airyscan FAST module. (**a, b**) Two acquisitions of HeLa cells (scale bar respectively at 20 and 10 μm) labeled with both 1-pHis mAbs and 3-pHis mAbs in green (488) and DAPI in blue (objective 63×, oil index 30). (**c**) One acquisition of HeLa cells (scale bar at 50 μm) labeled with rabbit IgG control in green (488) and DAPI in blue (objective 20×). (**d, e**) Two acquisitions of HeLa cells (scale bar respectively at 20 and 10 μm) labeled with both 1-pHis mAbs and 3-pHis mAbs in red (647) and DAPI in blue (objective 63×, oil index 30)

4. When making solutions, the pH is measured using a pH meter. pH stability after a period of storage can be checked using pH paper.
5. PBS is preferentially used without added Ca^{2+} and Mg^{2+} (PBS−/−), except if you are using adherent cells that are sensitive to detachment. Some pHis signaling has been related to ion channel activity. In order to limit the impact of PBS on ion channel function, we recommend avoiding the addition of Ca^{2+} and Mg^{2+} unless absolutely necessary. Should detachment of adherent cells be overserved, add 0.9 mM calcium chloride and 0.5 mM magnesium chloride to 1× PBS buffer to improve adhesion.
6. Adjust the pH with adequate buffering solution: sodium hydroxide 10 N (NaOH) to increase the pH value and formic acid 30% to decrease the pH. A few drops can be used initially to narrow the gap from the starting pH to the required pH. From then on, it is better to use a series of NaOH (e.g., 5 and 1 N) and formic acid (e.g., 10% and 5%) with lower ionic strengths to avoid a sudden increase/decrease in pH.
7. Paraformaldehyde used for fixation is free of methanol but complemented with glutaraldehyde and acrolein fixatives, more appropriate for preserving cellular structure and biomolecular distribution [6].
8. Select only PFA in pre-scored amber ampules under inert gas. This is important to protect the solution from both air oxidation and light as free formaldehyde oxidizes to formic acid with oxygen.
9. Aliquots frozen at −20 °C are stable for years. Do not re-freeze a thawed aliquot but keep at 4 °C if you plan to use it within a week. It is essential to use fresh PFA (a new aliquot or one that is less than 3–4 weeks old when stored at 4 °C) because PFA breaks down in solution (free formaldehyde oxidizes to formic acid with oxygen). It is important not to warm-up or thaw PFA at temperatures above 60 °C because it decreases the fixation efficiency.
10. Due to the viscosity of Triton X-100 and Tween 20, it can be challenging to obtain an accurate volume. Cutting about one third off the pipette tip with scissors makes pipetting easier, as the opening of the tip is larger. To be more accurate, Triton X-100 or Tween 20 can also be weighed to obtain the correct volume: the density is 1.065 and 1.095 g/mL, respectively.
11. Rabbit IgGs are used as a negative control, not only to check for nonspecific background but also to evaluate washing efficiency.

12. Fixation procedures (using PFA or formalin fixatives) cause autofluorescence in the green spectrum. We recommend using a fluorophore in the near-infrared range if you have an infrared detection system and/or to have a control condition to determine nonspecific background signal without using pHis primary antibodies, but rabbit IgG and secondary only.
13. DAPI is a known mutagen and should be handled with care. It has poor solubility in water; sonicate as necessary to dissolve. The 10 mg/mL DAPI stock solution may be stored at −20 °C for years. The intermediate diluted stock at 1 mg/mL can be stored at 4 °C for up to 6 months or at −20 °C for years.
14. Mounting media with DAPI can be used but is optional. We recommend using a mounting media without DAPI to keep the freedom of choice of different nuclear staining strategies or to limit potential background if you do not need to stain the nucleus.
15. Any standard mounting media can be used (e.g., Fluoromount-G from Southern Biotech, Vectashield from Vector Laboratories, Prolong Gold from Molecular Probes, Glycerol 90% (v/v), Mowiol 4-88 10% (w/v)). However, we recommend using mounting media with antifading, such as paraphenylenediamine, to maintain fluorescence and suppress quenching.
16. We do not recommend plastic slides but instead Quartz (UV transparency) or borosilicate glass (UV absorption), optically clear, for a better thermal and corrosion resistance.
17. Let the MeOH/HCl evaporate totally from the surface before using the coverslip/cover glass. Do not touch the coverslip with your fingers but use the forceps. Coverslips are fragile and must be manipulated with care. The sterile inside layer of the multiwell plate's lid can be used to position the coverslips for drying in a tissue culture hood.
18. If the coverslips are to be coated immediately prior to cell culture, coat the coverslips in the well plate designed to grow cells, then add 0.1–1 mL poly-L-lysine to cover the top side, where the cells will attach. Incubate, wash and dry the coverslips as described, then add cells in medium. If the coverslips are coated in advance, use a petri dish with 1 mL poly-L-lysine, immerse the coverslips in solution, then incubate, wash and dry as described; maintain at 4 °C.
19. Other negative controls can be envisaged, such as incubating with a specific pHis phosphatase, incubating a nonhydrolyzable pHis analogue peptide [15], or a histidine-phosphorylated peptide (*see* Chapter 13 on IHC) with the primary antibody prior to using it for IF. Degradation (hydrolysis) can also be induced by boiling the coverslips for 5–10 min in 0.01 M

citrate buffer (pH 3–5), which is generally used for heat-induced antigen retrieval. Citrate-based solutions are designed to break the protein cross-links; therefore, we recommend keeping the same buffer as the conservative condition-treated sample (not treated with citrate), such that only the effects of pH and temperature are being evaluated. In parallel, it is recommended that the global pHis signal is checked from the cell lines of your choice by immunoblotting prior any IF experiments (*see* Chapter 12).

20. If you plan to transfect the cells and select with antibiotics, it is recommended that you start with higher cell numbers to account for transfection efficiency.
21. If cells reveal significant autofluorescence, alternative compatible fixation methods can be used. Instead of PFA, methanol fixation can be performed by incubating cells at −20 °C for 15 min in cold 100% methanol. Methanol fixation is useful to limit pHis hydrolysis, but we do not recommend it in the first instance as it is generally used for cytoskeletal protein staining and cannot be adapted for staining lipid-associated proteins (hydrophobic bonds), or proteins localized in the nucleus or mitochondria because of a flattening effect (crenation) [18]. A photobleaching pretreatment step can also be implemented prior to immunostaining using a white phosphor light emitting diode (LED) to eliminate the natural background from tissue or due to aldehyde fixation processes [19].
22. Alkaline conditions used to help maintain stability of pHis are also known to improve fixation, as PFA decomposes faster at basic pH releasing formaldehyde which then fixes the sample. It also has been validated for in situ hybridization and immunohistochemistry using sodium phosphate and sodium borate [20, 21].
23. PFA results in chemical crosslinking of free amino groups which better preserves cellular architecture. Standardly, 2–4% of PFA is sufficient to fix cells, but higher concentrations (up to 25%) are used for “instant fixation.” Even if it has been reported that cross-linking is essential for proteome-wide localization studies [14], we still recommend using 8% PFA “instant fixation” to limit the over-crosslink of biomolecules because of increasing autofluorescence.
24. Collect the PFA solution and subsequent washes in a 50 mL Falcon tube for easy disposal of the PFA waste materials according to local regulations.
25. Alternatively, glycine 0.1 M (3.5 g/L in the appropriate PBS solution) can be used for 5 min instead of the first wash, to react with excess formaldehyde and stop fixation.

26. Depending on the need, other detergents like saponin can be used instead of Triton X-100 to increase ER signals. However, unlike Triton-X and Tween 20 that are nonselective and create stable pores in the membrane by interacting with both protein and lipids, saponin will not permeabilize nuclear membranes as it only removes membrane cholesterol. Furthermore, saponin can be totally washed out during the staining steps and permeabilization decreases if the saponin levels are not maintained in the subsequent washing buffer [7].
27. BSA (Bovine Serum Albumin) is typically used for blocking nonspecific binding of nonantigen molecule. Technically, milk or any serum can be used, but note that the serum has to be from a different species than the species in which the primary antibody was raised.
28. pHis antibodies for IF were used at 2.5 μg/mL but the concentration will need to be tested and adapted as a function of the cells/tissues used. When staining for multiple proteins or phosphorylation isomers, the primary antibodies can be combined. However, note that if you plan to distinguish them, the primary antibodies have to be derived from different species and distinct from the species being studied in order to avoid nonspecific background.
29. To limit the suction effect on coverslips and prevent the cells from drying, do not remove the previous solution. Instead, incline the plate slightly to catch sight of the edge of the coverslips and position the curved end forceps extremity. Once the coverslips hold out of the plate, absorb by capillary action the excess solution maintaining the coverslip vertically and the edge perpendicularly in contact with absorbing paper (Kimwipes) to limit extra-dilution of antibodies.
30. Be careful to limit light exposure as much as possible, keeping the samples in the dark during incubation times to avoid bleaching the fluorochrome. Use aluminum foil or an opaque box to protect from the light.
31. DAPI works at a concentrations range of 0.1–10 μg/mL. If the purpose is to visualize the chromatin compaction, do not exceed 1 μg/mL to distinguish chromocenters, interchromatin space, heterochromatin and euchromatin. If DAPI is used as an internal control for cell localization, a higher concentration is recommended to reduce incubation time for staining and laser intensity required for imaging, which will decrease bleaching of higher wavelength fluorophores. Hoechst (33342) can be used instead of DAPI.
32. Alternatively, the edges of the coverslip can be sealed with a small amount of clear nail polish. After drying, the slides can be kept at −20 °C to conserve fluorescence until imaging.

33. It is not recommended that DAPI is exposed first due to potential bleaching of signal. Always start from the higher wavelengths and decrease to short wavelengths.
34. Mounting media not only preserves the fluorescence but also increases the refractive index, allowing for oil immersion lens to be used for high-quality pictures.

Acknowledgments

This work was supported by the Waitt Advanced Biophotonics Core Facility of the Salk Institute with funding from NIH-NCI CCSG: P30 014195 and the Waitt Foundation.

References

1. Coons AH, Creech HJ, Jones RN (1941) Immunological properties of an antibody containing a fluorescent group. Proc Soc Exp Biol Med 47:200–202. https://doi.org/10.3181/00379727-47-13084P
2. Beutner EH (1961) Immunofluorescent staining: the fluorescent antibody method. Bacteriol Rev 25:49–76
3. Borek F (1961) The fluorescent antibody method in medical and biological research. Bull World Health Organ 24:249–256
4. Donaldson JG (2015) Immunofluorescence staining. Curr Protoc Cell Biol 69:4.3.1–4.3.7. https://doi.org/10.1002/0471143030.cb0403s69
5. Joshi S, Yu D (2017) Immunofluorescence. In: Basic science methods for clinical researchers. Elsevier, Amsterdam, pp 135–150
6. Hobro AJ, Smith NI (2017) An evaluation of fixation methods: spatial and compositional cellular changes observed by Raman imaging. Vib Spectrosc 91:31–45. https://doi.org/10.1016/j.vibspec.2016.10.012
7. Jamur MC, Oliver C (2010) Permeabilisation of cell membranes. Methods Mol Biol Clifton NJ 588:63–66. https://doi.org/10.1007/978-1-59745-324-0_9
8. Hermanson GT (2013) Fluorescent probes. In: Hermanson GT (ed) Bioconjugate techniques, 3rd edn. Academic, Boston, MA, pp 395–463
9. Sanderson MJ, Smith I, Parker I, Bootman MD (2014) Fluorescence microscopy. Cold Spring Harb Protoc 2014:pdb.top071795. https://doi.org/10.1101/pdb.top071795
10. Boyer PD, Deluca M, Ebner KE et al (1962) Identification of phosphohistidine in digests from a probable intermediate of oxidative phosphorylation. J Biol Chem 237: PC3306–PC3308
11. Stock AM, Robinson VL, Goudreau PN (2000) Two-component signal transduction. Annu Rev Biochem 69:183–215. https://doi.org/10.1146/annurev.biochem.69.1.183
12. Adam K, Hunter T (2018) Histidine kinases and the missing phosphoproteome from prokaryotes to eukaryotes. Lab Investig J Tech Methods Pathol 98:233–247. https://doi.org/10.1038/labinvest.2017.118
13. Kee J-M, Villani B, Carpenter LR, Muir TW (2010) Development of stable phosphohistidine analogues. J Am Chem Soc 132:14327–14329. https://doi.org/10.1021/ja104393t
14. Stadler C, Skogs M, Brismar H et al (2010) A single fixation protocol for proteome-wide immunofluorescence localisation studies. J Proteome 73:1067–1078. https://doi.org/10.1016/j.jprot.2009.10.012
15. Fuhs SR, Meisenhelder J, Aslanian A et al (2015) Monoclonal 1- and 3-phosphohistidine antibodies: new tools to study histidine phosphorylation. Cell 162:198–210. https://doi.org/10.1016/j.cell.2015.05.046
16. Khan I, Steeg PS (2018) The relationship of NM23 (NME) metastasis suppressor histidine phosphorylation to its nucleoside diphosphate kinase, histidine protein kinase and motility suppression activities. Oncotarget 9:10185–10202. https://doi.org/10.18632/oncotarget.23796
17. Hindupur SK, Colombi M, Fuhs SR et al (2018) The protein histidine phosphatase LHPP is a tumour suppressor. Nature

555:678–682. https://doi.org/10.1038/nature26140

18. Sonmez M, Ince HY, Yalcin O et al (2013) The effect of alcohols on red blood cell mechanical properties and membrane fluidity depends on their molecular size. PLoS One 8:e76579. https://doi.org/10.1371/journal.pone.0076579
19. Sun Y, Ip P, Chakrabartty A (2017) Simple elimination of background fluorescence in formalin-fixed human brain tissue for immunofluorescence microscopy. J Vis Exp. https://doi.org/10.3791/56188
20. Basyuk E, Bertrand E, Journot L (2000) Alkaline fixation drastically improves the signal of in situ hybridization. Nucleic Acids Res 28:e46
21. Berod A, Hartman BK, Pujol JF (1981) Importance of fixation in immunohistochemistry: use of formaldehyde solutions at variable pH for the localisation of tyrosine hydroxylase. J Histochem Cytochem 29:844–850. https://doi.org/10.1177/29.7.6167611

Chapter 15

High-Throughput Characterization of Histidine Phosphorylation Sites Using UPAX and Tandem Mass Spectrometry

Gemma Hardman and Claire E. Eyers

Abstract

Liquid chromatography (LC)-tandem mass spectrometry (MS/MS) is key for the characterization of phosphorylation sites in a high-throughput manner, and its application has proven essential to elucidate the phosphoproteome of many biological systems. Following proteolytic digestion of proteins extracted from tissues or cells, phosphopeptides are typically enriched by affinity chromatography using TiO_2 or metal-ions (*e.g.*, Fe^{3+}) coupled to solid-phase materials, prior to LC-MS/MS analysis. Separation of relatively low abundance phosphopeptides from nonphosphorylated peptides in these types of extremely complex mixtures is essential to maximize coverage of the phosphoproteome. Maintaining acidic conditions during these IMAC or TiO_2-based enrichment minimizes the concurrent unwanted binding of highly acidic peptides. However, while peptides containing phosphomonoesters, namely, phosphoserine (pSer), phosphothreonine (pThr), and phosphotyrosine (pTyr), are stable under these acidic binding conditions, phosphopeptides containing acid-labile phosphate group such as phosphohistidine (pHis), are not. Consequently, hydrolysis of these types of phosphopeptides occurs during standard phosphopeptide enrichment, and subsequent phosphosite identification by LC-MS/MS is severely compromised. Here we describe UPAX, unbiased phosphopeptide enrichment using strong anion exchange, for the separation of both acid-stable (pSer, pThr, pTyr) and acid-labile phosphopeptides (including those containing pHis) from nonphosphorylated peptides. We outline how implementation of UPAX prior to a minimally modified standard proteomics workflow can be used to identify sites of pHis as well as other acid-labile, as well as acid-stable phosphosites.

Key words Phosphohistidine, pHis, Phosphoproteomics, Mass spectrometry, Strong anion exchange, Enrichment

1 Introduction

Over the last two decades, mass spectrometry (MS)-based phosphoproteomics has proved invaluable for defining sites of phosphorylation on serine, threonine and tyrosine residues, in both low- and high-throughput studies. Although histidine phosphorylation is known to be important in relaying the extracellular signals that drive intracellular responses in a variety of organisms,

Claire E. Eyers (ed.), *Histidine Phosphorylation: Methods and Protocols*, Methods in Molecular Biology, vol. 2077,
https://doi.org/10.1007/978-1-4939-9884-5_15,

characterization of sites of phosphohistidine (pHis) by MS (or other analytical strategies) has remained a significant challenge, and interrogation of pHis lags far behind its canonical counterparts. The phosphoramidate bond of pHis is both heat- and acid-labile, with a -ΔG value of −12 to −14 kcal/mol compared to approximately −6.5 to −9.5 kcal/mol for the phosphoester bonds found in phosphoserine, or phosphothreonine [1]. Consequently the phosphate group of pHis is unstable under the acidic conditions typically used in standard phosphoproteomics workflows and the rapid hydrolysis hampers phosphosite mapping [2, 3].

A key stage in any phosphoproteomics workflow is the enrichment of phosphopeptides following proteolysis of a complex protein extract. Separation of phosphopeptides from the background of nonphosphorylated peptides which are present in vast excess, is essential for sensitive phosphopeptide identification and phosphorylation site identification by tandem mass spectrometry (MS/MS) due to both the relatively low abundance of phosphopeptides, where modification of an individual residues is typically sub-stoichiometric, and the compromising effect that addition of the negatively charged phosphate group can have on peptide ionization efficiency. Most of the currently available phosphopeptide enrichment approaches rely on acidic conditions to minimize unwanted binding of acidic peptides, which is not suitable for pHis due to the significant hydrolysis observed for both 1- and 3-pHis.

Strong anion exchange (SAX) chromatography can facilitate peptide separation based on electrostatic interactions of negatively charged groups with a positively charged stationary phase. Elution from the SAX column can be elicited by either a decreasing pH gradient [4–6], or increasing salt concentration at a constant pH [7]. At low pH, the negative charge of a single phosphate group is often not sufficient to overcome the electrostatic repulsion conferred by the N-terminus and the side chain of the C-terminal amino acid, resulting in poor retention of singly phosphorylated peptides. However, above ~pH 6 a second negative charge is acquired, making SAX chromatography feasible for separation of negatively charged phosphopeptides from the majority of peptides which do not contain a net negative charge [6]. Crucially, as the phosphate group of pHis peptides is stable above pH 6, SAX fractionation under these conditions is an attractive strategy for separation of acid-labile phosphopeptides, such as those containing pHis, from the majority of nonphosphorylated peptides, facilitating MS-based site identification. Here we describe a procedure termed UPAX (*u*nbiased *p*hosphopeptide enrichment by strong *a*nion e*x*change) for the separation and enrichment of phosphopeptides, including acid-labile pHis-containing peptides, permitting pHis phosphosite characterization by LC-MS/MS [8, 9].

2 Materials

Use HPLC-grade solvents and acids to prepare all solutions, and analytical grade buffers throughout. Low-bind Eppendorf tubes (or similar) should be used to minimize sample loss through adsorption to the vessel.

2.1 Sample Preparation

1. Dithiothreitol (DTT): 100 mM DTT in 50 mM ammonium bicarbonate (AmBic).
2. Iodoacetamide (IOA): 250 mM IOA in 50 mM AmBic.
3. Sequencing grade modified trypsin: reconstitute in 50 mM acetic acid at 0.5 mg/mL (*see* **Note 1**).
4. Heating block.
5. Benchtop centrifuge.

2.2 Strong Anion Exchange (SAX) Chromatography

1. Sonicating water bath.
2. SAX buffer A: 20 mM ammonium acetate, pH 6.8, 10% (v/v) acetonitrile (MeCN).
3. SAX buffer B: 300 mM triethylammonium phosphate, pH 6.8, 10% (v/v) MeCN (*see* **Note 2**).
4. SAX column: PolySAX LP column (PolyLC; 4.6 mm × 200 mm, 5 μm particle size, 300 Å).
5. HPLC system such as the Dionex U3000 HPLC instrument, equipped with a fraction collector and a UV detector capable of measurement at 280 nm (*see* **Note 3**).
6. Vacuum centrifuge.

2.3 Peptide Desalting

1. StageTips: 200 μL pipette tip containing 3 discs of C18 material (Empore™ Octadecyl C18, 47 mm) (*see* **Note 4**).
2. C18 StageTip elution buffer: 50% (v/v) MeCN in H_2O.
3. Methanol.

2.4 Liquid Chromatography (LC)-Tandem Mass Spectrometry (MS/MS) Analysis

1. MS dilution buffer: 3% (v/v) MeCN in H_2O.
2. C18 trap wash buffer: 2% (v/v) MeCN, 0.1% (v/v) TFA in H_2O.
3. LC-MS buffer A: 0.1% (v/v) formic acid in H_2O.
4. LC-MS buffer B: 0.1% (v/v) formic acid, 80% (v/v) MeCN in H_2O.
5. High resolution mass spectrometry system with online nanoUPLC system capable of high energy (beam-type) CID. For these studies we use a Thermo Orbitrap Fusion tribrid mass spectrometer (Thermo Scientific) attached to an UltiMate 3000 nano system (Dionex).

6. C18 trapping column: PepMap100, C18, 300 μm × 5 mm (Thermo Scientific).
7. C18 analytical column: Easy-Spray C18, 75 μm × 500 mm, 2 μm bead diameter.

2.5 LC-MS/MS Data Processing

1. Proteome Discoverer (PD) with *ptm*RS node (for these studies we used PD version 1.4).
2. Mascot (Matrix Science).

3 Methods

3.1 Sample Preparation

Proteins should be extracted from the required cell line or tissue according to sample specific protocols, based on the individual requirements of the biological material (*see* **Note 5**). To avoid heating of the sample or treatment under acidic conditions, proteins are typically extracted in a urea-based buffer (8 M urea) (*see* **Note 6**), and ~2 mg of protein per sample is digested (*see* **Note** 7).

1. Add 100 mM DTT to the protein lysate solution to achieve a final DTT concentration of 3 mM. Incubate at 30 °C for 20 min (*see* **Note 8**). Cool the sample to room temperature.
2. Add 250 mM IOA to achieve a final IOA concentration of 14 mM. Incubate at room temperature in the dark for 45 min.
3. Quench excess IOA by addition of sufficient DTT to achieve a final concentration in the protein sample of 7 mM. Dilute the sample by addition of 50 mM AmBic such that the urea concentration is at or below 2 M.
4. Add 2% (w/w) trypsin and incubate at 30 °C with shaking at 650 rpm overnight (~16 h).
5. Following overnight digestion, dry the samples to completion by vacuum centrifugation. Dried samples can be stored at −20 °C if necessary, or ideally subjected immediately to SAX fractionation.

3.2 SAX Fractionation

1. Wash the SAX column with 20 column volumes (CV) of SAX buffer B, then equilibrate with at least 20 CV of SAX buffer A (*see* **Note 9**).
2. Resolubilize the dried digested peptide samples in 180 μL SAX buffer A. Leave in a sonicating water bath for 5 min to aid recovery of peptides from the Eppendorf tubes (*see* **Note 10**).
3. Load the resolubilized peptide sample onto the column in 100% SAX buffer A using the same flow rate as for **step 1** (*see* **Notes 9** and **11**). Wash the column for 5 min with 100% SAX buffer A. Collect unbound peptides and store on ice.

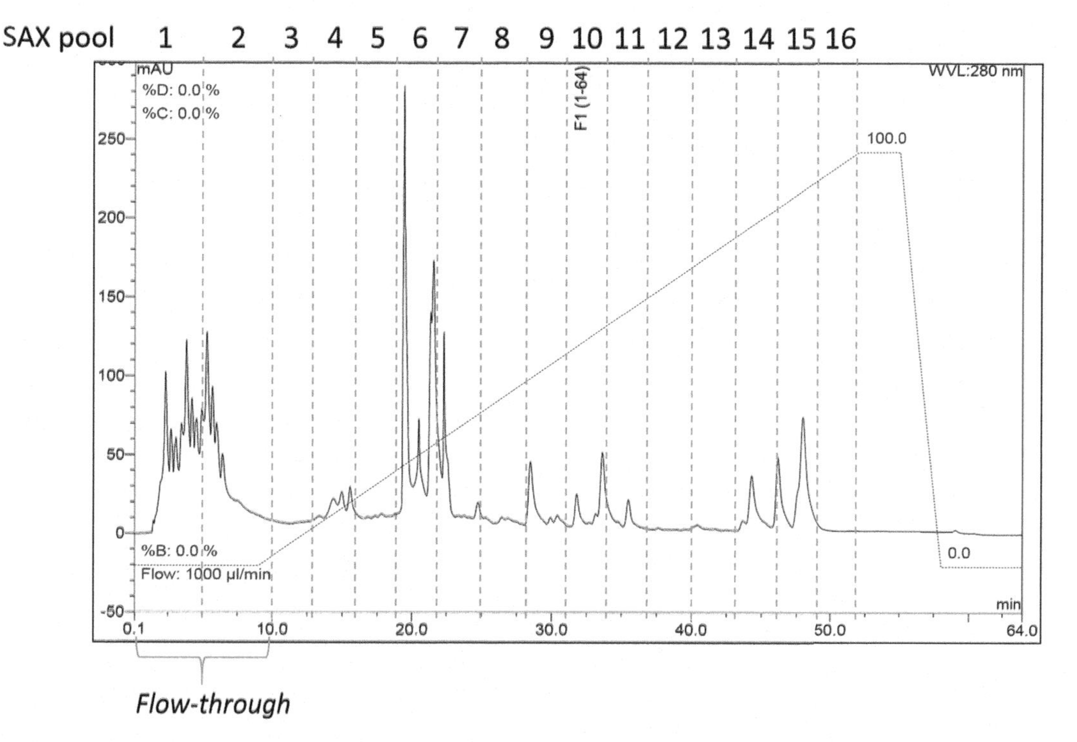

Fig. 1 Example SAX chromatogram (280 nm absorbance) following separation of 2 mg HeLa cell tryptic peptides on a PolySAX LP column. SAX buffer A: 20 mM ammonium acetate (pH 6.8), 10% MeCN; SAX buffer B: 300 mM triethylammonium phosphate (pH 6.8), 10% MeCN. Fractions are collected every minute (1 mL/min flow rate) and pooled as indicated

4. Elute bound peptides with a gradient of increasing concentration of SAX buffer B to 100% over 43 min. Maintain at 100% SAX buffer B for 5 min before equilibration to start conditions (100% SAX buffer A). Collect fractions every minute until the end of the gradient (48 min) and store them on ice. An example SAX chromatogram, with pooled fractions is shown in Fig. 1.
5. Pool the fractions to reduce the overall number of fractions for LC-MS/MS analysis to 16 (*see* **Note 12**). Reduce the volume of the pooled samples by vacuum centrifugation so that the volume of each pooled sample is ~500 µL (*see* **Notes 13** and **14**).

3.3 Peptide Desalting

1. Pack three discs of C18 material into a 200 µl pipette tip (*see* **Note 4**), repeat to create 16 StageTips, one for each set of pooled SAX fractions. Place the tips into Eppendorf tubes, held in place with a plastic stopper (*see* **Note 15**).
2. Condition each of the StageTips tips by sequential addition of 100 µL methanol, 100 µL 50% MeCN in H_2O and 100 µL H_2O, centrifuging for 2 min at 2000 × *g* between each step to pass the liquid through the tip (*see* **Note 16**).

3. Load 150 μL of sample onto each tip, centrifuge as in **step 2**, load the flow through back into the tip and centrifuge again (*see* **Note 17**). Ensure all the liquid has passed through the tip at each stage.
4. Wash each of the tips with 100 μL H_2O, then elute bound peptides with 50 μL 50% MeCN in H_2O into a fresh low-bind Eppendorf tube (*see* **Note 18**).
5. Dry eluents to completion by vacuum centrifugation. Samples are now ready for LC-MS/MS analysis and can be stored at −20 °C until needed.

3.4 LC-MS/MS Analysis

1. Resolubilize the dried desalted SAX pools in an appropriate volume of MS dilution buffer. Sonicate the fractions for 5 min in the sonicating water bath. Centrifuge in a bench-top at 15,000 × *g* for 15 min, then transfer to glass LC-MS vials.
2. Using a nanoUPLC system arranged in-line with the mass spectrometer, load the peptides onto a C18 trapping column using partial loop injection, for 7 min at a flow rate of 9 μL/min with C18 trap wash buffer. Resolve bound peptides at 300 nL/min using the C18 analytical column using an LC gradient from 3.8% LC-MS buffer B (96.2% LC-MS buffer A) to 50% buffer B (50% buffer A) over 90 min.
3. Acquire a full mass spectrum over *m/z* 350–2000 in the Orbitrap (120K resolution at *m/z* 200). Perform data-dependent MS/MS analysis using a top speed approach (cycle time of 3 s), with HCD (collision energy 32%, max injection time 35 ms) and neutral loss-triggered (Δ80 and Δ98 amu) EThcD (ETD reaction time 50 ms, max ETD reagent injection time 200 ms, supplemental activation energy 25%, max injection time 50 ms) for fragmentation. Detect all product ions in the ion trap (rapid mode) (*see* **Note 19**).

3.5 Data Processing

1. Convert .raw files to .mzml using ProteoWizard's msconvert tool in order to perform MS2-level deisotoping.
2. Using Proteome Discoverer (PD), split scans for each raw file into those arising from HCD and EThcD events using a collision energy (CE) filter (HCD: min CE 0, max CE 34; EThcD: min CE 35, max CE 1000) to generate two separate .mgf files. An example Proteome Discoverer workflow is shown in Fig. 2.
3. Search the .mgf files in PD using the MASCOT search algorithm against the appropriate database. For human derived cell extracts, we use the UniProt Human database. Set search parameters as follows: MS1 tolerance of 10 ppm; MS2 mass tolerance of 0.6 Da; enzyme specificity as trypsin, with two missed cleavages allowed; fixed modification: carbamidomethylation of Cys; variable modifications: phosphorylation of Ser, Thr,

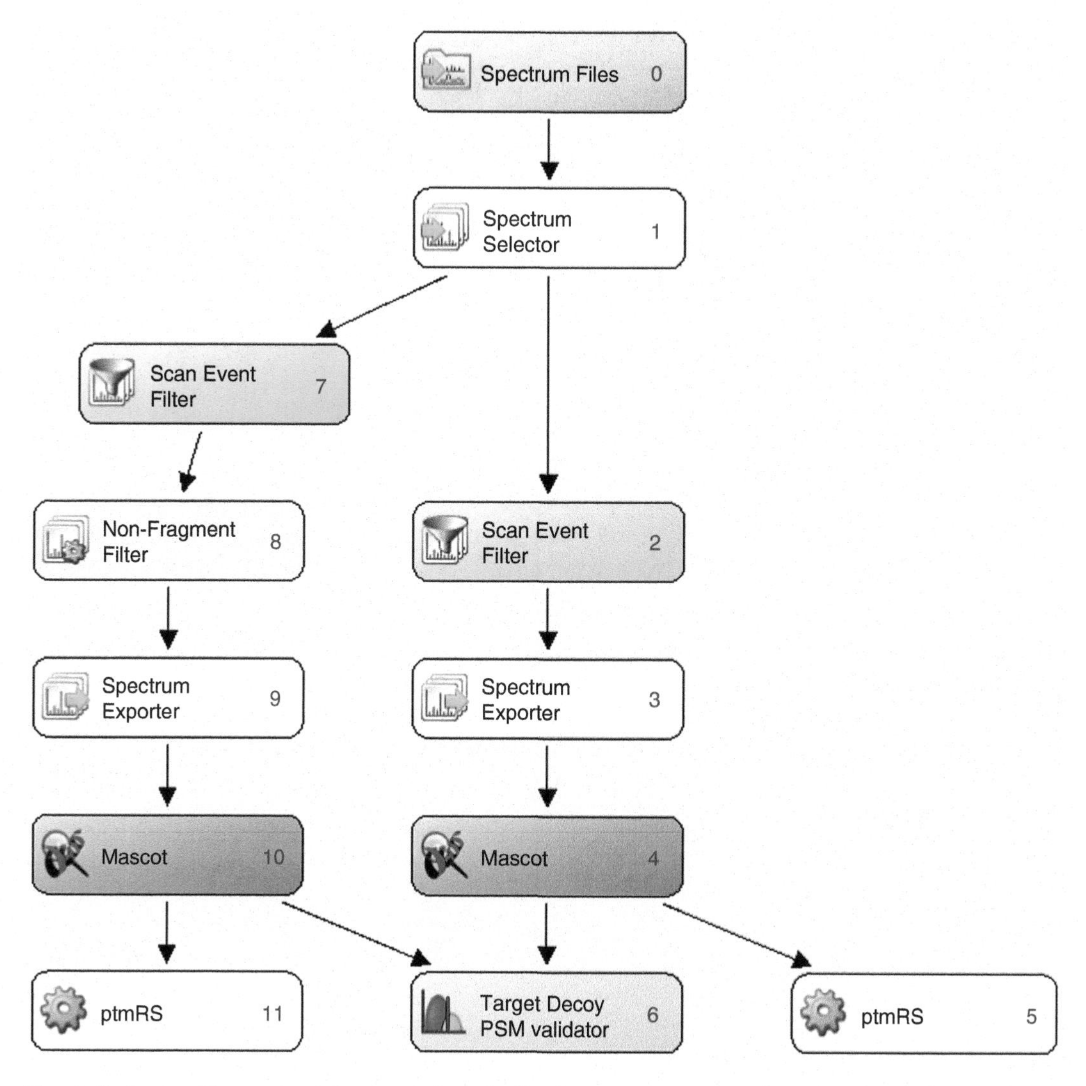

Fig. 2 Proteome discoverer data processing pipeline. Tandem mass spectra are separated according to fragmentation type prior to searching with MASCOT and application of ptmRS for phosphosite localization

Tyr, and His, oxidation of Met; instrument type set as ESI-Quad-TOF for HCD files and CID + ETD for EThcD files (*see* **Note 20**).

4. Analyze phosphopeptide spectra using the ptmRS node, with the "treat all spectra as EThcD" option selected for EThcD data (*see* **Note 21**).
5. Apply a peptide false discovery rate (FDR) filter (typically 1% or 5%) and export files to .csv for further data processing.
6. Process the exported .csv to identify phosphosites above desired ptmRS score cut-off. Scores above 0.75 are generally considered to be "localized," although we typically apply a

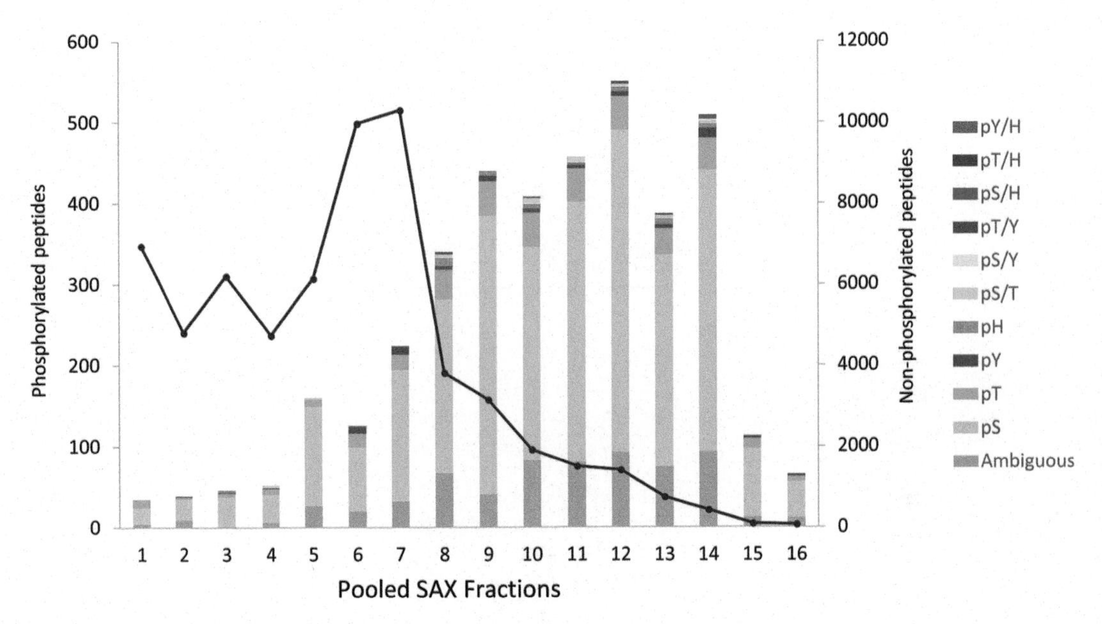

Fig. 3 Phosphopeptides identified in each pooled SAX fraction following separation of peptides from 2 mg HeLa cell lysate. Phosphosites with ptmRS score >0.75 are considered localized, whilst any peptide with an unassigned site is shown as "Ambiguous." The number of nonphosphorylated peptides in each fraction is also plotted, showing a proportional decrease in the number of non-phosphorylated peptides identified in the later SAX fractions, coinciding with an increase in the number of phosphopeptide identifications

more stringent 1% false localization rate (FLR) cut-off defined according to the mode of MS2 fragmentation and analysis [9, 10] (Fig. 3). More detailed strategies for analysis of this type of MS data are presented in Chapter 16.

4 Notes

1. Acidic stock solutions of trypsin can be aliquoted and frozen at 20 °C for up to 6 months.
2. Make up SAX buffer B by adding 30% volume of 1 M phosphoric acid to a graduated cylinder. Adjust the pH to 6.8 with trimethylamine – this is typically at least 10 mL per 250 mL total buffer volume. Add 20% volume of MeCN and make up to volume with water.
3. Any offline HPLC system can be used, ideally one that also has a fraction collector. If a fraction collector is not available, fractions can be collected manually, although this is more time consuming. Retain all fractions on ice.
4. Using a Plunger Assembly for a Hamilton Syringe, punch a disk of 3M-C18 material and carefully dispense into a 200 μl tip. Repeat twice so that for each sample, you have constructed a 200 μl C18-StageTip with three layers of C18 material [11].

5. pHis-containing samples should be kept on ice as much as possible. Temperatures above 30 °C should be avoided to minimize heat-induced loss of the labile pHis group. All sample processing procedures should be carried out as quickly as possible within minimal time between stages. Be sure to include phosphatase and protease inhibitors in cell/tissue lysis buffers. Care must be taken to avoid sample processing procedures that require acid-based treatment.
6. Mammalian cells grown in culture can be lysed in 50 mM AmBic, 8 M Urea with protease/phosphatase inhibitors. We have not tested any other lysis buffer compositions for compatibility with the SAX fractionation.
7. For large-scale phosphoproteome analysis we typically prepare 2 mg of protein and fractionate all of this by SAX. The amount of protein in the sample should be measured, for example, by Bradford assay or using a NanoDrop system, and used to calculate the amount of trypsin required (2% w/w).
8. The temperature of these samples must be carefully controlled — all incubation/digestion steps are performed at a maximum of 30 °C to minimize loss of pHis.
9. For the PolySAX LP column (4.6 mm × 200 mm) described here, we used a flow rate of 1 mL/min.
10. To minimize sample adsorption, we use low-bind tubes at all steps.
11. Sample injection onto the HPLC column will be dependent on the size of the injection loop. If the sample loop is at least 200 μL, all 180 μL can be injected in a single step. For smaller injection loops, two or three injections of smaller volumes can be made, that is, load 60 μL into the loop, switch the valve to wash sample onto the column for 90 seconds, switch the valve back to load a further 60 μL into the loop, switch again to load this portion of the sample onto the column then repeat again to load the final 60 μL. The column can then be washed with 100% SAX buffer A for 5 min as described in **step 3**. If the sample is delivered in multiple steps, then the fractions corresponding to this loading portion should be combined.
12. Fractions can be combined every 3, so that SAX fractions 1–3 become pool 1, fractions 4–6 pool 2 etc. Alternatively, as the latter fractions are more enriched in phosphopeptides, SAX fractions can be differentially combined so that more fractions in the earlier part of the gradient (which are less phosphopeptide rich) are combined (e.g., SAX fraction 1–6 combined to generate pool 1), while latter fractions are kept as separate pools. This can increase the phosphopeptide coverage.

13. Fractions are only dried to 500 μL (rather than to completion) as the high concentration of nonvolatile salt in the later fractions prevents total drying. Volumes smaller than ~500 μL have such a high salt concentration that C18 StageTip desalting becomes problematic; liquid does not easily pass through the tip.
14. Pooled SAX fractions can be stored frozen (−20 °C) after drying to 500 μL, although we recommend proceeding to the desalting step as soon as possible (i.e., within a day) to minimize sample degradation.
15. High concentrations of triethylammonium phosphate following SAX fractionation must be removed prior to LC-MS/MS analysis. C18 StageTips prepared in-house are a suitable choice for desalting [11], although commercial prepacked tip or column options are also available for peptide desalting.
16. When using StageTips be sure that the liquid in the Eppendorf tube does not reach the bottom of the tip. We typically use 2 mL low bind Eppendorf tubes to reduce the risk of this happening, although the 1.5 mL tubes also work.
17. For the workflow described here, desalting 150 μL of sample (out of the 500 μL volume remaining after drying) is sufficient based on a 2 mg starting amount of protein. However, if the amount of material in a given pool is low (based on A_{280}) or a different concatenation strategy is employed where fewer SAX fractions are pooled, it may be necessary to desalt more (or all) of each sample.
18. Following desalting (and prior to drying) peptide concentration can be estimated by NanoDrop. Peptide concentration estimates can be used to assist in determining the amount of each sample to load for LC-MS/MS analysis.
19. Suggested starting parameters for a Thermo Orbitrap Fusion mass spectrometer are described. Both HCD and ETD are implemented, with ETD fragmentation triggered by neutral loss of 80 or 98 amu from the precursor ion. It is recommended that sample loading and mass spectrometer parameters are optimised for a specific instrument set-up. While we describe low-resolution ion trap-based MS2 analysis, fragment ion analysis can also be performed in the orbitrap.
20. Search parameters should be adjusted according to the instrument set up: acquisition of MS2 data in the orbitrap will require a reduction to MS2 tolerance to 10 ppm. Alternative modifications may also be considered (e.g., Asn deamidation, N-terminal acetylation, phosphorylation of residues other than Ser, Thr, Tyr, and His).
21. For phosphopeptide identification it is important to have confidence in both the sequence and the site of modification.

Determining with confidence the position of modification when there is more than one possibility can be challenging. It is therefore beneficial to apply a localization score to indicate confidence in a given site of modification. A suggested pipeline for processing LC-MS/MS data utilises Proteome Discoverer with the MASCOT search engine [12] and ptmRS for phosphosite localization [13].

Acknowledgments

This work was supported by the Biotechnology and Biological Sciences Research Council (BBSRC) in the form of DTP funding to G.H. and grants to C.E. (BB/M012557/1, BB/R000182/1, BB/H007113/1), and the North West Cancer Research Fund (NWCR) (CR1157). We thank all members of the Centre for Proteome Research, particularly Dr. Philip Brownridge, for continued support.

References

1. Attwood PV, Piggott MJ, Zu XL et al (2007) Focus on phosphohistidine. Amino Acids 32:145–156
2. Hultquist DE, Moyer RW, Boyer PD (1966) The preparation and characterization of 1-phosphohistidine and 3-phosphohistidine. Biochemistry 5:322–331
3. Hultquist DE (1968) The preparation and characterization of phosphorylated derivatives of histidine. Biochim Biophys Acta 153:329–340
4. Dai J, Jin WH, Sheng QH et al (2007) Protein phosphorylation and expression profiling by Yin-yang multidimensional liquid chromatography (Yin-yang MDLC) mass spectrometry. J Proteome Res 6:250–262
5. Nie S, Dai J, Ning ZB et al (2010) Comprehensive profiling of phosphopeptides based on anion exchange followed by flow-through enrichment with titanium dioxide (AFET). J Proteome Res 9:4585–4594
6. Alpert AJ, Hudecz O, Mechtler K (2015) Anion-exchange chromatography of phosphopeptides: weak anion exchange versus strong anion exchange and anion-exchange chromatography versus electrostatic repulsion-hydrophilic interaction chromatography. Anal Chem 87:4704–4711
7. Han G, Ye M, Zhou H et al (2008) Large-scale phosphoproteome analysis of human liver tissue by enrichment and fractionation of phosphopeptides with strong anion exchange chromatography. Proteomics 8:1346–1361
8. Hardman G, Perkins S, Ruan Z et al (2017) Extensive non-canonical phosphorylation in human cells revealed using strong-anion exchange-mediated phosphoproteomics. bioRxiv 2017:202820
9. Hardman G, Perkins S, Brownridge PJ, Clarke CJ, Byrne DP, Campbell AE, Kalyuzhnyy A, Myall A, Eyers PA, Jones AR, Eyers CE, (2019) Strong anion exchange-mediated phosphoproteomics reveals extensive human non-canonical phosphorylation. The EMBO Journal 21:e100847. doi:10.15252/embj.2018100847
10. Ferries S, Perkins S, Brownridge PJ et al (2017) Evaluation of parameters for confident phosphorylation site localization using an orbitrap fusion tribrid mass spectrometer. J Proteome Res 16:3448–3459
11. Rappsilber J, Mann M, Ishihama Y (2007) Protocol for micro-purification, enrichment, pre-fractionation and storage of peptides for proteomics using StageTips. Nat Protoc 2:1896–1906
12. Perkins DN, Pappin DJ, Creasy DM et al (1999) Probability-based protein identification by searching sequence databases using mass spectrometry data. Electrophoresis 20:3551–3567
13. Taus T, Kocher T, Pichler P et al (2011) Universal and confident phosphorylation site localization using phosphoRS. J Proteome Res 10:5354–5362

Chapter 16

Proteome Bioinformatics Methods for Studying Histidine Phosphorylation

Andrew R. Jones and Oscar Martin Camacho

Abstract

In this chapter, we introduce the bioinformatics methods associated with studying histidine phosphorylation (pHis) by LC-MS/MS. We describe methods for converting and preprocessing raw data from MS instruments, the method of searching MS data against a sequence database and scoring the confidence associated with localizing the modification site on the peptide sequence. We also describe methods for performing pathway enrichment once a set of pHis-containing proteins have been identified to understand the putative functions of modified proteins. Several of the methods are relatively straightforward to run but require some theoretical knowledge to optimize parameters and correctly interpret outputs. We also describe some of the theory underpinning statistical considerations, to assist correct usage and interpretation of these bioinformatics methods.

Key words Proteomics, Phosphoproteomics, Database searching, Bioinformatics, Phosphohistidine, Site localization

1 Introduction

There are several methods under development for the characterization of phosphohistidine (pHis) in proteins using mass spectrometry (MS), including the use of anti-pHis antibodies for enrichment of proteins containing pHis for subsequent identification using standard proteomics methods [1]. However, in the absence of robust methods that can enrich labile pHis-containing phosphopeptides, localizing the site of modification can be problematic, and approaches that attempt to enrich for these acid-labile phosphopeptides using strong anion exchange (SAX) chromatography prior to LC-MS/MS analysis are showing great promise (*see* Chapter 15 for more information) [2]. Here we discuss the informatics associated with ascribing pHis site localization following phosphopeptide fractionation using, for example, SAX chromatography, from LC-MS/MS data where the sites of modification are potentially identifiable in the data generated.

Claire E. Eyers (ed.), *Histidine Phosphorylation: Methods and Protocols*, Methods in Molecular Biology, vol. 2077,
https://doi.org/10.1007/978-1-4939-9884-5_16,

This chapter introduces the key points related to the handling and processing of tandem MS data generated from mixtures of phosphorylated peptides, including those containing pHis. Specifically we discuss (1) the tools and approaches for converting raw MS data to a peak list file format ready for searching; (2) search engines for peptide (and protein) identification; (3) tools, approaches and statistics for modification site localization. We also highlight important considerations when analyzing LC-MS/MS data for pHis analysis, including common pitfalls in raw data processing, and the necessity to perform peak picking prior to undertaking a database search, as well as challenges in statistical power at the database search stage, arising due to the necessary inclusion of multiple sites of phosphorylation. Evaluation of phosphosite localization confidence is also considered in the context of different algorithms. Following the identification phosphorylation sites, it is common in workflows to perform steps to understand the functions of groups of modified proteins. We briefly cover some of the tools available for downstream functional analysis to help understand the pathways and interaction networks in which pHis containing proteins may be acting. Lastly, we introduce some theoretical considerations behind the methods to help readers to correctly interpret the results.

2 Materials

2.1 LC-MS/MS Data Processing and Analysis

1. Tandem MS data: product ion mass spectra (.raw files) generated following LC-MS/MS analysis of pHis-containing peptides derived from the biological sample of interest. Peptides are typically generated following tryptic digestion, although other enzymes can be used. Peptide ions may be fragmented using a number of different strategies (e.g., higher collision-induced dissociation (HCD), electron transfer dissociation (ETD), or a combination of the two (EThcD)). *See* **Note 1** for how instrument and fragmentation type influence data processing and analysis parameters.
2. Protein database: FASTA file containing the protein sequences from the species being analyzed.
3. Data processing software: the requirements for this will be dependent on which instrument (Vendor) was used to generate the MS/MS data.
4. Proteomics search algorithm.
5. Algorithm for phosphosite localization.
6. A high-performance workstation, server or compute cluster for running searches and postprocessing.

2.2 Phosphohistidine -Containing Pathway Analysis

1. Access to pathway analysis tools. These can be freeware including DAVID [3] and Panther [4], or commercial packages such as Ingenuity Pathway Analysis (IPA).

3 Methods

3.1 Preprocess the LC-MS/MS Data to Generate Peak Lists From .raw Files

Following LC-MS, the raw data that is generated by the instrument is usually in a vendor raw (binary) file format, which is specific to the manufacturer of the instrument (and sometimes the model). For data generated on Thermo instruments (LTQ-Orbitrap, Q-Exactive, Orbitrap Fusion, etc.), the instrument produces a ".raw" file. There are some search engines that can read these files directly, and perform internally the necessary preprocessing, but most search engines, including the majority of open source search engines require data to be converted to a peak list format for searching first, most commonly "Mascot Generic Format" (.mgf files).

An important consideration is whether data need to be centroided (peak picked) before exporting to MGF format. The process of centroiding is a data reduction technique. When instruments collect data, they are scanning all mass/charge (m/z) values, usually over a m/z range of 0–2000 Th. The data is recorded initially as pairs of points (m/z and intensity) across the whole range, capturing around tens of thousands of data points per scan. When a peptide (or fragment) ion is measured, the shape of the peak is thus captured by several, perhaps tens of data points. All noise is recorded, and depending on the manufacturer's encoding, regions with no signal can be explicitly encoded as having zero intensity at every m/z point measured. Peak picking greatly reduces the volume of data by selecting one single centre point (centroid) of each measured ion, where there is the highest intensity (Fig. 1).

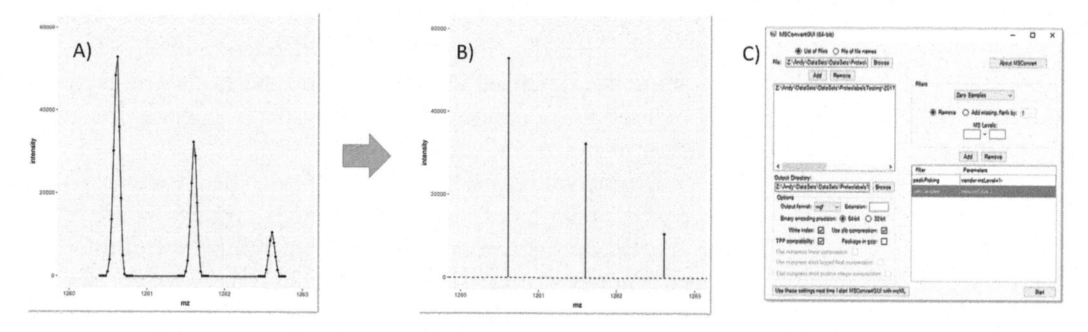

Fig. 1 (**a**) A small example of data points as represented in a raw file (Thermo QExactive HF), where the points are data values actually recorded (plotted in R: ggplot). Lines have been added to show the peak shape. (**b**) Following peak picking, data are only recorded for the apex of points (shown as dots), lines have been added to represent typical peak notation. (**c**) The MSConvert GUI from Proteowizard can be used to perform peak picking, removal of unnecessary zero values, and conversion to MGF format

Most search engine algorithms also expect that data is peak picked before the search, otherwise there can be a negative effect on results.

For workflows performing data-dependent acquisition (DDA), the only preprocessing necessary before database searching is to perform centroiding (where needed) and convert the file format such that it is ready for the database search software. For data-independent acquisition (DIA) methods, in which a wide window is applied to precursor peptide ions for fragmentation (e.g., from 25 Da up to the full mass window), several additional preprocessing steps are required. In DIA methods, the intrinsic link between a fragment ion in an MS^2 spectrum and the precursor ion(s) is lost, and must be inferred via one of several methods. DIA methods are not widely used for phosphoproteomics, since they make the challenge of modification site localization even more difficult, and thus detailed discussion of these methods is out of scope for this chapter.

Data acquired with older generation Orbitrap instruments (particular the LTQ-Orbitrap) is automatically centroided and thus there is no need to do this in another piece of software. However, centroiding of .raw files is actively required for data from Q-Exactive and later models before undertaking a database search. A reliable method to perform both centroiding and export . raw files to MGF format is to use ProteoWizard's MSConvert tool [5], which can be downloaded from http://proteowizard.sourceforge.net/. MSConvert contains internally the vendor's own libraries for file manipulations and peak picking, as well as some custom routines.

3.2 Search the Processed MS/MS Data to Define Peptide Sequence and Localize the Site of Histidine Phosphorylation

1. Search the peak lists (MGF files) against the UniProt database for the species of interest, using the search algorithm of choice.

   There are many different search engines that apply one of the following methods: sequence database search, *de novo* sequencing, hybrid (de novo followed by sequence database) or spectral library search. Full description of the various flavors of peptide identification are beyond the scope of this chapter, but sequence database search or hybrid methods would generally be used for phosphoproteomics, especially where the aim is to discover new sites, such as pHis.

   When deciding which search engine to use, there are often lab-based preferences, and some search engines work better with particular instruments. For a thorough review of different search engines, consult Verheggen et al. [6]. However, for the most common setups, there is not a great difference in performance between most search engines, especially for platforms that generate high-resolution fragment ions, making identification and site localization easier. Commercial search engines such as Mascot or Peaks offer intuitive graphical interface and some interactivity in terms of exporting results at a given FDR threshold. Importantly, both offer control over neutral loss settings (*see* **Note 6**), which is an important consideration when searching for phosphorylated peptides. In terms of open source search engines, there are many that run primarily from the command line, including MSGF+, X!Tandem, and MS-Amanda, among many others. Andromeda, the search engine embedded within MaxQuant can be run from either the command line or via the graphical user interface. Performance across different search engines is relatively similar, although for some instrument settings or search protocols differences can be observed. To our knowledge, MS-Amanda and Andromeda support customization allowing different precursor ion neutral loss masses to be considered. Both can be run from on Windows or Linux, although in both cases Linux was not natively supported originally, and these programs may not run on all forms of Linux without issues.

Once the search engine has been selected, almost all offer the ability to set the parameters summarized in Table 1. Users should understand that correct settings must be applied for the given instrument used (precursor/product ion tolerance), the digestion enzyme employed, the type of fragmentation used, and most importantly the selection of modifications and residues for the search engine to explore. *See* **Note 1** for how instrument and fragmentation type influence data analysis parameters. For a pHis search, users would typically search for pS, pT, and pY, as well as pH, although if phosphorylation of other residues are considered possible, then these would need to be added to the variable modifications list, albeit at a cost to the statistical power of the search (discussed in Subheading 3.3; *see* **Notes 2** and **3**). Typically users should employ a decoy database search, allowing estimates of false discovery rate (FDR) to be made.

2. Open the output file from the search engine. Typically this is either a report than can be visualized via a graphical user interface, or downloaded, for example, as a CSV file that can be opened in a Spreadsheet or statistical software package.
3. Decide on the statistical threshold to accept for peptide spectrum matches (PSMs) or peptides identified, that is, where multiple PSMs reporting on the same peptide have been collapsed to a single best scoring entry. The latter is generally more conservative since false positive identifications tend to be randomly distributed, with 1 PSM per false positive peptide, whereas true positives are more likely to be supported by >1 PSM (*see* **Note 4**). In general, 1% FDR is the recommended setting (for PSMs or peptides), although 5% FDR is used in some phosphoproteomics literature under the rationale that as phosphosites can be transient and often occur on low intensity

Table 1
A summary of the most important search engine settings

Setting type	Notes
Precursor ion tolerance	The optimal precursor tolerance is dependent on the instrument type and assumes the instrument is well calibrated. For optimal data interrogation, it can sometimes be worth performing a pilot search, checking the distribution of precursor tolerance values for true positives (high-scoring nondecoy hits), and setting a tolerance just wider than that distribution. In the latest version, the PEAKS search engine can perform recalibration automatically
Fragment ion tolerance	The main consideration is whether data were generated on a low-resolution trap (e.g., tolerance of 0.5 Da) or at high-resolution (5 ppm or 0.02 Da). For time-of-flight instruments, precursor and fragment tolerance values may be similar, such as 10 or 20 ppm depending on the model. Consult manufacturer's guidance if in doubt
Decoy search	A decoy search is generally recommended for calibrating the statistics for false discovery rate (FDR). The search engine will usually search both the actual sequences of interest (targets) as well as other sequences assumed to be incorrect (decoys), for example generated by reversing or shuffling amino acids within proteins
Enzyme	Trypsin is the most commonly employed enzyme used to generate peptides, although other enzymes can be used. The cleavage specificity of the proteolytic enzyme used must be defined
Variable modifications	This is a key setting when attempting to identify pHis (*see* **Note 3**)
Fixed modifications	In most protocols, iodoacetamide is used to prevent the formation of disulfide bridges on cysteine residues, resulting in a fixed mass Cys modification of +57.021464 Da (carbamidomethylation). If a different reagent is used, then this should be changed accordingly
Missed cleavages	This setting allows for the possibility of the proteolytic enzyme having missed a cleavage site. In the case of trypsin (which cleaves after K or R except when followed by P), a missed cleavage setting of, for example, 2, would permit peptides containing 0, 1 or 2 internal K or R sites to be considered
Database	For studies on humans or most model organisms, it is recommended to download a "proteome" set from UniProt (https://www.uniprot.org/proteomes/) [14]. For quantitative phosphoproteomics, it is sometimes recommended to follow the link "Download one protein sequence per gene," which makes data interpretation simpler. However, for discovery of pHis sites, downloading the full set of proteins (including alternative splice products) might be more appropriate
Instrument/ fragmentation type	All search engines allow users to specify directly or indirectly the ion series to search. Most common options include b- and y-ions for CID or HCD fragmentation; or c and $z + 1$ ions for ETD

peptides, it can be challenging to get high scoring phosphopeptides at scale. Nevertheless, peptides with weaker scores than the 1% FDR cut-off should be treated with caution, as there is a rapidly increasing probability that hits are false positives.

3.3 Compute Phosphosite Localization Confidence

A crucial step in phosphoproteomics analysis is determining the confidence at which a given site within a peptide has been modified, particularly considering that most peptides contain more than one residue that could have been phosphorylated. Site localization algorithms rely on the search engine having first accurately identified both the peptide sequence and the number (and type) of modifications on those peptides. The most popular localization algorithms Ascore [7] and *ptm*RS [8] (formerly called phosphoRS) use the peptide sequence, count of modifications and the number of theoretically possible sites, to generate theoretical fragment ions of all possible phosphopeptide isomers (i.e., all combinations of assumed possible modification sites on one given peptide), which are compared against experimental spectral data. The theory behind these methods and calculation of the statistics is summarized in **Note 5.**

*ptm*RS is available as a node within Proteome Discover (Thermo Scientific). Given that it is primarily parameterized by the search (e.g., the selection of which sites to score is driven by the variable modifications selected in the search), there are only a few settings that can be adjusted. Users can select whether to score neutral loss ions or not (*see* **Note 6**). As a rule, we would recommend not using neutral loss ions for detecting pHis modification sites. Ascore is available by default in Peaks, and does not appear to have any parameters that can be readily changed. An Ascore >20 is intended to give $p < 0.01$, thus only a 1% probability that the modification is incorrectly assigned. The *ptm*RS probability-like score is not straightforwardly converted to false localization rate (FLR), and can vary dependent on the fragmentation mode as discussed in [9]. However, as a rule of thumb, a *ptm*RS score > 0.99 would have approximately a 1% chance of false localization.

3.4 Determine Which Protein Groups or Pathways Are Enriched in pHis

In most phosphoproteomics studies, it is common to follow up the determination of phosphosites by LC-MS/MS (from quantitative or qualitative studies) using functional analysis software, to allow determination of key protein groups or pathways enriched in the phosphorylated proteins identified. Popular tools for pathway analysis include DAVID [3] and Panther [4], which take a list of (genes or) proteins, and map them onto functionally annotated groupings, such as Gene Ontology [10] terms or UniProt keywords, or pathways such as KEGG [11]. The tools then calculate which categories or pathways are significantly enriched with mappings from the protein set, compared with the frequency of mappings expected by random chance (*see* **Note** 7).

To identify the pathways that proteins containing pHis (or other types of modified residue) are enriched in:

1. Identify with confidence the proteins containing significant pHis peptides (i.e., with <1% peptide FDR and <1% site FLR).

2. Extract the UniProt accession numbers for these proteins (e.g., of type "sp|P12345"), and remove the preceding "sp|".
3. Upload the pHis protein accessions into the software tool. To perform pathway enrichment, the pHis protein set needs to be compared to a background distribution (*see* **Note 7**). An appropriate background distribution is all phosphorylated proteins identified in the current workflow, which then in effect asks the tool to determine the functions that are enriched in pHis proteins versus all phosphoproteins (taking into account any identification biases), that is, how are the functions of pHis proteins different from proteins carrying canonical phosphorylation sites.
4. Click the option for "Functional Annotation Tool" to discover the pathways or sets of annotation keywords that are enriched in the pHis protein set. Typically, a statistic should be used that corrects for the number of comparisons that have been made. In DAVID, this is the "FDR" statistic, where is it appropriate to select, for example, all pathways/sets with FDR < 0.1. Weaker thresholds can sometimes be applied if there is a lack of statistical power, arising due to analysis of relatively small protein sets.

4 Notes

1. When searching MS/MS data against a database, the search engine processes *each tandem mass spectrum in turn*, with the following steps. (a) The software first generates the list of peptide masses, including the possibility for modifications (*see* **Note 3**), that match the precursor mass +/− a small tolerance (generally considered as parts per million (ppm) difference between expected and theoretical m/z). The acceptable ppm is dependent on the instrument acquisition parameters; for example, <5 ppm is typically acceptable for high resolution Orbitrap-generated data. (b) The search engine then selects all peptides that have this mass and generates theoretical spectra for all candidates. For tandem mass spectra generated by collision-induced dissociation (CID) or higher energy collision dissociation (HCD), the search engine generates the complete set of prefix product ion sequences (called *b* ions), and suffix product ion sequences (*y* ions). For phosphopeptide analysis, it is also common to use ETD (or EThcD), under which scheme the peptide backbone breaks at different sites, generating *c* and $z + 1$ ions. ETD is sometimes preferred for phosphopeptide analysis as the mode of fragmentation means that the phosphate group is more likely retained at the site of modification; phosphate-derived neutral loss ions are not therefore observed (*see* **Note 5**). To explore the process of fragment ion

generation, the phpMS [12] fragmentation tool can demonstrate how search engines generate these masses [MS Fragment Ion Generator: http://pgb.liv.ac.uk/phpMs/?page=peptide_properties], as well as similar tools provided by Protein Prospector [MS-Product: http://prospector.ucsf.edu/prospector/cgi-bin/msform.cgi?form=msproduct] or the Institute for Systems Biology [ISB Fragment Ion Calculator: http://db.systemsbiology.net:8080/proteomicsToolkit/FragIonServlet.html].

2. Search engines compare an experimental spectrum against a theoretical spectrum, and counts the number of product ions that match between them. The count of matches is then applied to a statistical model to determine the likelihood of generating that *count of matches by random chance*, from which a *p*-value is calculated. To correct for the number of candidate spectra considered, the *p*-value is converted to an *e*-value (Expectation value), generally by multiplying the *p*-value by the number of possible peptides within the mass window that could be generated from the database. The e-value thus equates to the expected frequency of getting a match with a given *p*-value, based on how many peptides were considered for comparison. If a peptide-spectrum match (PSM) has a p-value of 0.01, but 100 peptides were considered against the spectrum, then the e-value will be 1 (a weak match, likely produced by chance). As one example of the importance of parameter selection in a search, if an instrument has a real tolerance whereby all true hits are within 5 ppm of the database peptide mass, and a user instead sets the search tolerance to be 10 ppm (instead of 5 ppm), this would have the approximate effect of doubling the *e*-values, for the same *p*-values and thus reducing statistical power.

3. When attempting to identify pHis sites on proteins from LC-MS/MS data, we recommend that variable modification (phosphorylation) of S, T, Y, or H be considered during database searching, since under most conditions the majority of phosphorylation sites will occur on these amino acids. If other amino acids are suspected to be phosphorylated, then the list of potential phosphorylated residues can be expanded. However, the search space expands combinatorially depending on the number of residues considered (Fig. 2), which is detrimental for two reasons. First, the ability to differentiate true positives from false positives is dependent on statistical power. Considering more candidates for each spectrum produces weaker *e*-values from the same *p*-values, and ultimately *e*-values are used for ranking all peptide-spectrum matches (PSMs). As a statistical side note, the formula for PSM *e*-value = PSM *p*-value × number of peptide candidates, is based on the

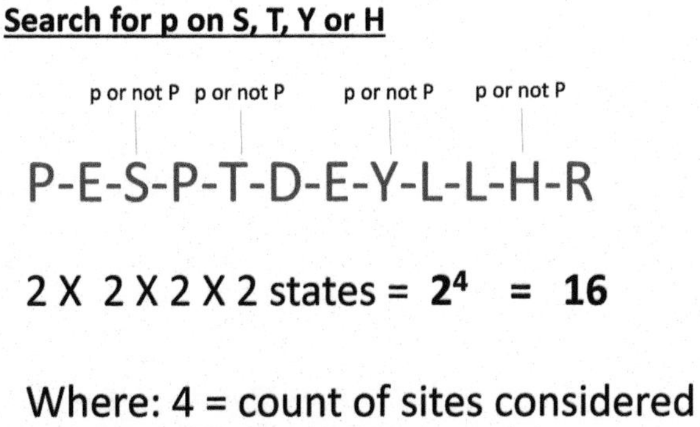

Search for p on S, T, Y or H

p or not P p or not P p or not P p or not P

P-E-S-P-T-D-E-Y-L-L-H-R

2 X 2 X 2 X 2 states = $\mathbf{2^4}$ = **16**

Where: 4 = count of sites considered

Fig. 2 A demonstration of the combinatorial expansion in search space when searching for phosphorylation on multiple sites

assumption that peptide candidates are independent, and randomly different. In general this assumption seems dubious (it is possible to have multiple peptides with short sequences in common), and it not supported at all for modification site assignment, when different phosphosites are considered as different peptide candidates yet share many fragment ions in common. Second, searches that consider larger numbers of potentially modified residues increases the computing power needed, and we would not generally recommend searching for a large number of variable modifications at the same time, at least using the standard search method.

4. FDR threshold can also be performed at the protein-level, although for peptide or modification site identification, this is less important than for protein discovery or quantification experiments.
5. The statistical model in both *ptm*RS and A-score assumes that correct matches between theoretical product ions (b- and y-ions from HCD) and ions in the MS^2 spectrum occur approximately homogeneously across the spectrum. Hence, to reduce the possibility of false matches in areas with greater peak density, only the most intense peaks are chosen for comparison within a given m/z bin width. Both algorithms divide the spectrum in 100 m/z windows and iterate through a set of predefined peak depths i where $1 >= i < 10$, that is, up to the ith most intense peaks per window. The cumulative binomial distribution $P(x)$ is used to calculate scores for all possible isoforms across all peak depths. The score represents the probability of obtaining at least a number of matches x just by chance.

$$P(x) = \sum_{k=n}^{N} \binom{N}{k} p^k (1-p)^{N-k}$$

where N is the number of trials or theoretical ions used in the comparisons, the number of successful matches is denoted as n, and p is the probability of matching a peak within a window. The probability p is directly linked to the search tolerance as this dictates the number of differentiating peaks per window. For example in the simplest case, a tolerance of ± 0.5 m/z (i.e., a 1 m/z match window) could have 100 possible peaks in a window then, for peak depth $i = 3$, the probability of obtaining a random match in a window is $p = 3/100 = 0.03$. The score for each peak depth and isoform is calculated as $-10 \log(P)$. By comparing scores at each peak depth, the optimal peak depth (defined as the value of i that maximizes the difference between the two highest scoring isoforms) is identified.

Up to this point the optimal peak depth is used to produce a final score for the isoforms but not for phosphosite localization discrimination. To score each potential phosphosite, Ascore simply takes the differences between the scores at the optimal peak level, and calculates a p-value. The Ascore is then a $-10 \times \log$ (p-value) transformation of this value, hence why an Ascore > 20, is derived from the equivalent of $p < 0.01$. In contrast, ptmRS postulates that one of the candidate isoforms must be correct and presents $1/P$ as a score of confidence about the correct assignment for each isoform sites. The *ptm*RS algorithm then converts the confidence score into a pseudo-probability, by summing the scores for all isoforms, and dividing each isoform score by the summed total. Through simulation, we have been able to show that having as few as two additional fragment ion matches between the top ranked and second ranked isoform is generally sufficient to produce a *ptm*RS "probability" > 0.98. Consequently, it is relatively likely in large datasets that a *ptm*RS value >0.98 can occur through random matches (particularly on low-resolution instruments), and hence is an important reasons that consideration should be given to the inclusion of neutral loss product ions when searching the data (*see* **Note 5**). Multiple observations of a given phosphorylation site are thus preferred to eliminate the probability of false localization.

In addition to the different methods employed for calculating phosphosite localization score, the range of the peak depths considered is different between these two algorithms. Also, significantly, *ptm*RS considers all theoretical peptide ions, while the Ascore statistical model only considers site determining ions. In *ptm*RS, these theoretical ions include water and ammonia mass losses as well as 2+ ions for all possible phosphoisomers. Consideration of neutral loss of phosphoric acid (−98 Da) as site determining is also optional in *ptm*RS. Neutral loss pathways are not well defined yet for pHis and loss of 98 Da is potentially misleading [2]; we therefore recommend not considering neutral loss ions for scoring of site localization, since they may potentially bias the results in favor of, for example, pSer or pThr where loss of 98 Da is prevalent (*see* **Note 5**).

Neither *ptm*RS nor *A*-score is able to generate a statistic of False Localization Rate (FLR) natively. Instead, the scores exported from these tools has been benchmarked and calibrated using synthetic peptide libraries, which although extremely useful, may not be truly representative of real, biological phosphopeptides. Importantly, since there is a current sparsity of synthetic peptide sets containing phosphohistidine (or other types of phosphorylated amino acid), it is not currently possible to know how well these tools perform for sites other than S, T or Y, especially considering lack of knowledge about fragmentation pathways for pHis (including neutral losses), and the wider search space through the inclusion of multiple phosphosites within many peptides.

6. A particular consideration for phosphoproteomics data searching is whether the software considers neutral losses within the fragmentation spectrum (*see* **Note 4**). A neutral loss occurs when a modification site (or amino acid side chain) loses a particular chemical group during the fragmentation process. Under HCD, it is relatively common to observe loss of phosphoric acid from pSer or pThr (H_3PO_4: 97.976896 Da). For pTyr, loss of phosphate is generally more prevalent (HPO_3: 79.966331 amu) with additional loss of a water molecule (−18.010565 amu) being sequence dependent. For pHis identification, the neutral loss pathways have not yet been well defined, and there is currently some debate over the prevalence of neutral loss ions [2, 13]. As such, to minimize the possibility of misassigning the phosphorylated residue, evidence for pHis site localization we recommend that only the presence of His +79.966331 Da containing ions be used as evidence for pHis localization.
7. Pathway enrichment methods such as DAVID or Panther apply a simple metric to determine which KEGG or REACTOME pathways, or which Gene Ontology/UniProt terms are enriched in the dataset of interest. For ease of implementation we highly recommend searching the LC-MS/MS data against the most up-to-date protein sequence database from UniProt. This is important to ensure that DAVID/Panther can map the proteins to its internal database of pathways.

Both methods map the input gene/protein list to each of their curated sets of terms and pathways to calculate a ratio:

- *a/b:* where *a* is the number of *proteins in the input list* mapping to the pathway (e.g., KEGG:Glycolysis), and *b* is all proteins mapped to the pathway set (all KEGG pathways).

The software then maps the background distribution (discussed below), that is, all proteins identified, to the same sets to calculate a second ratio:

- *A/B*: where *A* is the number of *background proteins* mapping to the given pathway (e.g., KEGG:Glycolysis), and *B* is the number of background proteins that mapped to all KEGG pathways.

The software then calculates the enrichment factor (EF):

- *EF = (a/b)/(A/B):* interpreted as how many times higher the proportion of proteins in the protein list is, compared to the background list (e.g., (12/56)/(60/500) = 1.79 EF)

A Fisher's exact test (or similar) is then used to calculate whether the counts mapping in the input list could have come from the same distribution as the background list. A small p-value thus indicates that the input list (for a given pathway or UniProt/GO term) is from a different distribution and thus significant. Global testing correction is then applied for the number of different pathways that the software has attempted to search for significance.

By default, the calculation of random chance mappings (the background distribution) uses the complete set of genes or proteins from the species being studied. It is important to select a suitable background dataset when analyzing phosphoproteomics data. As a hypothetical experiment, if a study had identified a set of *n* pHis containing proteins, it could be argued that the most statistically correct distribution for the background set of proteins should be all phosphoproteins identified in the study. Any significant pathways observed would thus indicate that there is enrichment for the presence of specifically pHis containing proteins over other phosphoproteins in these pathways. If the total theoretical proteome is used as the background distribution, it is almost inevitable that significant pathway enrichment would be observed for general phosphorylation containing proteins, and cell signalling-related categories, but it would not necessarily report anything about specific to the process of pHis-containing proteins, especially as it is likely that pHis-containing also have phosphorylation sites on S, T, or Y residues. Alternatively, a background distribution of all identified proteins could be used, which would ask the tool to determine the functions enriched in pHis-carrying proteins versus both phosphorylated and nonphosphorylated proteins. This comparison is likely to have greater statistical power, but if pHis proteins tend to carry canonical phosphorylation sites as well, then it is possible that pathway enrichment will discover known pathways/functional categories involving phosphorylation in general, rather than what is unusual about pathways just involving pHis-containing proteins.

Acknowledgments

We are pleased to acknowledge funding from BBSRC that supported this work [BB/M023818/1, BB/L005239/1].

References

1. Fuhs SR, Meisenhelder J, Aslanian A et al (2015) Monoclonal 1- and 3-phosphohistidine antibodies: new tools to study histidine phosphorylation. Cell 162:198–210
2. Hardman G, Perkins S, Brownridge PJ, Clarke CJ, Byrne DP, Campbell AE, Kalyuzhnyy A, Myall A, Eyers PA, Jones AR, Eyers CE (2019) Strong anion exchange-mediated phosphoproteomics reveals extensive human non-canonical phosphorylation. EMBO J e100847:21. https://doi.org/10.15252/embj.2018100847
3. Huang Da W, Sherman BT, Lempicki RA (2009) Systematic and integrative analysis of large gene lists using DAVID bioinformatics resources. Nat Protoc 4:44–57
4. Mi H, Poudel S, Muruganujan A et al (2016) PANTHER version 10: expanded protein families and functions, and analysis tools. Nucleic Acids Res 44:D336–D342
5. Chambers MC, Maclean B, Burke R et al (2012) A cross-platform toolkit for mass spectrometry and proteomics. Nat Biotech 30:918–920
6. Verheggen K, Ræder H, Berven FS et al (2017) Anatomy and evolution of database search engines—a central component of mass spectrometry based proteomic workflows. Mass Spectr Rev. https://doi.org/10.1002/mas.21543
7. Beausoleil SA, Villén J, Gerber SA et al (2006) A probability-based approach for high-throughput protein phosphorylation analysis and site localization. Nat Biotechnol 24:1285–1292
8. Taus T, Kocher T, Pichler P et al (2011) Universal and confident phosphorylation site localization using phosphoRS. J Proteome Res 10:5354–5362
9. Ferries S, Perkins S, Brownridge PJ et al (2017) Evaluation of parameters for confident phosphorylation site localization using an orbitrap fusion tribrid mass spectrometer. J Proteome Res 16:3448–3459
10. Ashburner M, Ball CA, Blake JA et al (2000) Gene ontology: tool for the unification of biology. Nat Genet 25:25–29
11. Kanehisa M, Araki M, Goto S et al (2008) KEGG for linking genomes to life and the environment. Nucleic Acids Res 36: D480–D484
12. Collins A, Jones AR (2018) phpMs: a PHP-based mass spectrometry utilities library. J Proteome Res 17:1309–1313
13. Oslund RC, Kee J-M, Couvillon AD et al (2014) A phosphohistidine proteomics strategy based on elucidation of a unique gas-phase phosphopeptide fragmentation mechanism. J Am Chem Soc 136:12899–12911
14. The Uniprot C (2017) UniProt: the universal protein knowledgebase. Nucleic Acids Res 45: D158–D169

Index

Claire E. Eyers (ed.), *Histidine Phosphorylation: Methods and Protocols*, Methods in Molecular Biology, vol. 2077, https://doi.org/10.1007/978-1-4939-9884-5, © Springer Science+Business Media, LLC, part of Springer Nature 2020

Printed by Printforce, the Netherlands